GROWTH OF CRYSTALS
FROM THE VAPOUR

Growth of Crystals from the Vapour

M. M. FAKTOR
Senior Principal Scientific Officer

and

I. GARRETT

Senior Scientific Officer
Post Office Research Department
Dollis Hill, London NW2 7DT

CHAPMAN AND HALL
London

A HALSTED PRESS BOOK
JOHN WILEY & SONS
New York

*First published 1974
by Chapman and Hall Ltd.
11 New Fetter Lane, London EC4P 4EE*

© *1974 M. M. Faktor and I. Garrett*

*Typeset by Preface Limited, Salisbury, Wilts
Printed in Great Britain by
T. & A. Constable Ltd, Edinburgh*

All rights reserved. No part of this book may
be reprinted, or reproduced or utilized in any
form or by any electronic, mechanical or
other means, now known or hereafter invented,
including photocopying and recording, or in
any information storage and retrieval system,
without permission in writing from the
Publisher

Library of Congress Cataloging in Publication Data

Faktor, M   M
   Growth of crystals from the vapour.

   "A Halsted Press book."
   1. Crystals – Growth. 2. Vapors. I. Garrett, I.,
joint author. II. Title.
QD921.F24   548'.5   74-9878
ISBN 0-470-25350-9

To Jill and Anne,
    Joanna and David,
    Jamie and Ben

Mercury to Ixion.
' "So! tread softly. Don't be nervous. Are you sick?".
"A little nausea; 'tis nothing".
"The novelty of the motion. The best thing is a beef steak. We will stop at Taurus and take one".
"You have been a great traveller, Mercury?".
"I have seen the world".
"Ah! a wonderous spectacle. I long to travel".
"The same thing over and over again. A little novelty and much change. I am wearied of exertion, and if I could get a pension I would retire".
"And yet travel brings much wisdom".
"It cures us of care. Seeing much we feel little, and learn how very petty are all those great affairs which cost us such anxiety" '.

IXION IN HEAVEN

by Benjamin Disraeli.

Contents

	Preface	page	ix
1	Introduction		1
2	Review of the Basic Science		7
	2.1 Thermodynamics of crystal growth		7
	2.1.1 Introduction		7
	2.1.2 Basic thermodynamics		8
	2.1.3 Reaction equilibrium		16
	2.1.4 The phase rule		20
	2.2 Description of crystals		24
	2.2.1 Elements of crystallography		24
	2.2.2 Structural crystallography		35
	2.2.3 Crystal imperfections		41
	2.3 Some aspects of inorganic chemistry affecting the choice of a transporting agent		48
	2.3.1 Thermodynamic considerations		48
	2.3.2 Volatility: estimation of entropy and enthalpy of sublimation		53
	2.3.3 Solubility of species in solids		60
	2.3.4 High temperature species and their stability		63
	2.3.5 Non-stoichiometric phases		67
3	The Vapour–Crystal Interface		71
	3.1 Adsorption		71
	3.2 Nucleation		82
	3.3 Surface roughness and surface rearrangement		87
	3.4 Surface diffusion		91
	3.5 Self-sustained growth without nucleation		96
4	Vapour Transport		101
	4.1 Diffusion		101

	4.2	Vapour transport in crystal growth systems	109
		4.2.1 Stefan's flow	109
		4.2.2 Example: sublimation and transport of silver	111
		4.2.3 Stefan's velocity in various systems	118
	4.3	Dissociative sublimation	120
	4.4	Chemical vapour transport	137
		4.4.1 General introductory remarks	137
		4.4.2 Example: Transport of nickel as nickel carbonyl	138
		4.4.3 Example: Transport of carbon with sulphur	149
		4.4.4 More complex chemical vapour transport systems	151
		4.4.5 Transport of gallium arsenide	152
		4.4.6 Transport systems involving several reactions	157
	4.5	Viscous flow of gas	161
	4.6	Laminar and turbulent flow	165
	4.7	Convection	170
	4.8	Heat flow	178
		4.8.1 Example: Isothermal growth of metanitroaniline	184

5 *Sequential Processes in Crystal Growth* 191
   5.1 Introductory remarks 191
   5.2 Formalism of sequential processes in crystal growth 194
   5.3 The effect of growth conditions on stoichiometry and electronic properties of crystals 202
   5.4 Morphological stability 206
   5.5 Separation, purification, and doping by chemical vapour transport 217

6 *Experimental Methods of Crystal Growth from the Vapour Phase* 226
   6.1 The sealed capsule method 226
   6.2 Almost sealed capsule methods 238
   6.3 Open flow systems 245

7 *Areas for Further Study* 259
   7.1 Exploration of concepts 259
   7.2 Maximisation of the growth rate 276
   7.3 Growth at high pressure 279

References 283

Index 293

Preface

In this monograph, we discuss growing crystals from the vapour. This is but one aspect of crystal growth. As is well known, crystals may be grown from the melt, from solution and by solid-state reaction.

Many books on crystal growth have appeared recently, following the heavy investment in crystal growing by the electronics industry and others. These books provide catholic coverage of the subject, with a wealth of experimental detail, of great value to the practical crystal grower. In launching another volume onto the scene, we have endeavoured to retain the practical aspects, while shifting the emphasis to the underlying concepts. We therefore limit our discussion to vapour phase growth, where we consider the concepts are simplest, though still far from simple. At least we avoid the very real complications of solution and melt chemistry. Although the composition of vapours is not necessarily straightforward, particularly at high temperatures, there are analytical techniques which go some way to providing a picture. By confining ourselves to systems containing gases and crystals only, we are dealing with states of matter successfully described by the theory of the solid state and by the kinetic theory of gases.

Crystal growth is a multi-disciplinary subject, and crystal growers come from a variety of fields — chemistry, physics, electronic engineering, chemical engineering. With this in mind, and to assist new workers entering the field, we have tried to include under one cover relevant concepts and information disseminated in the textbooks of a variety of disciplines: hydrodynamics, surface chemistry, thermodynamics, crystallography and inorganic chemistry. The levels at which these topics are covered varies considerably, not least because of the authors' shortcomings. It is hoped that these accounts will nevertheless be found useful. It is our experience that some of these topics are but hazily understood by many practising crystal growers. In some topics, such as thermodynamics and crystallography, we have given fairly long accounts starting at first principles, because these subjects are central to crystal growth and are well developed. Other subjects, such as surface chemistry, are no less important but are in a comparatively primitive state of development. Our coverage of such topics is necessarily scant. In all cases, references are given to recent reviews or to standard textbooks, as appropriate.

The task of presenting the subject matter of this book in an entirely logical manner defeated us. The unusual placing of the chapter on Experimental Methods may seem surprising; yet the principles and theory of vapour growth of crystals are best understood before attempting to design or modify apparatus, or to choose optimum conditions for growth.

It is a pleasure to acknowledge the help afforded us by our colleagues at the Post Office Research Department, and in particular, the constant encouragement and constructive criticisms of Dr. J. R. Tillman. We are grateful also to Dr. E. A. D. White, Dr. A. Finch, and the late Dr. P. Gross among many with whom we have had fruitful discussions.

M. M. F.
I. G.
February, 1974.

CHAPTER ONE
Introduction

A crystal is an array of atoms arranged in a three-dimensional structure which extends over distances which are very much larger than the size of an atom. Herein lies the distinction between crystals and other condensed phases — liquids, glasses and plastics, in which the ordering of the atoms is relatively short-range (although some solid polymers are partly crystalline). Large natural crystals weigh typically a few kilograms and synthetic crystals of a few tens of kilograms have been grown. Such crystals contain $10^{25}-10^{27}$ atoms. A very small synthetic crystal, such as the active region of a semiconductor laser, might be as small as $100 \times 200 \times 0.1$ μm, and would contain some 10^{14} atoms. Clearly there are forces in Nature which strongly favour ordering of atoms into crystals. Indeed, crystallinity is the rule rather than the exception in Nature among inorganic solids.

The perfection of the ordering varies considerably, and 'mistakes' of various kinds are observed. Very small concentrations of such defects give rise to many of the physical properties of crystals which have been exploited in the last twenty years. Here we would distinguish two types of properties: those that would be possessed by the pure, perfect solid, such as ferromagnetism, birefringence, and piezoelectricity, and those which exist only because the crystal is not pure or not perfect. In the latter category we again distinguish two groups: intrinsic properties, which are caused by defects in the pure crystal, such as some colour centres and intrinsic conductivity, and extrinsic properties which are caused by small or large quantities of impurities.

The rapid advances in solid state physics over the last few decades have uncovered many effects in crystals which have formed the basis of large industries. In developing the basic effect as discovered by the theoretical or experimental research physicist into a marketable device, the development engineer has been calling for materials with more and more stringent specifications. Because of the vast market for solid state devices, in the telecommunications and computing industries for example, a great deal of effort and money has been spent in recent years on producing single crystal materials. Rather less effort and money have been devoted to finding out how crystals grow.

The crystal grower is called on to produce materials with close control over the following properties:

(1) Defects: non-stoichiometry, dislocation density, etc.
(2) Impurity concentrations.
(3) Concentrations of foreign atoms added intentionally, i.e. dopants.
(4) Geometry, e.g. very thin epitaxial layers.

A discussion of the basic science behind the control of these properties will be found in Chapters 2 and 5. In these areas, the crystal grower is frequently asked to perform at the limits of science. For example, many of the present generation of semiconducting devices rely on material which must contain less than 1 part in 10^9 of certain impurities. Producing such material demands a high degree of experimental skill and a clear understanding of the purification processes. While it is inappropriate to try to treat these matters in depth in this monograph, we have tried to bring out the underlying chemical principles, and augment these by references to specialist textbooks. It is hoped that this information will save considerable expensive empirical investigation.

It is often forgotten that before the properties of a new material can be studied, the material itself has to be discovered (either on the basis of a theoretical prediction of which material should have the desired property, or more or less accidentally). It must then be prepared in a size and form on which measurements can be made. This almost invariably means that single crystals will be required. Discoveries of new materials, and of new effects and properties, are not made by research forecasters, planners or administrators. (This point has been stressed by Laudise.)[1] An excellent example of the unpredicted discovery of new materials with interesting properties is the niobium-oxygen phases found and studied by Schäfer and co-workers[2].

There is clearly great advantage to be gained in running pure or 'curiosity motivated' research in parallel with applied or 'goal-motivated' research. Pure research would cover the design of a material to secure some required combination of properties, and the design of a rational crystal growth system to produce the material. Successful materials could then be taken over by the applied research side for development, with the benefit of immediate liaison with the designers of the material and of the growth system. At the moment, there are very few institutions in this country which teach the concepts of chemical vapour transport, and the applied research in the industrial establishments, while successful in the main, has not been cheap because of a largely empirical approach. Collection and purchase of empirical 'know-how' is poor preparation for the future!

Not only must the crystal grower be aware of trends in properties of substances, such as solubilities and volatilities (it would be pointless to try to grow sodium chloride by chemical vapour transport, for example), but ideally he should have an understanding of the solid-state physics which goes into characterizing the grown crystals. This is a subject that we have not attempted to deal with at all in this monograph. Yet the only real test of the crystal grower's success, in many cases, is provided by measuring some physical property (such as current carrier mobility or electro-luminescence) and interpreting the results on the basis of solid state theory. Often, it is only by such techniques that the crystal grower can learn about the concentrations of defects, dopants and impurities that he is putting into the material or leaving there. Again, there should be rapid feedback between the grower and the physicist who characterizes the material.

Vapour phase crystal growth is, in principle at least, a flexible method of preparing materials. This potential flexibility has occassionally been exploited (see Chapter 6), but more often the principles have been poorly understood.

Preparation of crystals from the gas phase requires the general reaction:

$$\text{GAS (disordered)} \rightarrow \text{CRYSTAL (ordered)}$$

The gas phase may consist simply of molecules of the crystal substance, or of its separate constituents, if they are all volatile. Otherwise, the involatile constituents may be reacted with a transporting agent to provide volatile species. In all cases, an inert gas may be added to modify the transport kinetics. The basic reaction results in an increase in atomic order. The driving force for ordering is the substitution of strong bonds for disorder, with the consequent entropy decrease being more than compensated by transferring the heat of condensation or reaction to the surroundings.

We divide crystal growth from the vapour into various categories, depending on the way in which the vapour is obtained. These categories and discussed separately in detail in Chapter 4. The first category is sublimation or evaporation, in which the vapour is obtained from the pure condensed phase at an appropriate temperature. A compound may dissociate on sublimation or evaporation, and if the degree of dissociation is high, the growth system is then in the second category, namely dissociative sublimation. Here we have extra degrees of freedom in the vapour at our disposal. If a transporting reaction is used for one or more of the constituents of the crystal, we have either a chemical vapour transport (CVT) or a chemical vapour deposition (CVD) system. We distinguish between these two categories on the basis of the conditions over the growing crystal. If equilibrium is achieved fairly closely, the system is CVT, otherwise it is CVD. The distinction is more clear-cut in practice than in principle, and the transport phenomena in the vapour are the same in both systems. The industrial applications of CVD are growing explosively[3].

The outline of the pertinent, basic science in Chapter 2 starts with a section on thermodynamics. The emphasis here is on reaction equilibria and the phase rule, as these two topics are central to an understanding of vapour phase crystal growth. Although equilibrium is never achieved in a *successful* crystal growth experiment, it is often closely approached. Thus the equilibrium situation provides an excellent starting point for a discussion of the system, except perhaps in those extreme cases where massive departures from equilibrium have been engineered. Inclusion of the next topic, crystallography, occasioned the authors some misgivings. The information contained in this section is presented more lucidly in many standard textbooks, and is in any case not used directly in the subsequent discussion of crystal growth in the rest of the book. Yet there can be no doubt that the subject has much to teach us about crystal growth. For example, the kinetics of adsorption and growth are dependent on the reticular densities of crystal faces. Furthermore, this concise summary of crystallography is sufficient to enable crystal growers to report on the crystal structure and crystallographic faces of their specimens in the accepted manner. The present profusion of loose descriptive terms accompanied by indifferent photographs is to be deplored.

The section on crystallography is followed by a semi-quantitative discussion of the thermodynamic basis for choosing a transporting agent. Various kinetic aspects of this choice are discussed throughout much of the rest of the book. The final sections of Chapter 2 discuss

INTRODUCTION

ways of estimating thermodynamic parameters. A word of warning is required here. Thermodynamics, which provides a corner-stone for discussions and theories of crystal growth, has not enjoyed the status that it merits among crystal growers. It is perhaps helpful to see why this is. One can obtain accurate, quantitative information only if accurate data are available. Such data must be obtained from precise experiments. Estimated thermodynamic data, particularly when arrived at by circuitous routes, rarely have the necessary accuracy for such exercises as calculating the partial pressures, which are required for uncoupling sequential processes, since the data involved (e.g. the enthalpy and entropy change in some process) appear in an exponential in such calculations. Estimated values are nevertheless of great value in suggesting the right range of operation for the variables in a crystal growth system We suggest that it is failure to recognize these facts which has caused many crystal growers to turn a deaf ear to the teachings of this pre-eminently experimental science.

Chapter 3 is concerned with various effects at the vapour-solid interface: adsorption, migration or surface diffusion, and nucleation. The surface provides the communication link between the crystal and the vapour. It can confidently be stated that the continuing study of surface chemistry and physics will lead to an understanding of the atomic processes of crystallization. At present, this is a young field. Particularly for nucleation, erudite mathematical theories abound, with adjustable parameters to fit all observed nucleation phenomena. Their relevance to real situations remains to be demonstrated; therefore we give a mere outline of this topic.

Movement of gas and transport through the gas is the subject of Chapter 4. Obviously this is an essential part of crystal growth, since reactants must be brought to the crystal surface, and the gaseous products, if any, must be transported away. Here the picture is clear, and the theory is more tedious than difficult. We derive the flow equations, and illustrate their application with several worked examples. In these examples, we have obtained boundary conditions for the flow equations by assuming that there is equilibrium between the solid and the vapour. While this assumption is not always very accurate, it does represent an upper limit to the driving force for the gas transport, and hence allows us to calculate an upper limit to the rate of growth under given conditions.

In Chapter 5 we provide a formalism for decoupling vapour transport processes from surface reactions, so that the surface reactions may be studied and the effects of vapour transport taken into account. We also discuss the basis for controlling stoichiometry, doping, and morphology.

At this stage, the conceptual picture is complete, and we have the formalism for coupling or decoupling the sequential processes of crystal growth. We go on to discuss various experimental systems in Chapter 6. We make no attempt to cover this extensive subject comprehensively; our viewpoint is essentially personal. A number of excellent reviews have appeared recently, with which we would not wish to compete. Moreover, in all fields where experiment is ahead of theoretical understanding, ingenious experimental systems abound, and we could not attempt to do them all justice. We have selected some typical systems which can be discussed in the light of the foregoing chapters.

Finally, in Chapter 7, we discuss some of the difficulties in crystal growth from the vapour and explore some ideas. Here we no longer remain on firm ground covered in the literature, but enter the area of hypothesis and opinion. The reader may agree or disagree with us, but in either case, we hope that we will have provoked some thought about twilight areas of crystal growth.

CHAPTER TWO
Review of the Basic Science

2.1 Thermodynamics of crystal growth

2.1.1 Introduction

Thermodynamics is a logically constructed science based essentially on some very common observations:

(1) Two bodies in equilibrium have the same 'degree of hotness' or 'empirical temperature'.
(2) It is impossible to construct a perpetual motion machine.
(3) No natural process can be reversed in its entirety.

Because the results of thermodynamics are largely independent of any assumptions concerning the atomic and molecular structure of any particular substance, they may be applied very widely. The kind of results that may be obtained are:

(1) From the first law, relations may be established between heat and work. These relations are not restricted to systems at equilibrium.
(2) From the first and second laws, predictions may be made about the effect of changes in temperature, pressure and composition on a great variety of physico-chemical systems. These applications are restricted to systems at equilibrium.

The second law of thermodynamics is partly expressed by an inequality:

$$dS \geq 0$$

which only becomes an equality for reversible processes, that is, processes taking place under conditions only infinitesimally different from equilibrium For this reason, thermodynamics has most to say about *equilibria,* and it is with this aspect, and in particular, reaction equilibrium and phase equilibrium, that we shall be concerned here. An understanding of these subjects is of immense value in designing a crystal growth system and in analyzing its shortcomings.

While there are many predictions that can be made about a vapour transport system by applying the results of thermodynamics, there are certain areas where thermodynamics has nothing to say. In assessing any actual situation in crystal growth, it must be decided whether the essential features of the problem have to do with *equilibria* or with *rates* of various processes. If the rate at which a crystal grows, or the form that it grows in, is determined by the rate at which some process is taking place (e.g. nucleation on the surface) then the situation must be analysed using the appropriate rate theory, and thermodynamics may be of little value. If the processes are occuring very near to equilibrium, however, which is often the case at elevated temperatures, then thermodynamics can be of immense value in predicting the variation of yield, composition, etc., with temperature, pressure, gas composition and other experimental variables.

In the resumé of basic thermodynamic theory in the next section, we pass quickly over the first principles, as these are well discussed in the standard text books,[4,5] and a lengthy discussion here would be out of place. We devote a little more space to consideration of reaction equilibrium and the phase rule, as these are topics of central importance in understanding crystal growth from the vapour.

2.1.2 Basic thermodynamics

First law

The first law of thermodynamics expresses the equivalence of work and heat as means of increasing or decreasing the *internal energy, U,* of a body or a system. When a body absorbs an amount of heat q and has an amount of work w performed on it, the change in internal energy is given by:

$$U_2 - U_1 = q + w \tag{2.1}$$

where U_2, U_1 are the internal energies in the final and inital states of the body. The internal energy is defined by Equation 2.1, except for an additive constant that is arbitrary. The important statement of the first law is that U is a function of the state of the body only (that is, essentially, of its temperature and pressure, or other intensive variables), and the quantity $U_2 - U_1$ is consequently independent of the sequence of intermediate states that the body may pass through, whereas q and w are, in general, dependent on the sequence of intermediate states.

Equation 2.1 may be written in differential form for an infinitesimal change of state:

$$dU = đq + đw \qquad (2.2)$$

where the notation đ serves to remind us that the integrals of $đq$ and $đw$ are not uniquely defined, but depend on the path between the initial and final states.

Second law

The second law of thermodynamics defines an absolute temperature scale (apart from an arbitrary scaling factor) and introduces the extremely useful concept of *entropy*. We may state the law in three parts as follows:

(1) An absolute temperature T may be defined so that the ratio T_1/T_2 of the temperatures of two bodies is independent of the properties of any particular substance.

(2) The entropy S of a body or system, defined by:

$$dS = (dq/T)_{\text{reversible}} \qquad (2.3)$$

for a reversible change, is a function of the state of the body or system only.

(3) The entropy of a system in an adiabatic enclosure can never decrease: it increases in an irreversible process and remains constant in a reversible process.

Fundamental equation.

For a reversible change of state, therefore, $dq = TdS$. Also, the mechanical work done reversibly on the body or system (i.e. against zero friction and with no pressure difference between the system and its surroundings), is given by $dw = -PdV$. Hence we arrive at the *fundamental equation* for a reversible change of state in a closed system:

$$dU = TdS - PdV \qquad (2.4)$$

From this fundamental equation, many of the results of thermodynamics are derived. It is because this equation applies only to reversible changes that thermodynamics is largely concerned with equilibria.

Open systems – chemical potential

In a closed system, that is, one in which the quantity of matter present is constant, we may write dU as a complete differential as follows:

$$dU = \left(\frac{\partial U}{\partial S}\right)_V dS + \left(\frac{\partial U}{\partial V}\right)_S dV \tag{2.5}$$

In an open system, changes in internal energy may be brought about by matter entering or leaving the system, and we must add appropriate terms to Equation 2.5:

$$dU = \left(\frac{\partial U}{\partial S}\right)_{V, n_i} dS + \left(\frac{\partial U}{\partial V}\right)_{S, n_i} dV + \sum_i \left(\frac{\partial U}{\partial n_i}\right)_{V, S, n_j} dn_i \tag{2.6}$$

The quantity $(dU/dn_i)_{S, V, n_j}$ is called the chemical potential of species i. In Equation 2.6, n_i is the number of moles of species i present in the system. The chemical potential, usually denoted by μ_i, may be expressed in several other ways in terms of various derived functions, which will be described shortly.

Integrated form of fundamental equation

The fundamental equation for an open system may be written thus:

$$dU = TdS - PdV + \sum_i \mu_i \, dn_i \tag{2.7}$$

This equation may be integrated by the following process which is not purely mathematical: let the system be enlarged to k times its original size, keeping the temperature, pressure, and composition constant. Then

$$\Delta U = T\Delta S - P\Delta V + \sum_i \mu_i \Delta n_i$$

where $\Delta V = (k-1)V$, $\Delta S = (k-1)S$, etc. so that:

$$(k-1)U = T(k-1)S - P(k-1)V + \sum_i \mu_i(k-1)n_i$$

or

$$U = TS - PV + \sum_i \mu_i n_i \tag{2.8}$$

which is the integral form of the fundamental equation. The process of integration relies on an item of physical knowledge, namely, that the intensive variables, T, P, and μ_i do not depend on the size of the system, whereas the extensive variables are directly proportional to it.

Derived functions

We will now define three functions, the enthalpy, the Helmholtz free energy, and the Gibbs free energy, which are useful in practice for two reasons. They are more readily and directly determined experimentally than internal energy, and they provide us with criteria of equilibrium under conditions commonly prevailing in the laboratory.

The enthalpy H is defined by:

$$H = U + PV \tag{2.9}$$
$$= TS + \sum_i \mu_i n_i \tag{2.10}$$

In differential form, using Equation 2.7:

$$dH = TdS + VdP + \sum_i \mu_i dn_i$$

The enthalpy change of a system is equal to the heat absorbed under two restrictive conditions: (1) the pressure remains constant, (2) the only work done is in changing the volume of the system. These conditions are frequently met in the laboratory, and the term 'heat' is freely, though not always correctly, used as a synonym for enthalpy.

The Helmholtz free energy is defined by:

$$F = U - TS \tag{2.11}$$
$$= PV + \sum_i \mu_i n_i \tag{2.12}$$

In differential form, this becomes:

$$dF = -SdT - PdV + \sum_i \mu_i dn_i \tag{2.13}$$

The Helmholtz free energy provides a criterion for equilibrium in a system enclosed in a rigid container and kept at a constant temperature. Under these conditions, the heat q taken in by the system in any spontaneous change must be less than or equal to $T\Delta S$. But since

$$\Delta F = \Delta U - \Delta(TS)$$
$$= q + w - T\Delta S$$

if T is kept constant, we obtain:

$$\Delta F \leqslant w \tag{2.14}$$

If no work is done on the system, $\Delta F \leqslant 0$, i.e. the Helmholtz free energy can only decrease or remain constant during a spontaneous change. In equilibrium no spontaneous changes take place, so F has reached its minimum value.

The Gibbs free energy, otherwise known as the Gibbs function or the free enthalpy, or just the free energy, is defined by:

$$G = U + PV - TS \quad (2.15)$$

$$= H - TS = F + PV \quad (2.16)$$

$$= \sum_i \mu_i n_i \quad (2.17)$$

In differential form, this becomes:

$$dG = VdP - SdT + \sum_i \mu_i dn_i \quad (2.18)$$

The Gibbs free energy provides a criterion of equilibrium in a system maintained at constant temperature and pressure. Equation 2.17 reveals the chemical potential to be the molar free energy of a species, in which form it is perhaps most familiar. From Equations 2.10, 2.13 and 2.17 we can express the chemical potential in three ways:

$$\mu_i = \left(\frac{\partial H}{\partial n_i}\right)_{S, P, n_j} = \left(\frac{\partial F}{\partial n_i}\right)_{V, T, n_j} = \left(\frac{\partial G}{\partial n_i}\right)_{T, P, n_j} \quad (2.19)$$

The chemical potential μ_i of a species is an intensive variable and a function of state. We will demonstrate that it provides a criterion for chemical equilibrium between two phases. Consider a system of two phases α and β, in thermal and mechanical equilibrium ($T_\alpha = T_\beta = T$, $P_\alpha = P_\beta = P$). Let $dn_{i\beta}$ mols of species i pass from phase β to phase α. The change in the Helmholtz free energy of the total system is:

$$dF = dF_\alpha + dF_\beta = -PdV_\alpha - PdV_\beta + (\mu_{i,\alpha} - \mu_{i,\beta}) dn_{i,\beta}$$

since $dn_{i,\alpha} = -dn_{i,\beta}$. The first two terms are the work done on the system, dw, so that

$$dF = dw + (\mu_{i,\alpha} - \mu_{i,\beta}) dn_{i,\beta}$$

From Equation 2.14, $dF \leq w$, hence:

$$(\mu_{i,\alpha} - \mu_{i,\beta}) dn_{i,\beta} \leq 0$$

It follows that the sign of $(\mu_{i,\alpha} - \mu_{i,\beta})$ is opposite to that of $dn_{i,\beta}$, so that transfer of species i is from a phase where its potential is higher to one where it is lower. For a reversible change, i.e. one taking place at equilibrium $dF = dw$, so that $\mu_{i,\alpha} = \mu_{i,\beta}$.

The condition for chemical equilibrium between phases is therefore that each species shall have the same chemical potential in all phases between which it can pass freely.

Statistical interpretation of entropy

It is common observation that the high temperature form of a substance is less ordered than a low temperature form. Examples are melting and boiling, order – disorder transitions, and so on. The term $-TS$ in the Gibbs and Helmholtz free energies indicates that a system can lower its free energy (i.e. approach equilibrium) by adopting a high-entropy form at high temperature even though the increase in enthalpy in so doing would make the change unfavourable at lower temperatures. There is clearly a relation between entropy and disorder. The form of this relation is arrived at with the aid of Statistical Mechanics, using an atomic or molecular description of matter.

For each thermodynamic state of a system, there are in general a large number of quantum states, corresponding to different ways in which the atomic, molecular, and electronic energy levels may be occupied in a way consistent with the total energy of the system. We assume that all quantum states consistent with the total energy of the system are equally probable. It is possible to deduce from this assumption that the probability of any quantum state depends *only* on the total energy of that state. It can then be shown that [4], if a system has W possible quantum states, the quantity

$$S' = k \ln W$$

may be identified with the thermodynamic entropy, where k is Boltzmann's constant.

In a given system, the quantum states may be divided into those that arise from different geometrical arrangements of the microscopic components of the system, and those arising from different distributions of the total energy among the microscopic components. The corresponding contributions to the entropy of the system are referred to as the *configurational* and *thermal* entropies.

Heat capacity

The heat capacity or specific heat of a substance is defined by

$$C \equiv \frac{dq}{dT}$$

Measurements are usually performed at constant pressure or constant volume. Using Equations 2.2, 2.4 and 2.9 we obtain:

$$C_V \equiv \left(\frac{dq}{dT}\right)_V = T\left(\frac{\partial S}{\partial T}\right)_V = \left(\frac{\partial U}{\partial T}\right)_V \qquad (2.20)$$

$$C_P \equiv \frac{\mathrm{d}q}{\mathrm{d}T}\bigg)_P = \left(\frac{\partial H}{\partial T}\right)_P = T\left(\frac{\partial S}{\partial T}\right)_P \tag{2.21}$$

The two specific heats C_V and C_p are related by:

$$C_P - C_V = \left[\left(\frac{\partial U}{\partial V}\right)_T + P\right]\left(\frac{\partial V}{\partial T}\right)_P \tag{2.22}$$

The specific heat is one of the more readily measurable properties of a substance. It is clear from Equations 2.20 and 2.21 that the temperature dependence of the thermodynamic functions U, S, H etc. may be expressed in terms of the specific heats, so that if the values of these functions are known at one temperature, their values at other temperatures may be found by suitable manipulation of specific heat data. It is simple to derive the following relations:

$$H_2 - H_1 = \int_{T_1}^{T_2} C_P \mathrm{d}T + \int_{P_1}^{P_2} V(1 - \alpha T)\mathrm{d}P \tag{2.23}$$

$$S_2 - S_1 = \int_{T_1}^{T_2} \frac{C_P}{T} \mathrm{d}T - \int_{P_1}^{P_2} V\alpha \, \mathrm{d}P \tag{2.24}$$

where

$$\alpha = \frac{1}{V}\left(\frac{\partial V}{\partial T}\right)_P \tag{2.25}$$

is the volume coefficient of expansion. In particular, if $P_1 = P_2$, we have the familiar forms:

$$H_2 - H_1 = \int_{T_1}^{T_2} C_P \, \mathrm{d}T \tag{2.26}$$

$$S_2 - S_1 = \int_{T_1}^{T_2} \frac{C_P}{T} \mathrm{d}T \tag{2.27}$$

Third law
The third law may be stated as follows: 'for any process involving pure substances in internal equilibrium, the entropy change in any process tends to zero as the temperature tends to the absolute zero'. It seems likely that the entropy of even a pure crystalline solid is non-zero at the absolute zero of temperature, because although thermal and configurational contributions may be zero, there may be contributions from isotope mixing and from random occupation of quantum states within the nuclei. These last quantities are unlikely to alter in a chemical process, however, and so it is useful to adopt

the convention that when the entropy contributions from all sources except these are zero, the *conventional entropy* is zero.

Entropies, enthalpies, and free energies of formation

Standard entropies for pure substances are calculated by integration of specific heat data (Equation 2.27) from the lowest temperature at which measurements have been made up to the standard temperature of 298 K*. Extrapolation of specific heat data to absolute zero is done using some theoretical model, such as the Debye T^3 law for crystalline solids. Values of molar entropy of substances in the standard state $S°_{298}$ (i.e. in their natural stable form at 298 K, 1 atm pressure) are tabulated [6-10] (see for example the references in Section 2.1.3). The value of $S°$ at other temperatures can be obtained by integrating the specific heat as in Equation 2.27. Specific heat data are often available in the form of an empirical equation of the type:

$$C_P = a + bT + cT^2 \tag{2.28}$$

The entropy change accompanying any phase change which occurs in the range of integration must, of course, be included.

The absolute enthalpy of a substance is no more amenable to experimental or theoretical determination than is the absolute entropy. However, in all applications of thermodynamics, it is a *change* in enthalpy that is of interest, whether this change is brought about by mechanical work or heat or chemical reaction, etc. The enthalpy change that is usually tabulated and which is of most interest to chemists is the standard molar 'enthalpy of formation', ΔH^0_{f298}. This is simply the enthalpy change per mole which occurs when the substance is produced in its stable form at 298 K, 1 atm pressure (i.e. in its 'standard state') from its constituent elements in their standard states. Of course, it follows that the enthalpy of formation of an element in the standard state is zero.

The molar enthalpy change for a reaction may be found as the difference between the standard molar enthalpies of formation of the products and the reactants, in each case multiplied by the appropriate stoichiometric coefficient. If the enthalpy of reaction is required at some temperature T different from 298 K, the enthalpies of formation must be adjusted by the appropriate integrals of the specific heats, Equation 2.26.

---

*288 K in the Russian literature.

Standard molar free energies of formation (ΔG^0_{f298}) and of reaction (ΔG^0_{298}) may be found from ΔH^0_{298} and ΔS^0_{298}:

$$\Delta G^0_{298} = \Delta H^0_{298} - 298\, \Delta S^0_{298}$$

or, at any temperature T:

$$\Delta G^0_T = \Delta H^0_T - T \Delta S^0_T \tag{2.29}$$

The free energy change for a reaction may, of course, be measured directly, for example by electrochemical means, or by measuring the equilibrium constant for the reaction. Note that, because of the logarithmic dependence of ΔG on K_p (see Equation 2.39):

$$\Delta G = -RT \ln K_p$$

errors in measurement of K_p do not produce large errors in ΔG. On the other hand, if K_p is calculated from an inaccurate value of ΔG, large errors will result! We might wish to calculate ΔG_T at 1000 K for a reaction, using room temperature data for a combination of other reactions which gives the required reaction. If the error in ΔG_T is 40 kJ mol^{-1}, which is not unreasonable, the uncertainty in K_p is two orders of magnitude. Under these circumstances it would be too optimistic to expect close correlation between predicted and measured rates of growth or composition of crystals (see Chapter 4). A critical appraisal of the thermodynamic data is an essential first step in trying to establish such a correlation.

In this context it is worth noting that the approximate relation:

$$\Delta G_T = \Delta H_{298} - T\Delta S_{298}$$

is often very useful, as the temperature corrections to ΔH and ΔS tend to cancel out.

2.1.3. Reaction equilibrium

We consider firstly a single reaction between gaseous species e.g.:

$$aA + bB \rightarrow cC + dD \tag{2.30}$$

or, more generally, the reaction may be written:

$$\sum_{i}^{N} \nu_i M_i = 0 \tag{2.31}$$

where the ν's are the stoichiometric coefficients, coventionally positive for products and negative for reactants, and the M's are molecular weights. Let us consider an infinitesimal progression of the

REVIEW OF THE BASIC SCIENCE

reaction so that dn_i mols of species i take part. For mass to be conserved, we require that:

$$\frac{dn_1}{\nu_1} = \frac{dn_2}{\nu_2} = \ldots = \frac{dn_i}{\nu_i} = \ldots = \frac{dn_N}{\nu_N} \quad (2.32)$$

For chemical equilibrium, we require:

$$\sum_i \mu_i dn_i = 0 \quad (2.33)$$

whether we are minimizing Gibbs free energy, Equation 2.18 or Helmholtz free energy, Equation 2.13. Using Equation 2.32, all but one of the dn_i's in Equation 2.33 can be eliminated in terms of the ν_i's and, say, dn_1:

$$\left(\mu_1 + \frac{\nu_2}{\nu_1}\mu_2 + \ldots + \frac{\nu_N}{\nu_1}\mu_N\right) dn_1 = 0$$

Thus

$$\sum_i \nu_i \mu_i = 0 \quad (2.34)$$

is the condition for equilibrium in a reactive system. The quantity $-\sum_i \nu_i \mu_i$ has been called the 'affinity' of the reaction, and it can be shown that as a first approximation ($\sum_i \nu_i \mu_i < RT$) the reaction rate is proportional to the affinity [11].

Ideal gas mixture

An ideal gas mixture is defined as one in which the chemical potential of each species i is given by:

$$\mu_i = \mu_i^0 + RT \ln p_i \quad (2.35)$$

where μ_i^0 is a function of temperature only, and p_i is the partial pressure of species i. It can be shown that an ideal gas mixture obeys the ideal gas law, and that there is no heat of mixing of the components. Most gases, and many gas mixtures, obey the ideal gas equations at atmospheric pressure and below.

Equilibrium constant

Substituting Equation 2.35 in Equation 2.34, we obtain:

$$-RT \ln \prod_i p_i^{\nu_i} \sum_i \nu_i \mu_i^0 \text{*} \quad (2.36)$$

*Where $\prod_i x_i$ is the continued product $x_1 x_2 x_3 \ldots x_N$

or
$$-RT \ln K_p = \sum_i \nu_i \mu_i^0 \tag{2.37}$$

where we define

$$K_p \equiv \prod_i p_i^{\nu_i} \tag{2.38}$$

Now, since μ_i^0 is a function of temperature only, the same is true of K_p. That is to say, there exists a certain function of the partial pressures in a reactive gas mixture which has a constant value at a constant temperature if there is equilibrium. This constant is called the equilibrium constant for the reaction.

According to Equation 2.35, μ_i^0 is the chemical potential of species i at unit pressure, i.e. the standard molar free energy of species i. It follows that $\sum_i \nu_i \mu_i^0$ is the standard molar free energy change for the reaction ΔG_T^0. Thus:

$$-RT \ln K_p = \Delta G_T^0 \tag{2.39}$$

or

$$K_p = \exp(-\Delta G_T^0 / RT) \tag{2.40}$$

Vapour and condensed phases
The chemical potential of a pure substance in the solid or liquid phase is a function of temperature, but, as a good approximation, not dependent on the applied pressure.*

For substances in solution in a solid or liquid, the chemical potential is given by an equation analogous to Equation 2.35 for the chemical potential of a gas:

$$\mu_i = \mu_i^0 + RT \ln a_i \qquad 2.41$$

where μ_i^0 is a function of temperature but virtually independent of the applied pressure, and a_i is the *activity* of species i in the solution. In 'ideal solutions', the activity is equal to the mole fraction x_i. For the solute species in a dilute solution, a_i is proportional to x_i (Henry's law). For the solvent species, as $x_i \to 1$ (i.e. as the solution tends to a pure phase of species i), $a_i \to x_i$, so that we may speak of the activity of a pure condensed substance as being unity.

*But ignoring effects due to pressure-induced phase changes.

REVIEW OF THE BASIC SCIENCE

Our condition for equilibrium, corresponding to Equation 2.36, is now:

$$-RT \ln \underset{\substack{\text{condensed} \\ \text{phases}}}{\Pi a_i^{\nu_i}} - RT \ln \underset{\substack{\text{vapour} \\ \text{species}}}{\Pi p_i^{\nu_i}} = \underset{\substack{\text{all} \\ \text{species}}}{\Sigma \nu_i \mu_i^0} \qquad (2.42)$$

We define

$$K_p = \Pi a_i^{\nu_i} \, \Pi p_i^{\nu_i} \qquad (2.43)$$

So that

$$-RT \ln K_p = \Sigma \nu_i \mu_i^0 \qquad (2.44)$$

Again, the equilibrium constant K_p for the reaction is, as a good approximation, independent of pressure, and is essentially a function of temperature only.

Variation of K_p with temperature

Equation 2.39 may be written as follows:

$$\ln K_p = \frac{-\Delta G_T^0}{RT} \qquad (2.45)$$

Now, from Equation 2.29,

$$\Delta G_T^0 = \Delta H_T^0 - T \Delta S_T^0 \qquad (2.46)$$

and, using Equations 2.26 and 2.27, we can write:

$$\Delta H_T^0 = \Delta H_{298}^0 + \int_{298}^{T} \Delta C_P \, dT \qquad (2.47)$$

$$\Delta S_T^0 = \Delta S_{298}^0 + \int_{298}^{T} \frac{\Delta C_P}{T} \, dT \qquad (2.48)$$

where ΔC_P is the difference in specific heats between the products and reactants. On substituting in Equation 2.45 we obtain:

$$\ln K_p = -\frac{\Delta H_{298}^0}{RT} - \frac{1}{RT}\int_{298}^{T} \Delta C_P \, dT + \frac{\Delta S_{298}^0}{R}$$

$$+ \frac{1}{R}\int_{298}^{T} \frac{\Delta C_P}{T} \, dT \qquad (2.49)$$

When this is differentiated with respect to temperature, the derivatives of the two integrals cancel out and we obtain:

$$\frac{d \ln K_p}{dT} = \frac{\Delta H_{298}^0}{RT^2} + \frac{1}{RT^2}\int_{298}^{T} \Delta C_P \, dT = \frac{\Delta H_T^0}{RT^2} \qquad (2.50)$$

Equation 2.50 is known as van't Hoff's equation.

Equilibrium with several reactions
If there are many possible reactions that can take place in a system, we define the *number of independent reactions* R as the minimum number of reaction equations that must be written down so that all other reactions may be obtained by linear combination.

For equilibrium in a system of N species with R independent reactions, we again require that $\Sigma_i \mu_i dn_i = 0$. For each of the R reactions, we may write a conservation of mass equation such as Equation 2.32. In each case we can again eliminate all the dn's except one in terms of the ν's, as before, and obtain R equations such as Equation 2.34. There are consequently R independent equilibrium constants.

2.1.4 The phase rule

For thermal equilibrium between two phases, denoted by α and β, we require that $T_\alpha = T_\beta$. If the phases are not separated by rigid partitions or surfaces of appreciable curvature, then for mechanical equilibrium we require $P_\alpha = P_\beta$. Finally, if any species are able to pass from one phase to the other, for chemical equilibrium we require $\mu_\alpha = \mu_\beta$ for those species. Note that these relationships are all between intensive variables.

For a pure substance, fixing the temperature and pressure determines all other intensive properties of the substance, in particular, it determines the chemical potential. If the chemical potential is plotted in three-dimensions as a function of temperature and pressure, a surface is obtained, which is specific to the particular substance and to the phase of that substance. If a second surface is drawn corresponding to a second phase of the same substance, the two surfaces intersect along a line. This line relates the temperature and pressure in such a way that at all points on the line, the two phases have the same chemical potential. Fixing the temperature, for example, now fixes the pressure which must obtain for equilibrium between the two phases. If a third surface is drawn corresponding to a third phase, this surface cuts the line of intersection of the other two surfaces at a point. Thus there is a unique temperature and a unique pressure at which three phases of a pure substance are in equilibrium. This is called the triple-point.

Let us now consider a system containing C species which do not react among themselves, and P phases. Let us suppose, to begin with, that every species is present in every phase. The state of each phase is determined by the temperature, pressure and $C-1$ composition parameters (e.g. mole fractions), i.e. $C+1$ variables. For P phases, there are $P(C+1)$ variables required to specify the state (but not the

REVIEW OF THE BASIC SCIENCE

size) of the system.

For equilibrium, we require:

$T_\alpha = T_\beta = T_\gamma = \ldots$ ($P - 1$ equalities of temperature)

$P_\alpha = P_\beta = P_\gamma = \ldots$ ($P - 1$ equalities of pressure)

$\mu_{1\alpha} = \mu_{1\beta} = \mu_{1\gamma} = \ldots$ ($P - 1$ equalities for species 1)

$\vdots \quad \vdots \quad \vdots \qquad\qquad \vdots$

$\mu_{C\alpha} = \mu_{C\beta} = \mu_{C\gamma} = \ldots$ ($P - 1$ equalities for species C)

The number of equations is $(C + 2)(P - 1)$. The equalities between the chemical potentials may be expressed as relationships between the composition parameters. Since the $(C + 2)(P - 1)$ equations are independent, the number of variables must be at least as great as the number of equations. Consequently:

$$C + 2 - P \geqslant 0$$

The amount by which $C + 2 - P$ differs from zero is called the *variance* or the *number of degrees of freedom* of the system, and is denoted by F. For a system of non-reacting components,

$$F \equiv C + 2 - P \tag{2.51}$$

and is the number of variables which may be freely chosen, and *must* be chosen, for the state of the system to be determinate. Equation 2.51 is known as the *phase rule*. If one species is not present in one phase, there is one less equation; correspondingly, there is one less variable required to describe the state of the system. The same phase rule is therefore obtained. By repetition of this argument it can be shown that the same phase rule is obtained however many species are absent from any phases, provided, of course, that each phase contains at least one species and each species is present in at least one phase.

We will demonstrate that an equivalent phase rule is obtained in a system containing reactive components. Let there be N species, P phases, and R independent reactions. The number of variables in the system is, as before, $P(N + 1)$. In addition to the $(N + 2)(P - 1)$ equations relating these variables at equilibrium we must add R equations for reaction equilibrium. Consequently the variance of the system is:

$$F = N - R + 2 - P \tag{2.52}$$

If we now define the number of components C by

$$C \equiv N - R \tag{2.53}$$

we once more obtain the phase rule, Equation 2.51., as for a non-reacting system. It can be shown [4] that this definition of the number of components may be interpreted in practical terms as the number of substances that must be available in the laboratory to form all the species in the system in the required proportions.

A word must be said about non-stoichiometric compounds. In a system consisting of a binary solid AB_x and the vapours of its components, there are two degrees of freedom. At a given temperature, the ratio of the two components in the vapour may be altered over a range which is limited only by the appearance of a new condensed phase (e.g. liquid A saturated with AB_x), and which may be very wide. The chemical potentials of the vapour components may therefore be altered, and the chemical potentials of the components in the solid must be able to alter in the same way, by an alteration in the composition of the solid. For many compounds, the alteration in composition may be extremely small, for example, at 1100 K, the composition of gallium arsenide may be altered by about 10 p.p.m. either side of the stoichiometric composition [12]. For other compounds, the range of composition may be extensive (see Section 2.3.5.).

Evaporation of a single component

For a system of a single component solid or liquid in equilibrium with its vapour, there is only one degree of freedom, so that at a given temperature there is a unique vapour pressure in equilibrium with the solid or liquid. The variation of equilibrium vapour pressure with temperature is described by the Clausius–Clapeyron equation.

$$\frac{dP}{dT} = \frac{\Delta H}{T \Delta v} \tag{2.54}$$

where ΔH is the molar enthalpy change, and Δv the molar volume change, for the evaporation process. It is frequently a good approximation to ignore the volume of the condensed phase in comparison with that of the vapour.

Dissociative sublimation

The Clausius-Clapeyron equation can be applied only to systems of a single component, whether that component be a single element, or a compound which is essentially undissociated in the vapour. Many compounds, however, become significantly or almost completely dissociated in the vapour phase at quite moderate temperature, and here the situation is more complicated.

We will discuss the sublimation of a binary solid according to the reaction:

$$AB \text{ (solid)} \rightarrow A \text{ (gas)} + \tfrac{1}{2} B_2 \text{ (gas)} \tag{2.55}$$

for which the equilibrium equation is:

$$K_p = p_A p_{B_2}^{1/2} \tag{2.56}$$

Such a scheme is found in many IIB-VIB compounds [13], for example. We will make use of the phase rule, and consider two cases:

Case (a). The solid is pure AB, A and B being insoluble in it. There are two degrees of freedom, and so we could arbitrarily fix any two of the parameters T, P, p_A, p_{B_2}, etc. It is obviously inappropriate to speak of a 'vapour pressure' of AB. On the other hand, let it be supposed that the system was prepared *from pure AB only*. Then there is a stoichiometric relationship between the partial pressures of A and B_2:

$$p_A = 2 p_{B_2}$$

and consequently there is only one degree of freedom. It may readily be shown that the total pressure is a minimum in these circumstances, and indeed, this is quite general, that the vapour pressure over a compound is least when the vapour contains the chemical elements in the same stoichiometric proportions as in the solid. With this restriction, we may speak of a '(minimum) vapour pressure' of AB.

Case (b). Vapour phase plus solid AB in which both A and B are soluble. The phase rule again gives two degrees of freedom. The partial pressures p_A and p_{B_2} must satisfy the equilibrium constant for the dissociation reaction, but two parameters, e.g. temperature and total pressure, may be fixed arbitrarily. The amounts of A and B dissolved in the solid AB will depend on the solubilities of A and of B, and on the partial pressures p_A and p_{B_2}.

If we now consider the system to have been prepared from pure AB only, we find in this case that the stoichiometric restriction on the partial pressures no longer applies, since the amounts of A and B dissolved in the solid will not, in general, be equal. There are therefore still two degrees of freedom. The vapour pressure at any temperature is determined by the solubilities of A and B in the solid, and by the ratio of the total amounts of solid and vapour, i.e. by the volumes of the solid and the container. The minimum vapour pressure is obtained when the partial pressures p_A and p_{B_2} satisfy the stoichiometric relationship. In practice, this situation may be approached by subliming a small quantity of solid in a large free volume. Because of the different solubilities of A and B in the solid, however, stoichiometry of the vapour implies non-stoichiometry in the solid, and the converse. The vapour pressure over the stoichiometric solid will be different from the minimum vapour pressure, and may be very much greater if the solubilities are very different. In practice, this situation may be produced by subliming a large volume of stoichiometric solid into a small free volume, although there is a possiblity of precipitating a second condensed phase if the partial pressure of one of the compounds exceeds the equilibrium vapour pressure of that compound.

2.2 Description of crystals

2.2.1. Elements of crystallography

The earliest studies of crystals were no doubt prompted by the fascinating and often beautiful shapes of natural crystals. The first observations and measurements were confined to the external shapes, yet the first law of crystallography, the Law of Constancy of Angle (Nicolaus Steno, 1669) suggests a basic microstructure determining the external shape. The law may be stated: 'In all crystals of the same substance, the angles between corresponding faces have a constant value'. The sizes and shapes of crystal faces may be influenced by fortuitous external factors.

In this brief outline of elementary crystallography, we aim to do no more than provide the crystal grower with the language to discuss his problems and achievements, and to give some indication of how the physical properties of crystals are related to their symmetry. Much of what we have to say will already be familiar to many, though they may be glad of a resumé and reminder. For complete discussions of the subject we recommend the books by Phillips [14] de Jong [15] and Nye [16].

Crystal Symmetry

We begin with a discussion of crystal symmetry, that is, the symmetry of an ideal specimen or model, the crystal *point group.* The symmetry is discovered by examining actual speciments, or by X-ray methods, optical refraction and birefringence, pyroelectricity and other physical properties. The possible symmetry elements are:

A Centre. Each plane of the crystal is related to an equivalent plane on the opposite side of the centre. Similarly, physical properties

REVIEW OF THE BASIC SCIENCE

associated with any direction in the crystal are identical with those associated with the exactly opposite direction.

Mirror planes or planes of symmetry, symbol m. Such a plane divides the crystal into two parts which are mirror images.

Rotation Axes. A crystal possesses an n-fold rotation axis of symmetry if rotation of the crystal by $360/n$ degrees brings the crystal into a congruent position. The possible values for n are 1,2,3,4, and 6, corresponding to the unitary, diad, triad, tetrad, and hexad axes, symbols, 1,2,3,4, and 6.

Rotation-Inversion Axes. An n-fold rotation-inversion axis rotates each element of the crystal $360/n$ degrees and inverts it with respect to the centre. The values for n are the same as for rotation axes; the symbols for rotation-inversion axes are $\bar{1}, \bar{2}, \bar{3}, \bar{4}, \bar{6}$. The $\bar{1}$ axis is equivalent to a centre of symmetry, the $\bar{2}$ axis to a mirror plane, and the $\bar{6}$ axis to a triad axis plus a centre of symmetry.

The Seven Crystal Systems. On the basis of these symmetry elements, crystals are divided into seven crystal systems in the way shown in Table 2.1.

Table 2.1

System	Minimum symmetry
Triclinic	None
Monoclinic	One diad or inverse diad
Orthorhombic	Three orthogonal diads, some or all of which may be inverse.
Trigonal	One triad or inverse triad
Tetragonal	One tetrad or inverse tetrad
Hexagonal	One hexad or inverse hexad.
Cubic	Four triads or inverse triads, which have to be at 109.5° to each other.

Within each crystal system, there are several possible arrangements of symmetry elements, i.e. several point groups.

The Crystal Lattice. The diffraction of X-rays by crystals indicates that crystals are built up of great numbers of basic units arranged in a very regular way. The microstructure of a crystal may be described in terms of an extensive array or lattice of imaginary *lattice points*, each with identical surroundings. The *unit cell* of the lattice is a small unit which, when repeated indefinitely in all directions, reproduces the lattice (see Fig. 2.1).

The choice of such a cell is to some extent arbitrary, though it is usually convenient to choose a primitive cell, i.e. one which contains

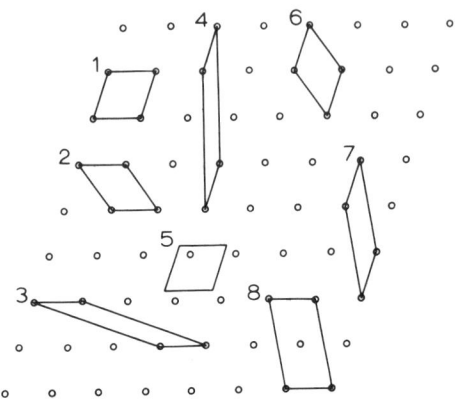

Fig. 2.1 Various possible choices of unit cell.

only one lattice point (not cell 8, for example, in Fig. 2.1), which has angles near 90°, and whose sides are as nearly as possible equal in length. Occasionally there are good reasons for choosing a non-primitive unit cell, for example, the face centred cubic lattice has a conventional unit cell as depicted in Fig. 2.2a, with a rhombohedral primitive unit cell also outlined and shown separately in Fig. 2.2b. The cubic unit cell is normally chosen for this lattice because it exhibits the full symmetry that the lattice possesses.

The lattice is defined by the repeat distances a, b, and c along three directions (*crystallographic axes*) parallel to the edges of the unit cell, and the three angles α, β and γ between the axes. If the unit cell is non-primitive, it is necessary to state the arrangement of lattice points within the unit cell also. The general arrangement for a primitive cell is shown in Fig. 2.3.

Miller Indices. The faces of crystals are parallel to *lattice planes*, i.e. planes passing through many lattice points. A lattice plane is described by its *Miller indices*, defined as follows: if the lattice plane makes intercepts on the x, y, and z crystallographic axes of a/h, b/k, and c/l units respectively, the Miller indices of that plane are (h,k,l). The indices are divided by any common factor they may possess. The choice of origin is arbitrary, and the Miller indices determine only the direction of the plane (i.e. the direction of its normal) and not its position in space. Fig. 2.4 illustrates the Miller indices. If the intercept on any axis is negative, the corresponding index is written with a bar over it, e.g.: $(1\bar{2}0)$.

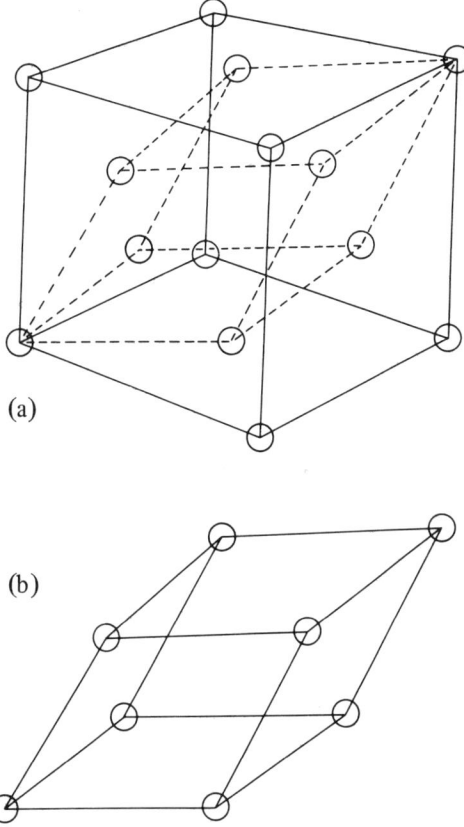

Fig. 2.2 Conventional unit cell (a) and primitive unit cell (b) of the face-centred cubic lattice.

Note that a plane parallel to one axis makes an infinite intercept on that axis: the corresponding index is zero. Thus the (100) plane is parallel to the y and z axes. It is not normal to the x-axis unless $\beta = \gamma = 90°$. Similar remarks apply to the other 'axial planes' (010) and (001).

Law of Rational Indices. It is observed that the faces occurring on real crystals have indices which are almost always small whole numbers, generally 0,1,2, or 3 and very rarely greater than 5. This observation led Haüy to posulate the Law of Rational Indices (1783), which may be stated in three stages:

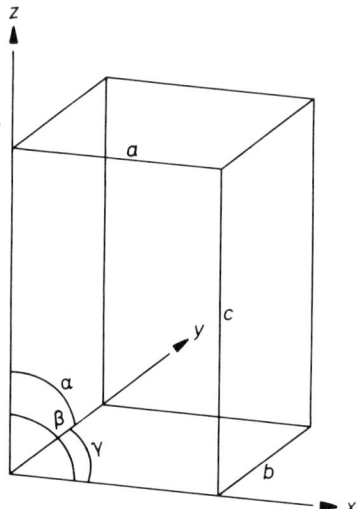

Fig. 2.3 General primitive unit cell.

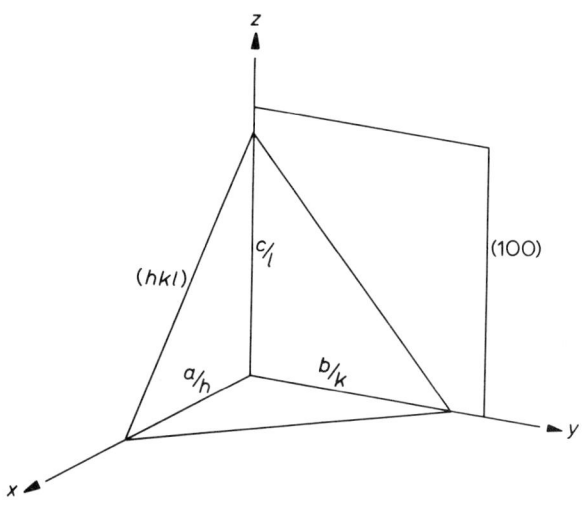

Fig. 2.4 Miller indices.

(1) Choose three faces (no two of them parrallel, nor necessarily perpendicular) on the crystal, and call these faces the axial faces, i.e. (100), (010), (001). The lines of intersection of these faces are then the crystallographic axes.
(2) Choose a fourth face, not parallel to any of the axial faces. Call this the (111) face. This defines the ratios $a:b:c$ (though not the absolute magnitudes), which is why the (111) face is sometimes called the parametral face.
(3) Index all other faces on the crystal. The Law of Rational Indices states that, however the axial faces and parametral face were chosen, all faces have rational indices. Here again we find strong observational evidence for the existence of an underlying microstructure or crystal lattice that determines the external shape of the crystal.

The observation that only planes with low indices are commonly found implies that those faces parallel to lattice planes with the highest density of lattice points per unit area are most stable. We can see qualitatively why this should be so. Consider a simple crystal with a basis of one atom per lattice point. In a plane containing a high density of atoms, most interatomic bonds lie within the plane, so that the number of bonds between one plane and the next is relatively low. It requires little energy to separate such planes, i.e. their surface energy is small, and they are therefore more stable than planes containing a low density of atoms. Crystals are usually bound by low-index, low surface energy faces. Crystals which cleave readily (mica, calcite, salt) do so along low-index planes for similar reasons.

Indices of a direction. We consider a line drawn through the origin in the direction in question. The indices of this direction are then the coordinates of an arbitrary point on it, expressed in units of the lattice repeat distances, and reduced to the smallest whole numbers. If the co-ordinates can be expressed as Ua, Vb, Wc, then the direction is the $[UVW]$ direction. Direction indices are written in square brackets to distinguish them from plane or face indices, which are written in round brackets. Note that in general, the (hkl) plane is not normal to the $[hkl]$ direction. Only in the cubic system is *every* plane normal to the direction having the same indices.

The Lattice, the basis and the structure. The lattice, as we have seen, is an imaginary geometrical construction. To build up a real crystal, we must arrange an atom or a group of atoms (the *basis*) round each lattice point to define the *crystal structure.* Thus the elements copper, silver, gold, and others have one atom associated with each lattice point in a face-centred cubic lattice. The hexagonal metals magnesium, zinc, cadmium etc. have two atoms to each lattice point.

Many binary compounds, e.g. the alkali halides have one atom of each species associated with each lattice point, and so on. A brief account of some structures is given in Section 2.1.2. Here we wish to emphasise that (a) it is not necessary to choose lattice points lying at the centre of atoms; (b) if the lattice points are chosen at the centre of some atoms, as is often convenient, it does not necessarily follow that *all* atoms are centred on lattice points, or even that all atoms of the same chemical element are centred on lattice points. For example, the atoms in the diamond structure and its derivatives occupy two distinct sets of sites (see Section 2.2.2).

Some definitions

A *Zone* is a collection of faces which meet (or would meet, if produced) in a set of parallel edges. The direction of the edges is the *zone axis*. According to the Weiss Zone Law, the face (hkl) lies in the zone [UVW] if $hU + kV + lW = 0$.

A *form* is a group of faces related by the symmetry of the crystal. The form containing the face (hkl) has the symbol $\{hkl\}$. A group of directions related by symmetry does not seem to have a name. It has the symbol $\langle UVW \rangle$.

A *unique direction* is one which is not repeated by operation of the symmetry elements, reversal being regarded as repetition here. Crystals possessing such a direction must be non-centro-symmetric, and may theoretically show vectorial pyroelectricity and piezoelectricity.

A *polar direction* is a direction which is not reversed by operation of the symmetry elements, though it may be repeated. Unique directions are therefore also polar, but the converse is not always true. A crystal with non-unique polar directions may show tensorial pyroelectricity.

The Thirty-two crystal classes

The thirty-two different point groups define the *crystal classes*, illustrated in Fig. 2.5. The stereographic projection method used to represent the symmetry elements will be familiar to many readers. A description will not be given here, but can be found in the standard texts, e.g. Phillips.[14] The symbols for the various classes (Hermann-Mauguin notation) are explained in Table 2.2

Note that 'associated with' a particular direction means symmetry axes parallel to that direction, and mirror planes normal to it.

Miller-Bravais axes

In the hexagonal and trigonal systems, a redundant fourth axis, the u-axis, is used, lying in the xy plane at 120° to x and y. The Miller indices then consist of four figures ($hkil$), and it can easily be shown that $i = -(h + k)$. The indices for the faces of a form bear a more obvious relation to one another in four figure indices than in three, as illustrated below for the form $\{11\bar{2}0\}$:

$11\bar{2}0$	110
$\bar{1}2\bar{1}0$	$\bar{1}20$
$\bar{2}110$	$\bar{2}10$
$\bar{1}\bar{1}20$	$\bar{1}\bar{1}0$
$1\bar{2}10$	$1\bar{2}0$
$2\bar{1}\bar{1}0$	$2\bar{1}0$

In four-figure notation, the first three indices consist of two 1's and a 2. In three-figure notation, no such obvious pattern can be seen.

Table 2.2 *The Hermann-Mauguin notation*

System		Meaning
Triclinic	One figure only	A centre ($\bar{1}$) or no symmetry (1)
Monoclinic	One symbol only	Symmetry associated with the diad axis direction: $2, m$ or $2/m$
Orthorhombic	Three symbols	Symmetry associated with the $x, y,$ and z crystallographic axes.
Trigonal	First symbol	Main symmetry axis: the z-direction.
Tetragonal	Second symbol	Symmetry elements associated with the x and y axes
Hexagonal	Third symbol	Symmetry elements associated with directions bisecting the angles between the x and y axes
Cubic	First Symbol	Symmetry elements associated with the $x, y,$ and z axes.
	Second symbol	Always 3, refers to the triad axes which are along the $\langle 111 \rangle$ directions.
	Third Symbol	Symmetry elements associated with the bisectors of the angles between the x, y and z axes.

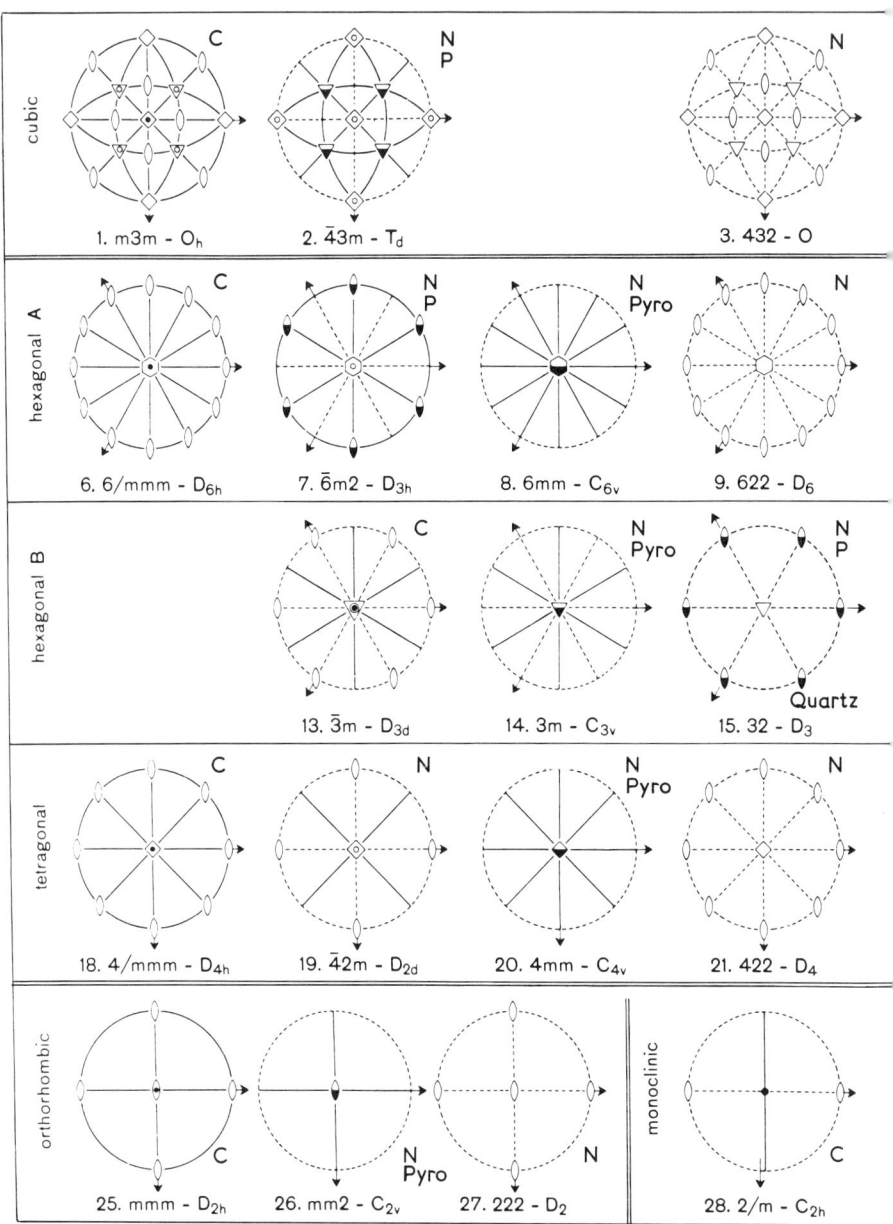

Fig. 2.5 Stereographic projections of the 32 crystal classes.

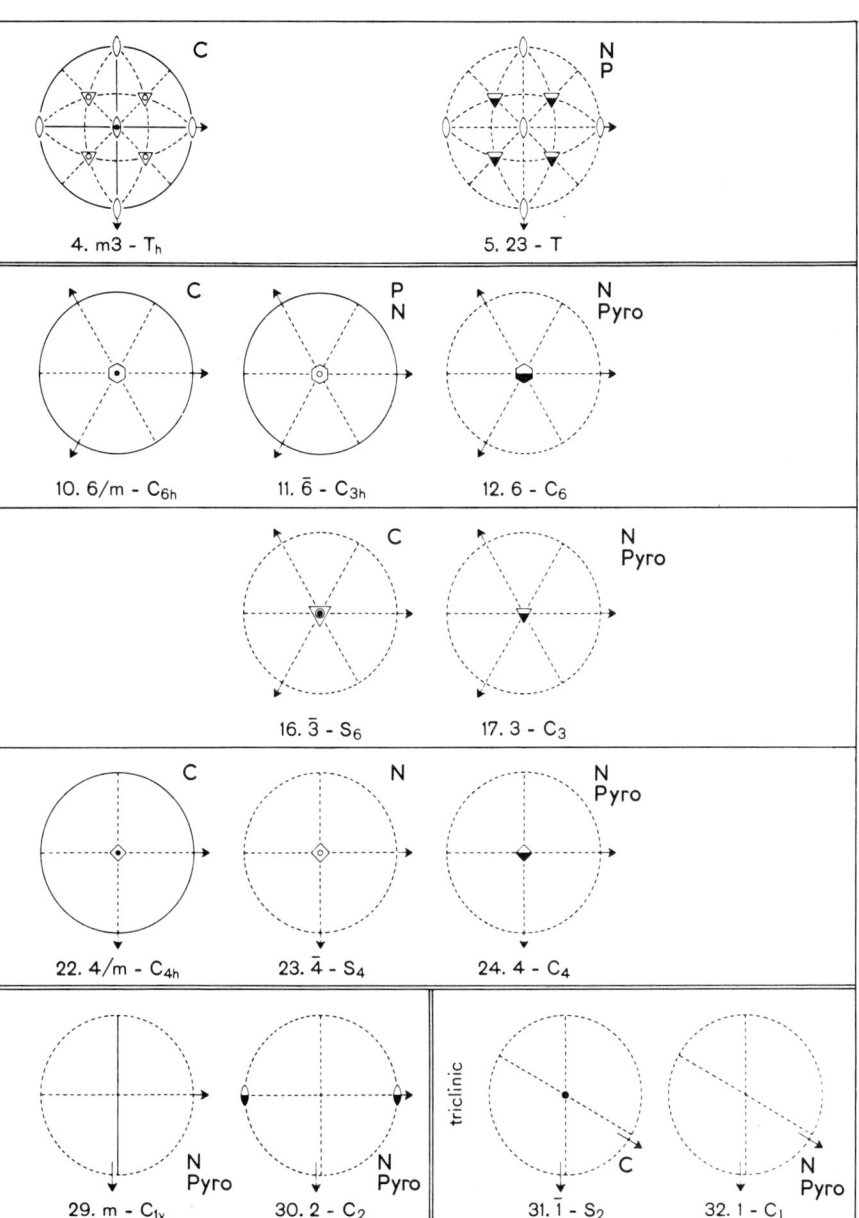

Effect of symmetry on crystal properties

We give only a few general statements here. The subject is well covered in the standard texts.[16, 17, 18]

Firstly, we observe that many physical properties of interest may be represented by second rank tensors (II-tensors). Such properties include those relating a vectorial effect to a vectorial cause (permittivity, permeability, electrical or thermal conductivity, etc), and those relating a II-tensorial effect to a scalar cause (thermal expansion, piezocaloric effect, strain under hydrostatic stress). It can be shown that the tensors appropriate to these properties are symmetrical, and the crystal behaves as if it had a centre of symmetry. For such properties, all crystals behave as if they belonged to one of the eleven centrosymmetric classes (Friedel's classes or Laue groups). See Table 2.3.

Table 2.3 *The Friedel classes*

Friedel class	Classes with identical II-tensor behaviour			
$\bar{1}$	1			
$2/m$	2	m		
mmm	222	$mm2$		
$\bar{3}$	3			
$\bar{3}m$	$3m$	$\underline{32}$		
$6/m$	6	$\bar{6}$		
$6/mmm$	$\bar{6}m2$	$\underline{6mm}$	622	
$4/m$	4	$\bar{4}$		
$4/mmm$	$\bar{4}m2$	$4mm$	422	
$m3$	$\underline{23}$			
$m3m$	$\bar{4}3m$	432		

A second-rank tensor property may be represented by a *quadric* which is a geometrical surface (in general an ellipsoid or hyperboloid) the dimensions of which in any direction are proportional to the magnitude of the physical property in that direction.

To see how the shape of the quadric depends on the symmetry elements of the crystal, we invoke *Neumann's principle,* which may be stated as follows: the symmetry elements of any physical property of a crystal must *include* the symmetry elements of the point group of the crystal. In the cubic system the quadric must possess four triad axes, and can therefore only be a sphere. Cubic crystals are isotropic as far as second-rank tensor properties are concerned. In the *uniaxial* systems, i.e. the trigonal, tetragonal and hexagonal systems, the quadric must possess the major symmetry axis and is therefore a figure of rotation about the major symmetry axis. The quadric is therefore completely specified by two numbers: the dimension in the direction of the major symmetry axis, and the dimension in a direction at right angles. In the biaxial systems – orthorhombic, monoclinic, and triclinic – the quadric is a general ellipsoid or hyperboloid and requires three numbers to specify its dimensions. In the orthorhomic system, the three principal axes of the quadric must lie along the three diad axes of the crystal. In the monoclinic system one of the three principal axes of the quadric must lie along the diad axis, but apart from that, the orientation of the quadric with respect to the crystal is not fixed by any symmetry requirements, and may alter in response to a change in temperature, state of strain, applied electric field etc. In the triclinic system, there is no symmetry restriction on the orientation of the quadric.

Tensors of higher rank than 2, representing such physical properties as the piezoelectric effect, the Pockels effect (third-rank tensors), the elastic moduli, and the piezo-and elasto- optical constants (fourth-rank tensors) must also comply with Neumann's principle. For a detailed treatment of the effect of crystal symmetry on such properties, we refer to the reader to Nye[16] or Mason[18].

We have not discussed the optical properties of crystals – double refraction (birefringence) and rotary polarization (optical activity). These effects are treated in many text books, and no doubt most of our readers will have some knowledge of these effects.

2.2.2. Structural crystallography

In this section, we consider the arrangement of atoms or groups of atoms in the crystal, or more particularly, in the unit cell.

Atom co-ordinates

The co-ordinates of an atom in the unit cell are given as fractions of the repeat distances a, b and c along the x, y and z axes. Thus, for example, an atom occupying the geometric centre of the unit cell has co-ordinates (½, ½, ½) in any crystal class.

Co-ordination

The co-ordination of an atom is a description of its immediate environment, that is, a statement of the number of adjacent atoms, their arrangement in space, and their separation from the central atom. Thus in the NaCl structure, the sodium atoms have sixfold, octahedral co-ordination to the chlorine atoms (see fig. 2.8).

Co-ordination polyhedra

The co-ordination polyhedron is the geometrical figure defined by the co-ordinating atoms. For example, in the NaCl structure, the polyhedra are octahedra.

Crystal structures may be divided into groups according to the type of structural unit which can be identified in the crystal. In many inorganic crystals, the largest unit which may conveniently be identified is a single atom or ion. Such structures are referred to as co-ordination structures, and may be thought of as being built up of co-ordination polyhedra fitting together to fill space partly or entirely. In some inorganic compound crystals, radical groups, such as CO_3 in calcite, together with ions, form convenient structural units, and crystals of these compounds have *radical* structures. Other common structural units are long chains (e.g. silicates such as pyroxenes, asbestos, polymer crystals) and layers (e.g. graphite, $CaCl_2$, CdI_2, silicates such as mica, clay minerals). Many organic compounds form molecular crystals consisting essentially of individual molecules within which the chemical bonds are relatively strong, but between which there are only weak bonds. Finally, several silicate materials such as feldspar form *framework* structures, built up of $[SiO_4]^{4-}$ tetrahedral groups in which the silicon may be replaced by aluminium to some extent, with the corners of the tetrahedra joined by sharing of oxygen atoms between two tetrahedra. Three-dimensional infinite complexes may be built up in this way.

In the co-ordination structures, the arrangements of the atoms are determined by the relative sizes of the different types of atoms and the charges on them i.e. the degree of ionicity of the bond. The simplest type of co-ordination structure consists of one type of atom only, and we will describe the four most important structures. Elements with metallic, non-directional bonding crystallize in *close-packed* structures, where each atom has twelve nearest neighbours, this being the maximum for spheres of a uniform size. There are two common arrangements: cubic close packing (also known as the face-centred cubic (f.c.c.) structure) and hexagonal close packing (h.c.p.). These structures are illustrated in Fig. 2.6. Each may be considered as being built up of close-packed layers. There are two possible ways in which one close-packed layer may be fitted on top of another. If we denote the first layer as an A-layer, and the second layer as a B-layer, the atoms in the third layer may lie directly above those in the first, and thus be an A-layer, resulting in hexagonal close-packing, or they may sit in the other possible set of positions, forming a C-layer, resulting in cubic close packing. The face-centred cubic unit cell is outlined on Fig. 2.6; in this structure all atoms occupy equivalent positions. The close-packed planes are the {111} planes. In the hexagonal close-packed structure, the atoms in the

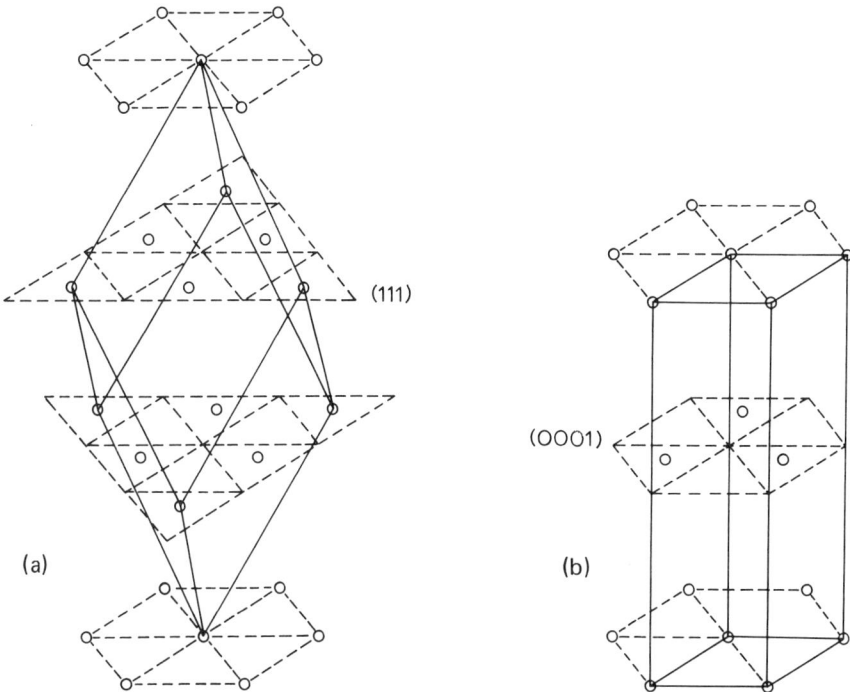

Fig. 2.6 Cubic close-packing (a), and hexagonal close-packing (b).

A-layers have different surrounding from the atoms in the B-layers, so that there are two sets of sites. The close-packed layers are the {0001} planes. Many metals crystallize in one or other of these structures (e.g. Cu, Au, Ag, Ca, Sr, Pt, Al are f.c.c, Mg is the best example of h.c.p) and several more (Zn, Hg) form crystals with a structure very close to hexagonal close-packing, but with a lattice

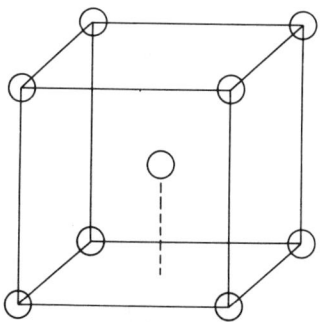

Fig. 2.7 The body-centred cubic structure.

parameter ratio c/a a little different from the value of 1.667 appropriate to true hexagonal close-packing.

The other common structure found among metals is the body-centred cubic structure (Fig. 2.7). All atoms occupy equivalent positions and have eightfold coordination. Examples are the group Ia, Va, and VIa metals, Fe, Eu Ba. The structure is fairly close-packed, with the $\langle 111 \rangle$ directions being close-packed.

The fourth structure of one type of atom that we describe is the diamond structure, shown by silicon, germanium, and grey tin besides diamond. This structure, shown in Fig. 2.10 has a face-centred cubic lattice, with atoms at (0,0,0) and (¼,¼,¼). Each atom has fourfold tetrahedral co-ordination. The packing is not close and the bonds are strongly oriented. This structure, the related sphalerite structure (Fig. 2.10) and a more distant relation, the wurtzite structure (Fig. 2.11) are the structures of a large number of commercially useful semi-conducting elements and compounds.

When there are two types of atoms in a co-ordination structure, the bonding is partly ionic, often largely so. The structure is determined by the conflicting requirements that (1) each ion shall have as many oppositely charged ions around it as possible, (2) each ion shall have as few similarly charged ions around it as possible. The *radius ratio*[19] rule gives a rough guide to the likely structure based on the number of the larger ions (usually the anions) which can be fitted around the smaller ions (usually the cations) without too much overlapping of the electron clouds of the large ions. If we denote the radii of the cation and anion by r_A and r_X respectively, and assume for the moment that $r_A < r_X$, then the possible range of values of r_A/r_X for each co-ordination number is given in Table 2.4.

These ranges of r_A/r_X are worked out using simple geometry and assuming that the ions are hard spheres.[20] The radius ratio rule is a rough guide only, although its predictions may be improved by

Table 2.4

Co-ordination number	Co-ordination polyhedron	r_A/r_X
12	Close packing	1
8	Cube	1 −0.73
6	Octahedron, trigonal prism	0.73−0.41
4	Tetrahedron	0.41−0.22
3	Equilateral triangle	0.22−0.15
2	Straight line	0.15−0.

The NaCl structure (halite)

This structure has a face-centred cubic lattice, with A-atoms at (0,0,0) and X-atoms at (½,½,½). Crystal class $m3m$. The co-ordination for both types of atom is sixfold, octohedral. A third of all AX compounds examined crystallize in this structure[15] including all the alkali halides except (Cs, Rb)(Cl, Br, I). The {100} cube faces are electrically neutral and of low surface energy: crystals with this structure frequently cleave well along these planes. The {111} planes consist of one type of atom only.

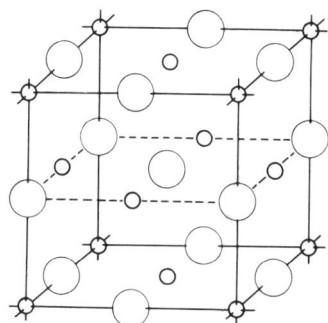

Fig. 2.8 The NaCl structure
Na ∘, Cl ○

CsCl structure

A cubic structure of class $m3m$, consisting of two inter-penetrating primitive lattices, one of A-atoms (0,0,0) and the other of X-atoms (½,½,½). The co-ordination for both types of atoms is 8, and the co-ordination polyhedra are cubes. Several halides of Cs and Tl crystallize in this structure.

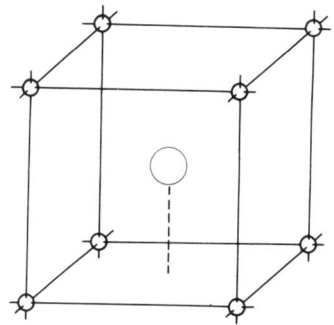

Fig. 2.9 The CsCl structure.
Cs ∘, Cl ○

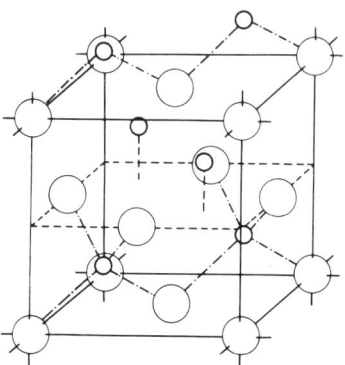

Fig. 2.10 The α – ZnS (sphalerite) structure.
Zn ∘, S ◯

The α-ZnS structure (sphalerite)
A cubic structure of class 43*m*, consisting of two inter-penetrating face-centred lattices, one of A-atoms at (0,0,0), and the other of X-atoms at (¼,¼,¼). This structure is related to the diamond structure: if both A and X atoms are replaced by carbon, diamond results. The co-ordination is fourfold, tetrahedral, for both types of atom, with strongly oriented bonds. Many semiconducting compounds of the type IIIb–Vb crystallize in this structure. The neutral planes are {110}, which are the cleavage planes. The {111} planes consist of one type of atom only, alternate planes being all A-atoms and all X-atoms. We can build up this structure by stacking double layers in the same manner as for face-centred cubic close packing. The double layers consist of a layer of A-atoms with a layer of X-atoms directly in register above it. Each atom is surrounded by six similar atoms within the layer, but these atoms are not nearest-neighbours, being further away than the atoms of the opposite type in the layers above and below. The layers are therefore not close-packed. We now arrange these double layers in the same sequence as the single, close-packed layers in cubic close-packing to obtain the sphalerite structure.

The β-ZnS structure (wurtzite)
A hexagonal structure, class 6*m*, with two inter-penetrating primitive hexagonal lattices. The co-ordination is tetrahedral as in sphalerite. This structure can be built up by stacking double layers, as described for the sphalerite structure, but with the hexagonal close-packing sequence. Many important II-VI semiconductors crystallize in this structure, as do the monovalent copper halides.

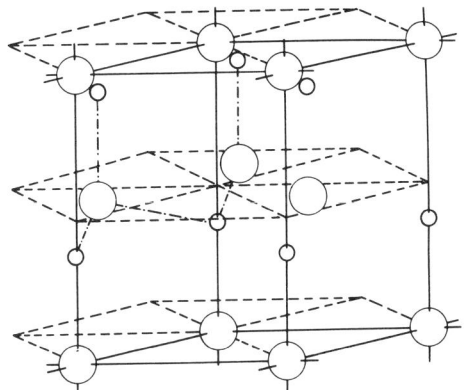

Fig. 2.11 The β-ZnS (wurtzite) structure.
Zn ∘, S O

There are, of course, many other structures of two types of atom, and countless structures of more than two types. We will end our description of the structures here, however, as the few that we have covered include those most frequently encountered in vapour phase crystal growth, and serve adequately to illustrate the concepts that will be introduced later in this chapter and in Chapters 3 and 5. Further information on structural crystallography may be found in the classic texts.[21, 22]

2.2.3 Crystal imperfections

In the preceding section, crystal lattices and crystal structures were described, and it was tacitly assumed that the periodic structure of the crystal extended infinitely without interruption. For one reason or another, real crystals contain defects which disturb the perfect lattice. Some types of defect (point defects) are thermodynamically stable, and are present in all crystals. Other defects, (e.g. dislocations, stacking faults, etc) while being unstable thermodynamically, are formed during the crystallization process when the crystal is in a non-equilibrium state, and cannot readily be removed because of kinetic factors. Very nearly all real crystals contain such defects.

Although generally speaking the defects occupy only a very small fraction of the volume of the crystal, they may have a very great influence on many of the crystal's properties. For example, the presence of certain types of point defects is responsible for the electrical conductivity of crystals of many ionic compounds, and for the wide range of intrinsic conductivity in compound semiconductors. Again, the mechanical properties of metal crystals are often determined by the presence of dislocations.

Point defects

There are two basic single point defects: the vacant lattice site (vacancy) and the interstitial atom. The concentration of each present in a crystal depends on the free energy of formation of the defect.

The enthalpy of formation of a vacancy may be estimated from the number and strength of the bonds to each atom in the solid. There is usually some relaxation of the lattice around a vacancy which modifies its enthalpy of formation. The enthalpy of formation of an interstitial atom contains a strain energy term, which may be very large in a fairly close-packed structure. In metallic and closely packed ionic structures, there are essentially *no* interstitial atoms (except possible interstitial *impurities,* which may be very small). Interstitial atoms have low enthalpies of formation (say 40 kJ mol^{-1}) in the open sphalerite and wurtzite structures discussed in the previous section. In ionic crystals, the metallic cations are normally very much smaller than the non-metal anions, and we expect a much lower enthalpy of formation for interstitial cations than for anions. The 8–4 co-ordinated fluorite structure is unusual in this respect, in being able to accommodate interstitial anions more readily than interstitial cations.[23]

The configurational entropy of the point defects may be calculated assuming a random distribution. The molar vibrational entropy will be of about the same magnitude as the lattice vibrational entropy. For small concentrations of point defects, the molar entropy is high, and at any temperature above absolute zero, there is a certain concentration of vacancies and interstitials that is in equilibrium with the crystal lattice.

In ionic crystals, the requirement of charge neutrality implies that vacancies on one or other sub-lattice, or interstitial ions, cannot exist without some other defects which compensate for the charge of the defect. There are two defect pairs which can occur so as to leave the ratio of atoms of different types unchanged. The Frenkel defect consists of a vacancy and an interstitial atom or ion, and is formed by an ion leaving its normal site and taking up an interstitial position. The Schottky defect consists of a vacant anion site and a vacant cation site. The Schottky defect affects both sub-lattices equally. Frenkel defects usually occur predominantly on one or other sub-lattice since the enthalpy of formation of interstitial atoms is usually very different on the different sub-lattices. Because ionic size is important in determining the enthalpy of formation of an interstitial ion, Frenkel defects are usually confined to the sub-lattice of the smaller ion, which is often the cation.

A vacancy or interstitial ion which is not one half of a Schottky defect or a Frenkel defect can have its charge compensated for by a change in oxidation state of an equivalent number of ions on one sub-lattice. Thus ferrous oxide ('FeO') contains several percent of vacant sites on the metal sub-lattice,[24] which are compensated for by twice the number of trivalent iron ions. The composition of the compound is no longer exactly represented by the simple chemical formula, i.e. the compound is non-stoichiometric. The important topic of non-stoichiometric compounds is discussed in Section 2.3.5.

In crystals of intermetallic compounds and covalent compounds, atoms of one type may change places with atoms of the other type on a certain fraction of the lattice sites ('place exchange'). The enthalpy of formation of this substitutional defect is low if the atoms have similar covalent radii and electronegatives. Some of the III–V covalent compound semiconductors behave in this way[23].

Line-defects – dislocations[25]

There are two elementary types of dislocation: the pure edge dislocation and the pure screw dislocation. The two types are illustrated in Fig. 2.12.

The pure edge dislocation may be thought of as the edge of an extra half-plane of atoms inserted in the crystal. The screw dislocation results from shearing the lattice.

Along the centre of both types of dislocation is a highly-strained region called the core. The enthalpy of formation of a dislocation is largely associated with the strain energy in the core. If the bonds in the crystal have appreciable ionic character, the dislocation core will

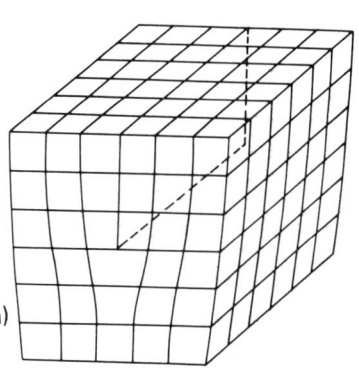
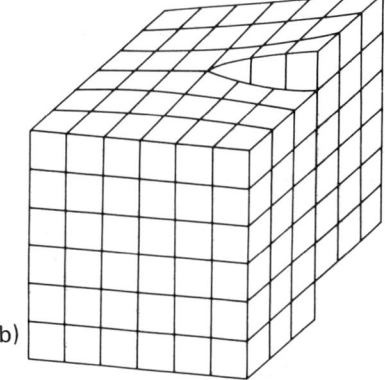

Fig. 2.12 Edge dislocation (a) and screw dislocation (b).

also be a region where the periodic charge distribution of the lattice is disturbed, and this will give an electrostatic contribution to the enthalpy of formation of the dislocation as well.

It can be shown that a dislocation line cannot end inside a crystal.[25] It must reach either the surface or a grain boundary, or form a closed loop, or meet other dislocation lines at a dislocation *node*. A dislocation line intersecting the surface of a crystal produces a disordered region on the surface, which may be a favourable site for nucleation. A screw dislocation emerging at a surface produces a step. This step is not removed when more atoms are added to it, and can therefore assist the crystal to grow by avoiding the necessity for two-dimensional nucleation on the crystal surface[26] (see Chapter 3). This is the famous mechanism of Frank.

A dislocation need not be pure edge or pure screw, but may often be of an intermediate type, and may change its type along its length. For a fuller discussion of dislocations see Cottrell.[25]

The entropy of a dislocation per unit length (i.e. per inter-atomic spacing) is small compared with that of a point defect. This is because the number of possible configurations for the two-dimensional defect of, say 10^6 inter-atomic spacings in length, is not as great as the number of configurations of 10^6 point defects. The enthalpy of formation per inter-atomic spacing for the dislocation may be as large as the enthalpy of formation of an interstitial atom, however. The free energy of formation of a dislocation is thus generally large and positive, typically 1 eV per atomic spacing, or 10^{-4} erg cm^{-1}. Nevertheless, nearly all crystals contain dislocations. In cold-worked metals[27] the dislocation density may be as high as 10^{12} cm^{-2}. At the other end of the scale, good single-crystals of silicon may contain only of the order of 10 dislocations per square centimeter. Some carefully prepared crystals are completely free of dislocations.

Dislocations are produced in a crystal in various ways. Mechanical strain can be relieved by the movement of dislocations which may start from a free surface or a grain boundary. Mechanical strain may be caused by lattice mismatch between an epitaxial deposit and a substrate, or by rapid cooling of a large single crystal, or by sawing, slicing, lapping and polishing operations. Dislocations also arise during the growth of a crystal. For example, if an epitaxial layer is deposited on a substrate, it may, during the initial stages of nucleation, take the form of many unconnected islands of deposit. Sometimes, each of these islands is exactly oriented with respect to all the others.[28] When the orientation is not exact, a small-angle grain boundary is formed between the misoriented islands as they

REVIEW OF THE BASIC SCIENCE

coalesce. Such a grain boundary is, in effect, an array of dislocations.[25]

Once dislocations have been produced in a crystal, it may be impossible for them to move out. In covalently bonded crystals, dislocations cannot move easily, because the strongly directional bonding makes the activation energy for movement of the dislocation very high. Even when the dislocations are mobile, as in many metals, they can be 'pinned' in several ways, e.g., by precipitates, or by intersecting other dislocations. Perhaps the most important 'pinning' mechanism, so far as the electronic properties of many crystals are concerned, is 'decoration' of a dislocation. The strain energy around a dislocation core can be relieved if impurity atoms accummulate along the dislocation line. Such segregation of impurities may impair the electronic properties of a crystal.

Related to segregation of impurities on dislocation lines is the rapid diffusion of species along the cores of the dislocation. Because of local disturbance in the atomic ordering of the crystal, the activation energy for diffusion of impurities (see Section 4.1) may be very much lower along the dislocations than in the bulk of the crystal, so that material diffuses faster along the dislocation cores at moderate temperature. If dopant species diffuse during the epitaxial deposition of $p-n$ junctions for example, the distribution of the dopant may be made very irregular by this phenomenon.

Planar Defects

We include under this heading:

(a) Stacking faults
(b) Grain boundaries
(c) Twinning and twin planes.

Stacking faults may best be illustrated by reference to the face centred cubic structure (Section 2.2.2), in which the perfect lattice is built up by stacking {111} planes in a regular order *ABCABCABC* . . . A stacking fault is produced if the sequence is disturbed: either side of the fault the sequence is correct, e.g.: *ABCA CABC*, or *ABCA CBCABC*. In the first case, a *B*-plane is missing; in the second, an extra *C*-plane has been inserted. Small areas of stacking fault may be produced in the body of a crystal by, for example, segregation of vacancies or interstitial atoms into a planar cluster.[29] During the growth of a crystal, a misoriented island may form on the surface and be incorporated and covered with further material before it has time to correct its orientation.

A grain boundary separates two regions of a single piece of material which differ appreciably in orientation. Across the grain boundary there is a more or less abrupt change in orientation over a distance of a few atomic spacings.[30,31,32] If the amount of misorientation is small, the grain boundary is composed of an array of dislocations (Fig. 2.13). For example, a misorientation of 9.4° can be taken up by dislocations spaced six layers apart. As the misorientation increases the dislocations have to be closer and closer together, till eventually the amount of disorder at the grain boundary is so great that the idea of dislocations cannot be applied usefully. Mott[33] has proposed a model for grain boundaries in which islands of atoms at the grain boundary fit the lattices on either side fairly well, and are separated by regions where the fit is poor.

If the lattices on either side of the grain boundary are in a special orientation to one another, the crystal is said to be twinned, and the boundary is a twin boundary. The special orientation relationship may be reflection in some plane (but not a symmetry plane of the lattice) or rotation by 180° about an axis (but not a symmetry axis of even degree). Twinned crystals are frequently observed in nature in many substances. Two criteria may be used to decide whether a given structure may be expected to twin. Firstly, if the atoms at the twin boundary have the correct nearest neighbours, even though second nearest neighbours and more distant atoms are misplaced by twinning, the energy increase due to twinning is small. Secondly, if the crystal has an axis or plane of pseudo-symmetry (e.g. potassium

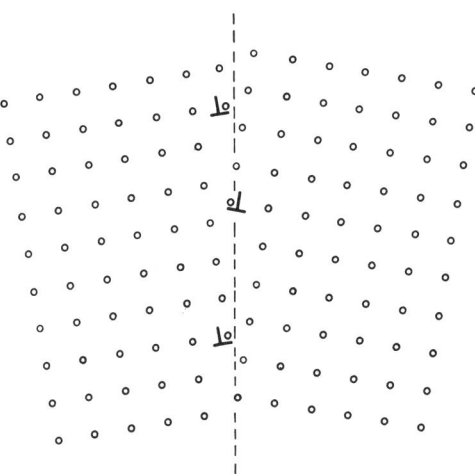

Fig. 2.13   Dislocation array at a small-angle grain boundary.
⊥ indicates a dislocation line.

sulphate, class *mmm*, has pretensions to hexagonal symmetry), the atomic displacements at the twin boundary are small, and the energy of the crystal is increased by only a small amount by twinning.

The free energy of two-dimensional defects contains a large, positive enthalpy and a negligible entropy term. Two-dimensional defects are thermodynamically unstable, and whenever possible, a crystal will try to rid itself of them. Usually the kinetics of the removal process are such that complete removal is not achieved. Stacking faults may be removed by migration of vacancies or interstitials.[34] Grain boundaries may be very mobile at high temperature, and recrystallization (i.e. the formation of a few large crystallites from a great number of small crystallites) of polycrystalline material can be rapid initially. At a certain stage, the grain boundaries have arranged themselves into a meta-stable network, and further annealing produces little change. Twins can normally be removed only by displacing all the atoms of one twin. Such rearrangements of large numbers of atoms can take place under mechanical stress, as when a crystal of tin is bent.

The energies of two-dimensional defects span a wide range. Cottrell[25] has calculated that grain boundary energies are about one third of the surface energy, e.g. 600 erg cm^{-2} in copper, as compared with 1650 erg cm^{-2} for the surface energy.[34] Stacking fault energies are considerably lower[36] e.g. 100–300 erg cm^{-2} in many metals, and may be below 30 erg cm^{-2}, as in gold, α-brass, and some other alloys. Twin boundary energies are generally low if the boundary is coherent, i.e. if it coincides with the twin plane. Values of 25 erg cm^{-2} for copper and 187 erg cm^{-2} for iron have been reported.[35] If the twin boundary is incoherent, its energy corresponds roughly to the grain boundary energy.

To the crystal grower, these two-dimensional defects are almost always undesirable. The disruption of atomic order at the defect causes scattering of electrons and photons and so impairs the electronic and optical properties of the crystal. The problem of avoiding the production of high-angle grain boundaries (i.e. growing a single crystal) is familiar to crystal growers, and such techniques as seeding and imposing a steep temperature gradient are common, in the attempt to prevent spurious nucleation and the growth of more than one crystal. Low angle grain boundaries and mechanical twinning may be caused by the stresses resulting from thermal mis-match between the crystal and its container. Stacking faults may be produced during growth by nucleating a misoriented island of atoms on the growing surface. These atoms will usually have the correct nearest neighbours but incorrect second-nearest neighbours.

The energy available for re-orienting the island is quite small, and if the crystal is being grown quickly, the stacking fault may be trapped (buried) before it can re-orient.

A word need saying about 'substrate-induced twinning'.[37] If an epitaxial layer of a substance of low symmetry is deposited on a substrate of higher symmetry, there are two or more equivalent orientations for the first nuclei. The epitaxial layer is almost always twinned, unless strenuous precautions are taken, such as imposing a steep temperature gradient across the substrate so that the layer nucleates at one corner or edge and spreads across the substrate.

Two-dimensional defects intersecting the surface of the crystal produce line defects on the surface which can be shown up by etching. These line defects may act as centres for nucleation. Thus the morphology of an epitaxial layer may be determined by the defect structure of the substrate, though in many cases, contamination on the surface may be the dominant factor controlling morphology.

2.3 Some aspects of inorganic chemistry affecting the choice of transport agent

2.3.1 Thermodynamic considerations

In this section we discuss the factors that determine whether a given chemical reaction will be of likely use as a transport reaction. Consider, for instance, the transport of an involatile element according to either of the following schemes:

$$\text{M (solid)} + \frac{n}{2} X_2 \text{ (gas)} \rightarrow MX_n \text{ (gas)} \tag{2.57}$$

$$\text{M (solid)} + n\text{HX (gas)} \rightarrow MX_n \text{ (gas)} + \frac{n}{2} H_2 \text{(gas)} \tag{2.58}$$

M is the element to be transported, and X is a halogen. Such schemes would be appropriate to the transport of an involatile metal with a unique valency, with the volatile halide being the transporting species.

The phase rule (Section 2.1.4) tells us that in the first system we have two degrees of freedom (e.g., temperature and the total amount of halogen present) and in the second system, three degrees of freedom (we could, for instance, add the total quantity of hydrogen to the variables under our control.) That is the situation so long as we have one solid phase and the vapour phase. Suppose, however, that we exceed the vapour pressure of MX_n, so that a third phase, solid or liquid, starts to form. We now lose one degree of freedom. Clearly we have no control over the amount of halogen present in the vapour, as the pressure of MX_n is now fixed at the equilibrium vapour pressure for the temperature we have chosen. Thus the transport kinetics in the vapour phase may be made slower by a reduction in the partial pressure of the transporting species, and hence of the transporting agent also, in order that the equilibrium constant relation remain satisfied. The presence of a second condensed phase on the growing crystal would modify the growth kinetics. In general, therefore, we are concerned to ensure that only one condensed phase can form. We must therefore choose our transporting agent so that all species which will form by reaction with the solid to be transported have adequate volatility.

There are, of course, occasions when it is desirable to have two or more condensed phases present in the crystal growth system. We will mention two. One may want to work with a fixed partial pressure of one species, in which case this species could be present as solid or liquid in a side reservoir, for example, so that the partial pressure of that species is pegged at the equilibrium vapour pressure. Another time when a liquid phase may be introduced on purpose is when crystals are grown by the vapour-liquid-solid (VLS) mechanism.[38] The surface of the crystal is covered, partly or wholly, with a liquid. Material is readily dissolved in the liquid from the vapour phase, as the adsorption kinetics are frequently very much more rapid on the disordered surface of a liquid than on a solid. This material eventually saturates the liquid (material is also dissolved from the solid crystal in the early stages), at which point deposition takes place on the solid surface. Probably the best-known example of this process is the growth of silicon whiskers via a film of gold.[38] Gold is sparingly soluble in solid silicon, while silicon dissolves in liquid gold to an appreciable extent. These are obvious requirements for the process. A similar process is thought to occur in the vapour growth of thin films of gallium arsenide and similar compounds. Droplets of liquid gallium may be deposited on the surface of the film and cause rapid, localized growth.[39] In this case the VLS mechanism produces unwanted effects, which illustrates the need for control over the number of phases present in the system.

There are other thermodynamic conditions that we must fulfill to obtain efficient transport. For instance, the equilibrium constant for the reaction must not be extreme. The equilibrium constant relations for our two examples are:

$$K_p = \frac{p_{MX_n}}{p_{X_2}^{n/2}} \tag{2.59}$$

and

$$K_p = \frac{p_{MX_n} p_{H_2}^{n/2}}{p_{HX}^n}$$

In the first case, we expect most efficient transport when $p_{MX_n} \simeq p_{X_2}$, so that if $n = 2$, we require $K_p = 1$. When $n \neq 2$, the optimum value of K_p depends on the total pressure, but clearly we are unlikely to obtain efficient transport if K_p is extremely large, when the vapour consists almost entirely of MX_n, or extremely small, when X_2 predominates. Under most normal circumstances, we require K_p to be not too far from unity,[2,40] but observe that this 'rule of thumb' is not inviolable. If n is large (e.g. hexa-halides of uranium) or if some polyatomic transporting agent is used (e.g. SF_6, S_8, $AsCl_3$, PCl_5 etc), and if the total pressure is kept very low, we require K_p to depart greatly from unity. If $p_{MX_n} = p_{X_2} = 10^{-6}$ atm, and $n = 6$, we have $K_p = 10^{12}$. The requirement is, of course, that no one component should be dominant everywhere. The value of K_p which satisfies this requirement depends on the total pressure and the stoichiometry of the reaction (see the worked examples in Chapter 4). We may look at this in another way. In the first of our two examples of transport reactions, the volatile halide is decomposed at the hot end of the transport system to yield the metal plus halogen. For such a system to provide transport at all, we require the halide to have limited stability at the operating temperature. In the second example, we require the halide to be reduced by hydrogen. We will discuss the stability and volatility of species in the following sections.

Of course, in several applications, the efficiency of the transport process may not be an important factor; for depositing very thin layers of material on substrates, too fast a rate of deposition may in fact be an embarrassment. In such circumstances, the choice of transporting agent may be governed by such factors as its solubility in the growing crystal. The stoichiometry of a growing compound is controlled by the composition of the vapour over it. The transporting agent may be chosen so as to obtain the required vapour composition.

Besides generally requiring that the equilibrium constant be not extreme, we also need an appreciable variation of K_p with temperature, i.e. ΔH for the reaction must not be too small

(Equation 2.50). The *sign* of ΔH indicates the direction of transport; the *magnitude* of ΔH gives an idea of the maximum rate of transport that can be obtained with a given temperature difference.[41] Actual rates may fall short of the maximum by many orders of magnitude, because of excess of one component in the vapour, inert gas, residual gas, volatile impurities etc. The influence of these factors on transport is examined in detail in Chapter 4.

The general 'rule' that K_p be near unity means that the free energy change for the reaction must be small, i.e.:

$$\Delta G = \Delta H - T\Delta S \simeq 0 \tag{2.61}$$

We have already seen that ΔH must be fairly large, a few tens of kilocalories, at least (see example on carbon+sulphur in section 4.4.3.) The temperature range that can be used in chemical vapour transport is limited by the slow rate of reaction kinetics at low temperatures, and by technological problems, particularly with container materials, at very high temperatures. We see that the entropy change for the reaction must also lie within workable limits. The entropy change is due largely to a change in translational entropy accompanying a change in the number of gas phase molecules, though the contribution from rotational entropy may be large also if the gas phase species on one side of the reaction are all monoatomic or diatomic, and those on the other side are non-linear triatomic or polyatomic molecules. The entropies of gas molecules[42] are shown in Fig. 2.14, as functions of molecular weight, and we see that, as a rough guide, the translational entropy contributes two-thirds of the total entropy of diatomic molecules, and half of the total for polyatomic molecules. Usually, then, if there is no change in the number of gas phase molecules as the reaction proceeds, we expect the entropy change to be a few entropy units only, unless there is a change from light, diatomic molecules to heavy, polyatomic molecules. Then if the enthalpy change for the reaction is large enough to give a sensible change in K_p with temperature — say 120 kJ mol^{-1} — we find that in order to have $K_p \simeq 1$, we require a temperature of several thousand degrees. Reactions with no change in the number of gas molecules are therefore not usually efficient transport reactions. The change in the number of gas molecules also determines the magnitude and direction of the Stefan velocity (see Chapter 4) which may make a significant contribution to the transport rate.

We have already mentioned the possibility of the transporting agent being dissolved or incorporated in the growing crystal. Most of the materials prepared by chemical vapour transport are required for

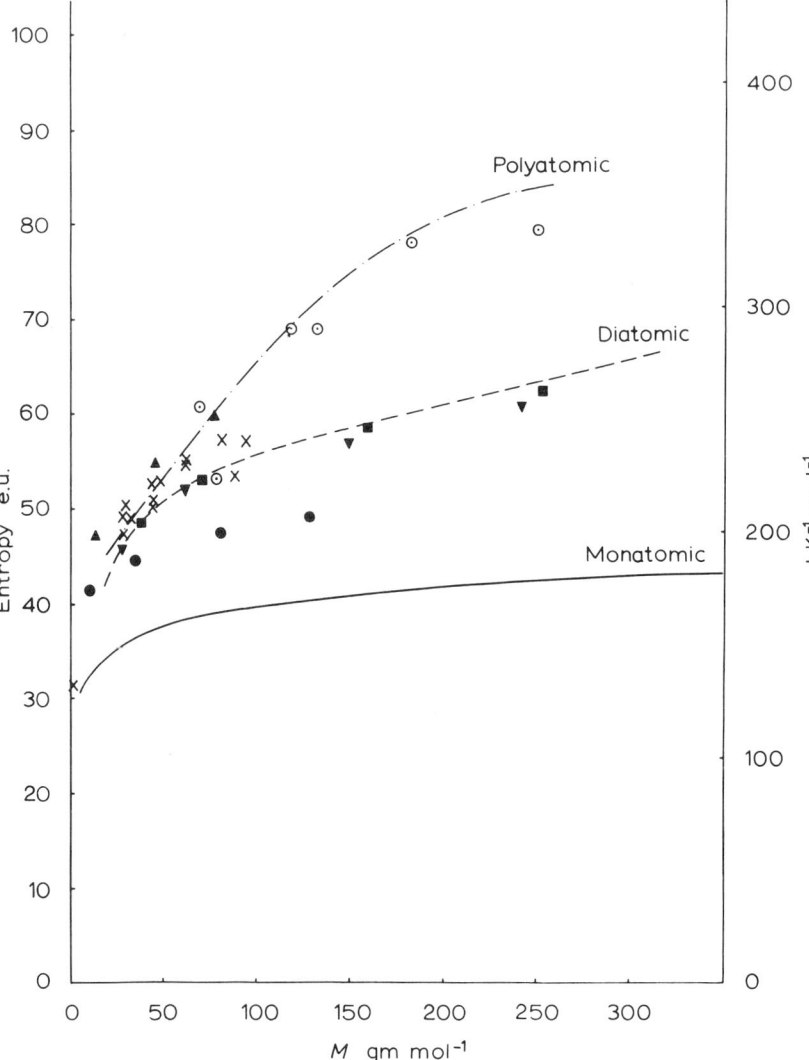

Fig. 2.14 Entropy of gas molecules versus molecular weight.
- ■ Halogens
- ● Hydrogen halides
- ▲ Alkali metals
- ▲ Pnictides
- X Other molecules.

manufacturing sophisticated electronic devices. These devices exploit properties of the crystals, such as semiconductivity, which are extremely sensitive to the presence of defects and impurities or dopants. The transporting agent may dissolve in the crystal and may act as a dopant; this could possibly be useful, or it might be an embarrassment, and in either case we should want it to be under our control. We also have to consider the interaction between the chemical vapour transport system and any impurities present in the source material. If the impurities are transported relatively slowly (ΔH and ΔS for their transport unfavourable) considerable purification may be achieved. We have a little more to say on this subject in Chapter 5. A further important consideration is the interaction of the chemical transport system with the container or crucible material. Silica, for example, may be transported in H_2/HCl gas mixtures at commonly used temperatures (700–800°C), though the rate of transport can be made very slow.[43,44]

To summarize the factors involved in choosing a transporting agent, we require that the equilibrium constant should be of suitable magnitude at a convenient temperature ($K_p \simeq 1$ is often a good guide), and we require knowledge of the volatility and stability of all the vapour species formed in the reaction, and of their solubility in the growing crystal. Sometimes the relevant thermodynamic data are available, and they should of course be used in the rational design of a chemical vapour transport system. Often, the data are not available or not reliable. We discuss the estimation and reliability of data in Section 2.3.2. Independent measurement of the data, though laborious, may well be the quickest way of selecting a successful system. If that is not feasible, the theoretical background of volatility, stability and solubility may be helpful, and a short treatment is presented in the following sections.

Experimentation in growing crystals is, of course, the final arbiter of the usefulness of the transport system, but is also the least enlightening avenue of investigation, as it rarely reveals the reasons for failure, nor does it point to conditions under which success would be achieved. Exploring these conditions in full by *ad hoc* empiricism is uneconomical, though remaining popular.

2.3.2. Volatility

Our interest in the volatility of solids and liquids is twofold. For the purpose of choosing a suitable transporting agent, we need to know what compounds of the various components of the substance to be transported have sizeable vapour pressures in the temperature range selected for growth. Here we may be concerned with the volatility of both solids and liquids, as we do not usually want to exceed the saturated vapour pressure of the particular volatile compound and precipitate another condensed phase (but see the comments in Section 2.3.1 on VLS growth). On the other hand, we will also need to know to what extent the substance to be transported exists as undissociated, uncombined molecules in the vapour phase. For example, many compounds formed of elements from groups IIB and VIB, such as ZnS, CdS, CdTe, etc., may be transported by dissociative sublimation (see Section 4.3 for a worked example) according to the reaction:

$$\text{AB (solid)} \rightarrow \text{A (gas)} + \frac{1}{n} \text{B}_n \text{ (gas)} \tag{2.62}$$

Undoubtedly, the sublimation reaction:

$$\text{AB (solid)} \rightarrow \text{AB (gas)}$$

also takes place, although the equilibrium vapour pressure of AB molecules may be extremely small. However, if the AB molecule can more readily be accommodated on the surface of the crystal and incorporated as part of the solid than can the B_n molecule (e.g. S_2, S_4, etc.) the small partial pressure of AB molecules may provide a significant contribution to the overall growth rate. In discussing this question, we are, of course, concerned with the volatility of the solid. The experimental determination of molecular species in a predominantly dissociated vapour is fraught with pitfalls, and the results are ambiguous.[45–48] Sometimes it is possible to estimate the fraction of molecular species that may be expected.

The equilibrium between a solid and its vapour is determined, like all equilibria, by the molar free energies (chemical potentials) of the solid and vapour phases. At equilibrium, the change in free energy, δG, of the system solid plus vapour, accompanying the sublimation or condensation of an infinitesimal quantity of the substance is zero, and $\mu_{\text{solid}} = \mu_{\text{vapour}}$. From Equations 2.35 and 2.41 we can express μ_{solid} and μ_{vapour} in terms of the activity of the solid and the partial pressure of the vapour:

$$\mu^0_{\text{solid}} + RT \ln a = \mu^0_{\text{vapour}} + RT \ln p \tag{2.63}$$

If the solid is pure, its activity is unity, so we obtain:

$$\mu^0_{\text{vapour}} - \mu^0_{\text{solid}} = - RT \ln p \tag{2.64}$$

But $\mu^0_{\text{vapour}} - \mu^0_{\text{solid}}$ is the standard molar free energy of sublimation:

$$\mu^0_{\text{vapour}} - \mu^0_{\text{solid}} = \Delta G^0_{\text{subl}} = \Delta H^0_{\text{subl}} - T \Delta S^0_{\text{subl}} \tag{2.65}$$

where ΔH^0_{subl} and ΔS^0_{subl} are the changes in enthalpy and entropy on sublimation of one mole of solid to form vapour in the standard state (25°C, 1 atm pressure). The corresponding changes in enthalpy and entropy at some other temperature will differ from ΔH^0 and ΔS^0: the variations of these quantities with temperature are given by Equations 2.26 and 2.37, so that suitable corrections can be made, using data for the specific heats of the solid and the vapour. If these specific heats are not known, the corrections may be estimated by, for example, Kubaschewski's method.[6] When ΔH and ΔS are combined to find ΔG, the corrections, which may be quite small, tend to cancel out.

If ΔH^0_{subl} or ΔS^0_{subl} are unknown, it is often possible to estimate their values, and we will discuss some of the possible ways of doing this.

Entropy of sublimation

Trouton's rule[49] gives the entropy of boiling of 'normal liquids' as 88.7 J mol^{-1} K^{-1}. For sublimation a corresponding rule due to Le Chatelier and Matignon[50] gives the value 134 J mol^{-1} K^{-1}, but this rule is not so closely obeyed as is Trouton's. The entropy of sublimation may be estimated in two ways: either as the sum of the entropies of fusion and evaporation, or as the difference between the entropies of the vapour and the solid. Taking the first way, the entropy of fusion may be estimated by a method given by Bondi[51] which takes into account the smaller entropy of liquids in which the molecules have symmetry. The entropy of evaporation is given approximately by Trouton's rule so long as there is no ordering mechanism in the liquid (for example, polar forces and hydrogen bonding in water). If we want to use the second way of estimating the entropy of sublimation, we can find the entropy of the solid using the excellent semi-empirical methods of Latimer and Buffington[52] and a particularly accurate one due to Drozin.[53] The entropy of the vapour is made up of translational, rotational, and vibrational contributions. The translational contribution may be calculated exactly using the Sackur-Tetrode equation.[54]

$$S^0_{trans} = \frac{3}{2} R \ln M + \frac{5}{2} R \ln T - R \ln P - 9.669 \text{ J mol}^{-1} \text{ K}^{-1} \quad (2.66)$$

for 1 mole of gas of molecular weight M at P atm. pressure. For monatomic gases, the other contributions are negligible at accessible temperatures. In diatomic gases, the rotational degrees of freedom yield a significant contribution to their entropy. The entropy may be estimated using the empirical formula of Kubaschewski, Evans and Alcock:[6]

$$S^0_{298} = 225 + 0.18 M - \frac{1004}{M} \text{ J mol}^{-1} \text{ K}^{-1}. \tag{2.67}$$

which is fairly good for gases, other than the hydrides, of molecular weight between 20 and 300. For polyatomic gases the empirical formula:

$$S^0_{298} = 163 + 1.42 M - 2.6 \times 10^{-3} M^2 \text{ J mol}^{-1} \text{ K}^{-1} \tag{2.68}$$

holds to within a few entropy units up to mass 250. As a very rough guide, the translational contribution accounts for two-thirds of the entropy of diatomic gases and half the entropy of poly-atomic gases. Certain types of compound, for example the carbonyls, have exceptionally high entropies.

Enthalpy of sublimation

The estimation of enthalpies of sublimation is a difficult exercise, and we can expect only limited success. It is essentially an attempt to describe inter-atomic forces in the gas and in the solid, which can be done relatively easily only for substances in which the bonding is largely ionic. The enthalpies of formation of solids and vapours in which the bonding is covalent may be estimated[20,42] using 'bond strengths'. Where the bonding is mixed to a significant extent, the errors in estimating enthalpies become large.

In the case of ionic solids, we consider the Born-Haber cycle[55] Fig. 2.15.

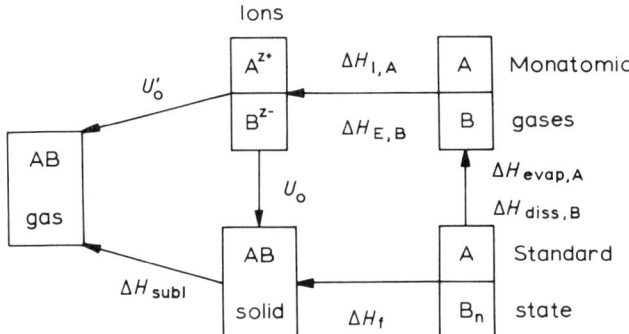

Fig. 2.15 The Born–Haber cycle, with extension to the gaseous molecule.

The various enthalpies around the cycle are as follows:

$\Delta H_{I,A}$ = Ionization energy to form the A^{Z+} ion

$\Delta H_{E,B}$ = Electron affinity to form the B^{Z-} ion

ΔH_{subl} = Enthalpy of sublimation

ΔH_f = Enthalpy of formation of solid AB
$\Delta H_{evap,A}$ = Enthalpy of vapourisation of solid A
$\Delta H_{diss,B}$ = Enthalpy of dissociation of B_n, assumed to be gaseous in the standard state.
U_0 = Internal energy of the solid AB
U_0' = Internal energy of the gas molecule AB

Strictly speaking, we should not equate the internal lattice energy U_0 to the sum of the enthalpies around the cycle without adding a small term $P\Delta V$ to U_0. Furthermore, calculations of U_0 using the ionic model give the energy of a lattice of stationary ions. The zero-point energy and thermal energy may be added if the extra accuracy is required. These corrections are usually small.

For the solid, we may use the Born-Landé model[56] for the inter-atomic binding energy:

$$U_0 = -\frac{Z^2 NAe^2}{r_0}\left(1 - \frac{1}{n}\right) \qquad (2.69)$$

where r_0 is the equilibrium separation of the ions in the solid, A is the Madelung constant, and n is the exponent of the repulsion potential (typically 5 to 10 for ionic solids). More elaborate expressions may be derived for the lattice energy, to include dipole-dipole and dipole-quadrupole interactions and the zero-point energy. We will not pursue these refinements.[57,58]

The energy of the gas molecule may be expressed a similar way:

$$U_0' = -\frac{Z^2 e^2 N}{r_0'}\left(1 - \frac{1}{m}\right) \qquad (2.70)$$

where r_0' is the equilibrium separation of the ions in the gas molecule, m is the exponent of the repulsion potential. The Madelung constant is, of course, unity. We now have:

$$\Delta H_{subl} = U_0 - U_0'$$
$$= Z^2 e^2 N \left[\frac{(n-1)A}{nr_0} - \frac{(m-1)}{mr_0'}\right] \qquad (2.71)$$
$$= U_0\left[1 - \frac{(m-1)n}{(n-1)m}\frac{r_0}{r_0'}\cdot\frac{1}{A}\right]$$

A similar calculation has been performed by Verwey and de Boer[59] for the alkali halides. Experimentally, it is found[60,61] that r_0' is less than r_0 by 15–20%. For the alkali halides, $A = 1.75$, and if $m \simeq n$, we find:

$$\Delta H_{subl} \simeq 30\% \text{ of } U_0$$

This is in good agreement with experimental data for the halides[6] of Li, Na and K, being in error by a few kJ mol^{-1} only.

For covalent substances, it is often possible to calculate the enthalpy of formation of the vapour molecule from its constituent atoms, using the known or estimated thermochemical bond strengths.[20] A good discussion of this topic is given in Dasent's book.[42] When the 'bond strength' is not known, it can be estimated by, for example, Pauling's method.[20] Suppose we want to find the bond energy E_{AB} of the A–B bond. According to Pauling, we may write:

$$E_{A-B} = \frac{1}{2}(E_{A-A} + E_{B-B}) + 96 \Delta \text{ kJ mol}^{-1}$$

where E_{A-A} and E_{B-B} are the energies of the A–A and B–B bonds, and Δ is the (positive) difference between the electronegativities of atoms A and B. Pauling's table of electronegativities has been worked out so as to fit this relationship as well as possible, though this method does not provide a very accurate estimate in many cases.

For compounds in which the bonding is mixed, it is not possible to estimate enthalpies of sublimation with any accuracy using the relatively simple methods that are applicable to ionic and covalent compounds. As an example, let us consider cadmium sulphide, for which Δ is 0.8 eV, indicating 15% ionicity. We will attempt to estimate ΔH^0 for the process:

CdS (solid) → CdS (gs)

Firstly, let us take the ionic model approach. The Born-Haber cycle gives the lattice energy of cadmium suphide as 3360 kJ mol^{-1}.[61] In going from the solid to the vapour, we do not expect such a large percentage change in energy as the 30% which we found for the alkali halides. The atomic separation is unlikely to decrease by anything like the 15% observed in the alkali halides, since this contraction of the bond indicates an increase in covalent character in going from solid to vapour and has, as it were, already taken place in the solid cadmium sulphide. On the other hand, the Madelung constant is 1.64 in the wurtzite structure[62] instead of 1.75. Without taking into account the polarization forces which probably contribute significantly to the binding energy of the CdS vapour molecule, we will use Equation 2.71 to obtain an estimate for ΔH^0_{subl}:

$$\Delta H^0_{subl} \cong 39\% \text{ of } U_0 = 1260 \text{ kJ mol}^{-1}.$$

This is undoubtedly too large.[47] The corresponding enthalpy change for dissociative sublimation:

$$\text{CDS (solid)} \rightarrow \text{Cd (gas)} + \frac{1}{2} S_2 \text{(gas)}$$

is around 330 kJ mol^{-1} [13,47,63], so that one would expect dissociative sublimation to far outweigh sublimation as molecular cadmium sulphide. Let us now see what the bond-strength method predicts. We need to know the energy of the Cd–S bond. Using Pauling's method, we have:

$$E_{Cd-S} = \frac{1}{2}(E_{Cd-Cd} + E_{S-S}) + 96(2.5 - 1.7) \text{ kJ mol}^{-1}$$

Since there is no evidence for the existence of Cd dimers in cadmium vapour[64], we know that E_{Cd-Cd} is small, so we will ignore it. The S–S double-bond energy is -426 kJ mol^{-1},[65] so that E_{Cd-S} comes to -290 kJ mol^{-1}. Our cycle is now as in Fig. 2.16:

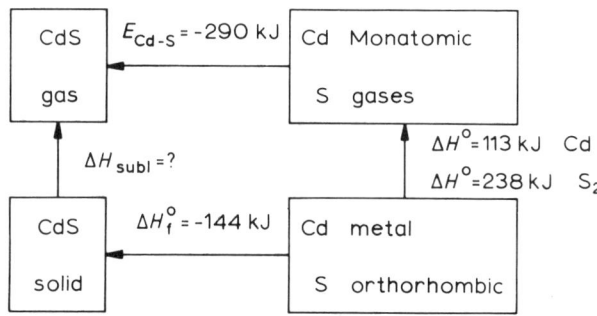

Fig. 2.16 The Born–Haber cycle for cadmium sulphide.

The enthalpy of sublimation is 210 kJ mol^{-1} by this method, as compared with 1260 kJ mol^{-1} using the ionic model. The best we can do is to take some sort of weighted mean. The electronegativity difference between cadmium and sulphur, 0.8 eV, indicates that in the solid, the bonding has about 15% ionic character. If we weight our two estimates in the ratio 0.15 to 0.85, we reach a figure of

364 kJ mol^{-1}. This is startlingly near a recent estimate based on torsion-effusion measurements of the vapour over cadmium sulphide,[47] which is not doubt largely fortuitous. We do not place any great confidence in this crude estimate, but we have included this example to show what may be done by way of informed guesswork in the absence of experimental data or reliable theory.

In conclusion, we will list some further general points:

(1) Volatility decreases with increasing charge on the ions, since the lattice energy is then greater.

(2) Because the Madelung constant is generally higher for structures with high co-ordination, the lattice energy is higher and the volatility lower.

(3) If the bonding in the solid is appreciably ionic in character, the bond length in the gas molecule may decrease as a result of polarisation forces and an increase in covalency on going from solid to vapour. The vapour molecule is thus stabilized. This effect cannot, of course, occur in covalent compounds.

(4) The stability of the associated vapour molecule is decreased if it can dissociate to form stable molecules. For example, zinc and cadmium chalcogenides dissociate to form chalcogenide molecules and metal atoms.

(5) The large entropy change on going from solid to vapour is mainly due to the acquisition of translational and rotational entropy. Both these entropy contributions increase with the molecular weight, so that heavy molecules and polymeric molecules are more volatile than lighter ones.

(6) The possible molecular weight that may be achieved in producing a volatile compound of an element which is to be transported is governed by the maximum co-ordination number round the central cation. The co-ordination number is a function of ligand size and central ion size, and is governed by the Radius Ratio rule. If the co-ordination is to non-bridging ligands (e.g. CCl_4, WF_6) the volatility is increased.

2.3.3 Solubility of species in solids

We are concerned here with the chemical equilibrium between species present in the vapour and the species dissolved in the growing crystal. The criterion for chemical equilibrium is that each species shall have the same chemical potential μ_i in both phases, i.e. that the free energy change dG accompanying the removal of an infinitesimal quantity dn_i mols of species i from the vapour and dissolving it in the solid should be zero. This free energy change is, of course, the difference of an enthalpy term dH_i, and an entropy term TdS_i. The quantities $(\partial H/\partial n_i)_{T,P,n_j}$ and $(\partial S/\partial n_i)_{T,P,n_j}$ are the molar enthalpy and molar entropy changes for the dissolution process, which we will denote by ΔH_i, ΔS_i.

The enthalpy of solution is a combination of several factors. The dissolved species may be incorporated on interstitial sites, or it may replace atoms of the solvent lattice, i.e. be dissolved substitutionally. In either case, there will in general be an elastic strain set up in the host lattice around the solute species to account for the size difference between the solute and the lattice site on which it is incorporated. Then there will normally be a contribution to ΔH_i from the altered chemical bonding around the solute. For example, if a divalent metal replaces some of the monovalent metal atoms in an alkali halide (with some additional defects, e.g. vacancies on the metal sub-lattice to preserve charge neutrality) the Coulombic forces between the doubly charged cation and the surrounding anions will be greater than the original cation-anion forces. There may be a change in the character of the bond, too. For example if one cation is replaced by another which is less electropositive its bonding to the surrounding anions becomes more covalent in character. Pauling's scale of electronegativities[20] provides an indication of the effect, and for a fuller discussion of this topic, we refer the reader to the literature.[20,42] The solubilities of various elements in metals, and in particular, of metals in metals, have been extensively analysed by Hume-Rothery,[66] who from rational principles has formulated some simple guiding rules.

The entropy of the solute contains two contributions: a configurational contribution arising from the random distribution of solute in the solvent, and a vibrational contribution arising from the relaxation of the host lattice around each solute atom. The configurational entropy may readily be calculated for dilute solutions (i.e. no interaction between the solute atoms and hence no tendency either to cluster or form a more or less regular distribution or superlattice). A full discussion of this topic may be found in Swalin's book.[35]

As a general rule, the solubility of a species in a solid decreases monotonically with decreasing temperature, being greatest at the highest temperature at which the solid is stable. However, many semiconductors have a maximum solubility for several elements at some lower temperature, for example, copper in silicon and in germanium[67] and gallium in gallium arsenide.[12] This phenomenon is known as 'retrograde solubility'. It occurs when the enthalpy of solvation is large and positive, for example, about 80 kJ mol^{-1} for germanium.[35] It has some practical application in semiconductor technology, in that it is possible to prepare material at a temperature above the solubility maximum, so that on cooling, the material becomes understaurated and may remain so down to a temperature where solid diffusion is extremely slow. In this way, precipitation of excess solute can be avoided.

If the solute atoms can become ionized in a solid solvent (extrinsic semiconductor) the situation becomes more complex and more interesting. We then have the following equilibria:

$$M \text{ (gas)} \rightarrow M \text{ (in solution)}$$

$$K_1 = \frac{[M]}{p_M} \text{ in solution}$$

$$M \text{ (in solution)} \rightarrow M^+ + e \text{ (in conduction band)}$$

$$K_2 = \frac{[M^+]n}{[M]} \text{ in solution.}$$

where n is the total concentration of electrons in the conduction band. There are two contributions to n; the intrinsic electrons and those contributed by ionized impurities or dopants. In equilibrium, the product np of the electron and hole concentrations is determined by the energy gap E_g and the temperature, and this fact has two effects. Firstly, the ionization of strong donors (i.e. those donors having a small ionization energy in solution) suppresses the ionization of weaker donors, and thereby reduces their solubility in the solid. Secondly, the interaction of donors with acceptors enhances the solubility of each, in a way somewhat analogous to the formation of salts in aqueous solution. For example, the presence of lithium in germanium enhances the solubility of gallium, and vice versa.[68]

Similar arguments may be applied to native defects (vacancies and interstitials, and to misplaced atoms in solid compounds). The subject is fairly complex, and has been treated rigorously at length by Kröger.[69] There is also a very readable and helpful monograph by Van Gool.[70]

We must now consider the solubility of the vapour species in the surface of the growing crystal. Most models of surface processes in crystal growth invoke a population of mobile adsorbed atoms or molecules on the surface (a two-dimensional gas) with more or less stationary islands, clusters, or steps, whose peripheries represent the surface of a two-dimensional liquid or solid at which the two-dimensional gas is adsorbed and incorporated as part of the growing solid. The density and effective pressure of the two-dimensional gas may well be so high that its behaviour departs significantly from ideality.[71] In general we cannot assume a linear relation between the partial pressure of a species in the vapour and its concentration on the surface. Instead, we have to use adsorption isotherms as a basis for an analysis.

Obviously, we would like any impurities that may be in the crystal growth system to be insoluble in the surface of the crystal. This ensures that the growing crystal is relatively free from impurities, and also that the surface does not become covered with an appreciable fraction of a monolayer of impurities which would block the approach of the components of the crystal from the vapour (poisoning of the surface). The situation is different for the transporting agent. Usually, some reaction must take place on the surface in order for the transporting agent to be liberated in the form in which it travels away from the growing crystal. Such a reaction might be, for example, the formation of a halogen molecule X_2 from two X atoms. This reaction will be assisted by a significant concentration of mobile X atoms on the surface, i.e. by the presence of a reasonable amount of species X in the two-dimensional surface gas. However, if species X is strongly adsorbed at the surface, it will not only be less mobile, but may be more readily incorporated in the growing solid than a species which is not strongly adsorbed, unless some other factor, such as atomic size, outweighs the chemical bonding contribution to the enthalpy of solution.

A good review of the theory of semiconductor surfaces and of the experimental methods (low energy electron diffraction, Auger spectroscopy, proton back-scattering, field-emission and field-ion microscopy, etc.[72]) available to study them has been published recently.[73]

2.3.4 High temperature species and their stability

For many years, identification of the molecules in a gas at high temperature depended largely on the interpretation of spectroscopic data. Only simple spectra can be interpreted easily, and the high-temperature species that were identified unambiguously were mostly diatomic. Coupled with the fact that at sufficiently high temperature, all molecules eventually dissociate, because the loss of the enthalpy of association is offset by the increase in translational entropy, the impression has arisen that high temperature gas molecules are very simple – single atoms or diatomic molecules mainly.

With the development of mass spectrometric techniques that could be applied to gases at high temperature came a flood of interesting but confusing data about unexpected species that exist at high temperatures,[74] for example, Si_5, Al_2C_2, MoV_4O_{12}, $(CuCl)_3$ planar rings.[75] It is found that oxygen, hydrogen, and the halogens form stable diatomic molecules with elements of all groups except the inert gases. Furthermore, the valency states of the atoms in molecules at high temperature may be very different from the observed room-temperature states.

It may be shown[76] that the minor species in a saturated vapour should become relatively more important as the temperature increases, since they generally have higher heats of sublimation (e.g. polymeric species). The vapours of many substances at their normal boiling points consist largely of polymeric species, e.g. C_2, C_3, $(BeO)_4$.

For an overall picture of high-temperature inorganic chemistry, the reader is referred to the excellent review by Searcy.[77] This field of knowledge is rapidly expanding but unfortunately lacks systemization.

In chemical vapour transport, we are using temperatures that may be well below the normal boiling points of many of the species formed in the vapour. Hence the assumption of simple monomeric species must be regarded with some suspicion, though in the absence of mass spectrometric or spectral data, there may be no better assumption available. In Section 2.3.2 we discussed the formation of CdS molecules in the vapour over solid cadmium sulphide, and concluded that it is likely to be present only at very low concentration in a largely dissociated vapour. We can say the same about molecules such as ZnO, AlN, etc, where the atoms are light and where the group V or VI element forms a stable vapour molecule, thus favouring dissociation. With compounds of Sb and Bi, on the other hand, dissociation is far less favoured, and we may expect to find molecules of compounds of these elements with group III elements in the vapour.[45]

There are, at present, three main ways of investigating the composition of a vapour at high temperatures, all of which are of limited use for detecting low concentrations of relatively unstable species. Spectral methods may be used if the molecules are fairly simple. Standard thermochemical methods (effusion and torsion-effusion methods, for example) are not selective and usually not sensitive to species present at concentrations of 0.1% or less. The mass spectrometer may well not reveal the true partial pressure of species which are more easily dissociated than ionized by electron impact, or which have short life-times as molecular ions.

Our interest in the stability of vapour species at high temperature is threefold. Firstly, we may consider the stability of the transporting species. We do not usually require the equilibrium constant for the transporting reaction to be extreme (see Section 2.3.1). In other words, the transporting species should not be very stable nor very unstable at the temperature used. Secondly, the presence of polymeric or undissociated species in the vapour will affect the Stefan flow velocity (see Section 4.2). Both of these factors affect the rate of transport of material through the vapour phase. Lastly, polymeric and undissociated species may have a very great effect on the kinetics of growth processes on the crystal surface. Growth may be hindered by the need to dissociate diatomic or polyatomic molecules before the component atoms can be incorporated as part of the growing crystal. On the other hand, undissociated molecules of the compound which is being transported, though present perhaps at very low partial pressure only, may be very readily incorporated at the growing interface. This aspect of the stability of vapour species is probably the most important for the crystal grower: it is regrettably the least well understood. We may suppose that there is an appreciable energy barrier (activation energy) to dissociation of a molecule such as As_4 on a surface of a crystal such as GaAs, and the rate of growth of the crystal may be limited by the rate at which this dissociation takes place. We expect the activational energy to depend on the crystallographic orientation of the surface in general, and to be sensitive to impurities and defects on the surface, as well as to the surface stoichiometry; that is, in the case of gallium arsenide, the ratio of gallium atoms to arsenic atoms in the surface. Such conditions of growth do not favour the development of smooth surfaces, but rather, they favour the formation of small facets, hillocks, etc. Furthermore, if the surface is largely covered with polymeric species with a long residence time before dissociation, the monomeric species, which may be much more active in the surface processes, are prevented from adsorbing on the surface. We may overcome this problem in two ways. If the total pressure is reduced, the ratio of monomers to polymers is increased. Alternatively we may employ a transporting agent which reacts with the substance to be transported to form molecules in which there are only single atoms of the substance to be transported. This aspect of chemical vapour transport, which is perhaps akin to the choice of a suitable solvent for a reaction or crystallization in more conventional chemistry, has not been emphasized, and little in the way of concrete experimental knowledge is available. Yet we would hope that research in this direction would yield rewarding results, both from the point of view of growing good crystals under controlled conditions and of understanding growth processes at the surface.

Turning now to the stability of the transporting species, let us consider the relatively simple example of the transport of a single element by means of a diatomic gaseous transport agent, e.g. a halogen:

$$M + \frac{n}{2} X_2 \rightarrow MX_n$$

The equilibrium relationship is:

$$K_p = \frac{p_{MX_n}}{p_{X_2}^{n/2}}$$

In section 2.31, we mentioned that for the transport rate to be large, we require MX_n to be the dominant species at the source end of the crystal growth apparatus, and X_2 to predominate near the growing crystal. Roughly half way in between, then, the partial pressures are equal, and each is ½P, where P is the total pressure. We therefore require that the mean value of K_p, appropriate to the mean temperature at which the apparatus is operated, should be given by:

$$\bar{K}_p = \left(\frac{2}{P}\right)^{n/2 - 1}$$

If $n = 2$, $\bar{K}_p \equiv 1$ for all values of total pressure. If n does not have the value 2, $\bar{K}_p$ depends on the total pressure, and may differ from unity by many orders of magnitude if n is large (say 4 or 6) and P is small. However, for many reactions taking place at or near atmospheric pressure, $\bar{K}_p \sim 1$ for maximum transport rate. This implies that:

$$\Delta G_{reaction} \sim 0$$

and so

$$\Delta H_{reaction} = T \Delta S_{reaction}$$

In this simple example, we may obtain a rough guide to the value of ΔS from the value of n:

Table 2.5 *Entropies of formation of gaseous halides*

n	ΔS^0, J mol^{-1} K^{-1}
1	80 to 100
2	20
3	−60
4	−125 to −145
5	−200
6	−290 to −340

As mentioned before in Section 2.3.1, the dihalides do not often make efficient transporting species, because of the small entropy change for the transport reaction. If $\Delta H_{\text{reaction}}$ is small, so that $K_p \sim 1$ at an accessible temperature, very large temperature differences are required for appreciable transport (see Section 2.3.1 and the example on carbon transport in sulphur in Section 4.4.3). Or else, if $\Delta H_{\text{reaction}}$ is not small, the dihalide is too stable ($\Delta H \ll 0$) or too unstable ($\Delta H \gg 0$), so that one species greatly predominates over the other throughout the growth apparatus.

In the case of the monohalides, we are mainly interested in those that form endothermically ($\Delta H > 0$). These species are unlikely to be observed at room temperature. If the enthalpy of formation is +80 kJ mol^{-1}, so that $K_p \sim 1$ at about 1000 K, then at 300 K, $K_p \simeq 10^{-10}$. As the temperature is raised, species which form endothermically become progressively more important.

With the trihalides and higher halides the reverse argument applies. The argument may be extended also to hydrides (but for them rather different entropy changes are observed[42] than those given in Table 2.5), oxides, hydroxides, oxyhalides, etc.

For many of the species with which we are concerned, thermochemical data are not available, or not accurate. We then have to estimate the required enthalpies and entropies, and we refer the reader to our previous remarks on this subject, in Section 2.3.2. Again, the estimation of entropies is easier than the estimation of enthalpies. In particular, if the ionic model is used to estimate enthalpies via the Born-Haber cycle, it is usually necessary to take into account all the terms involving dipole-dipole and other interactions. This is because the enthalpy of formation of the molecule comes out as the difference of two large numbers, one being a sum of ionization potentials, electron affinities, heats of atomization, etc. (all of which are usually well known) and the other is the molecular internal energy, calculated using the ionic model, which is subject to errors that may be a large fraction of the required heat of formation.

2.3.5 Non-stoichiometric phases

In Section 2.2.3 we considered various types of defects which could be present in a crystal of a compound. Disorder due to Frenkel and Schottky defects, and substitutional disorder (misplaced atoms) do not alter the ratios of the different atoms present in the solid compound. On the other hand, vacancies on one sub-lattice which are not compensated by interstitial atoms or by vacancies on the other sub-lattices result in the compound having a composition differing from that represented by the simple chemical formula.

Thus, for instance, a binary compound AB is better described as AB_x, where x may take values over a range near unity.

The composition of a compound depends on the composition of the vapour over it. In general, the stoichiometric compound is not in equilibrium with the stoichiometric vapour, but with some other vapour composition, which may well be unattainable in practice, so that the stoichiometric solid is never seen. In any case, the stoichiometric compound is something of an exception, and departures from it are the rule.

The change of stoichiometry accompanying changes of partial pressures in the vapour is governed by the free energy of formation of the defects responsible for the non-stoichiometry. Provided that the defect concentration is not so large that the interaction between the defects becomes significant, the configurational entropy of the defects can be calculated on the assumption that they are randomly distributed. When the interaction between the defects is significant, it may be that a new phase appears. We discuss this point later in the section.

We can obtain a qualitative estimate of the enthalpy of the defects from a few simple considerations:

(1) The strain energy associated with the interstitial defect. This energy will be smallest in structures with large interstices, such as the tetrahedrally co-ordinated sphalerite and wurtzite structures. The [8:4] co-ordinated fluorite structure can also accommodate a large excess of the non-metal on interstitial sites.

(2) The breaking of bonds required to form a vacancy. If the bond strength is not known, it must be estimated, using the methods discussed in Section 2.3.2. The relaxation of the lattice around the vacancy reduces the energy of formation to some extent.

(3) The requirement of charge neutrality implies that if anions are accommodated interstitially, an equivalent number of cations must adopt a higher oxidation state. Except in cases where the cation has well-established higher oxidation states (e.g. the transition metals, copper, lead), the extent to which excess anions can be accommodated is very small. On the other hand, if some of the cations are replaced by a higher valent species, a large percentage of excess anions can be accommodated (e.g. YF_3 in CaF_2).[78]

(4) In a similar way, vacancies on the anion sub-lattice require some cations to adopt a lower oxidation state (e.g. $Fe_{2+x}O_3$) or become neutral ($Na_{1+x}Cl$).

(5) The change in cation size when it changes its oxidation state produces a strain energy.

REVIEW OF THE BASIC SCIENCE

A wide range of stoichiometry is thus only possible if the strain energy is small and if one of the component atoms possesses two or more common oxidation states. If these criteria are not met, the range of stoichiometry will be small by the standards of normal chemical analysis, though it may be accompanied by large variations in defect-sensitive properties.

As a general rule, deviations from stoichiometry take place in the direction of another stable oxidation state of the element concerned. If there is no second stable valence for either component, and if the lattice is so closely packed as to exclude interstitial atoms, there may be no detectable variation in stoichiometry e.g. MgO.[23]

The replacement of atoms on one sub-lattice by atoms of the other species (place exchange) is energetically unfavourable in ionic crystals, though it is common in intermetallic compounds, and also observed in covalent compounds if the covalent radii and electronegativities of the two species are similar. This behaviour is observed in some III-V compounds, for example GaSb.[23]

Small deviations from stoichiometry, which we have so far considered, have been studied at length, and readers are referred to the works of Anderson,[79] Greenwood,[23] Van Gool,[70] and Swalin[35] for the detailed exposition of the theory. For large deviations from stoichiometry, we can envisage various possibilities:

(a) Quasi-random homogeneous distribution of defects. This is essentially the same picture as that already discussed for small deviations from stoichiometry, except that the theory must be modified to take into account the interaction between defects. Examples are the high-temperature forms of the monoxides of titanium and vanadium.[80]

(b) Ordering of defects on a super-lattice. If the interaction between the defects is such as to produce ordering of the defects on a super-lattice, we may properly say that a new phase has formed. An excellent example is the series of oxides of praesodymium, $Pr_n O_{2n-2}$, where ordered intermediate phases occur for $n = 4,7,9,10,11,12$ and ∞.[81] In general, each of these intermediate phases will have a range of composition.

Since the driving force for ordering is the enthalpic term in the free energy of the phase, opposed by the entropic term, a rise in temperature favours disordering. Unless the substance melts or decomposes on heating, the high temperature form is often the quasi-random homogeneous distribution of defects described previously.

(c) Intermediate phases based on shear structures. The lattice defects which give rise to non-stoichiometry can be removed not only by ordering onto a super-lattice, but also by clustering into groups which are self-eliminating by a shear or collapse of the original lattice into a new structure. The oxides of molybdenum and tungsten provide good examples of this behaviour. (Magneli phases).

The subjects of grossly non-stoichiometric compounds and the shear structures are discussed at length in a particularly readable and perceptive account by Greenwood.[23] We would also direct interested readers to the specialist review by Libowitz.[82]

It is clear that chemical vapour transport is a route to the preparation of these somewhat exotic non-stoichiometric compounds,[2] as the composition of the solid which is prepared depends on the experimental conditions in a way that can be predicted if the relevant thermochemical data are available. Equally clear is the need for some careful thought and calculations in designing a crystal growth system. The composition of the vapour over the growing solid may be dependent on the transport rate in the vapour phase, as well as on the solid/vapour equilibrium that is obtained. Thus in an open flow crystal growth system, a change in flow rate will alter the thickness of the boundary layer over the growing crystal through which the vapour components travel by Stefan flow and diffusion (see Chapter 4). This brings about a change in the growth rate of the crystal, and also changes the composition of the vapour in equilibrium with it. This change may be sufficient to cause a different phase to be stable.

Besides the possibility of controlled preparation of the more exotic non-stoichiometric compounds by chemical vapour transport, those compounds showing a narrow range of composition can, in principle, be prepared to the required degree of non-stoichiometry by suitable control over the vapour composition. The electronic properties of the prepared substance can thus be controlled, at least in principle, and often in practice.[83] A further detailed discussion of the composition of the vapour in the chemical vapour transport system is given in Chapter 4.

CHAPTER THREE
The Vapour-Crystal Interface

3.1 Adsorption

When a gas molecule strikes the surface of a solid or a liquid, one of two things may happen. It may either bounce back from the surface elastically, or it may stick to the surface for a certain length of time (be adsorbed) and then fly away, with a velocity, and in a direction, unrelated to its original velocity and direction. In the latter case, there is energy transferred between the gas and the solid or liquid, and this is by far the most usual case, as evidenced by the normal heating of a cold gas by a warm surface.

The mean length of time τ spent by the molecules adsorbed on the surface varies with the nature of the surface and of the molecule, and the temperature. For hydrogen adsorbed on many surfaces at room temperature, $\tau \sim 10^{-12}$ s. For oxygen atoms on a tungsten surface, at room temperature, $\tau \sim 10^{85}$ centuries[71]. The residence time τ is thus one of Nature's more variable quantities.

When a molecule is adsorbed on a surface, a certain amount of energy Q is given up (for example, the surface gets a bit hotter). The relationship between τ and Q was proposed by Frenkel[84] in 1924:

$$\tau = \tau_0 \exp(Q/kT) \tag{3.1}$$

where τ_0 is the period of oscillation of the adsorbed molecule in a direction normal to the surface.

In most discussions of adsorption, a distinction is drawn between physisorption and chemisorption. A molecule which is physisorbed on a surface is held to the surface by van de Waals forces, so that Q is fairly small, say $1 - 20$ kJ mole^{-1}. Such a molecule undergoes no major rearrangement of its electron clouds, and does not split up into its constituent atoms. A chemisorbed molecule, on the other hand, has rearranged its electron clouds so as to obtain a more or less strong bond to one or more surface atoms. This may involve dissociation of the molecule, and often involves passing through an intermediate state of higher energy, so that chemisorption is frequently an activated process. However, because the strengths of chemical bonds cover the complete range from the weakest van de Waals bonding to the strongest covalent and ionic bonds, the distinction between physisorption and chemisorption is more apparent than real, though often useful.

Since adsorbed molecules spend an appreciable time τ on the surface, the concentration of molecules on the surface will be greater than in the gas. This concentration of molecules on the surface is available to undergo any necessary processes to become incorporated as part of a growing crystal. Such processes might be migration to surface steps or the peripheries of clusters, or chemical reaction with other adsorbed molecules to produce the elementary building blocks of the crystal. The concentration of molecules on the surface is thus a very important parameter in a discussion of crystal growth mechanisms. This concentration is a function of the heat of adsorption of the molecules on the surface, temperature, pressure of the molecules in the vapour, etc.

Several theoretical approaches to adsorption have been given, and because of the importance of the subject in crystal growth, we give a brief review of the better-known theories below.

The Langmuir isotherm

In 1918, Langmuir[85] derived an expression for the fraction θ of a surface covered by adsorbed molecules as a function of the pressure of those molecules in the vapour. His treatment was as follows: for the simplest case of adsorption when all surface sites are equivalent, consider a flux μ moles hitting unit area of the surface in unit time. From the kinetic theory of gases:

$$\mu = \frac{p}{(2\pi MRT)^{\frac{1}{2}}} \qquad (3.2)$$

where M is the molecular weight of the impinging molecules. Imagine that some fraction θ of the surface is covered by adsorbed molecules. Langmuir argued that, for simple adsorption, a molecule landing on top of a previously adsorbed molecule would be very weakly bound, and would have such a short residence time that we can suppose it to be reflected immediately. Of the molecules impinging on the remaining fraction $(1 - \theta)$ of the surface, only a fraction α are adsorbed, where α is the condensation coefficient. The value of α may range from 0 to 1, being less than 1 if adsorption is activated. Thus the rate of adsorption is $\alpha\mu(1 - \theta)$. The rate of desorption is

THE VAPOUR CRYSTAL INTERFACE

$\nu\theta$, where ν is the rate of desorption from a complete monolayer. When the adsorbate is in equilibrium with the vapour,

$$\alpha\mu(1-\theta) = \nu\theta$$

If we let

$$\frac{\alpha}{\nu} = \sigma$$

we obtain the Langmuir isotherm:

$$\theta = \frac{\sigma\mu}{1+\sigma\mu} \tag{3.3}$$

Since μ is proportional to the pressure of the molecules in the vapour, this equation relates coverage θ to pressure at constant temperature. If the pressure is very low, we can ignore $\sigma\mu$ in comparison with unity, so θ is proportional to pressure. At the other extreme, if the pressure is high, $\sigma\mu \gg 1$, and θ tend to unity. The general shape of the isotherm is illustrated in Fig. 3.1.

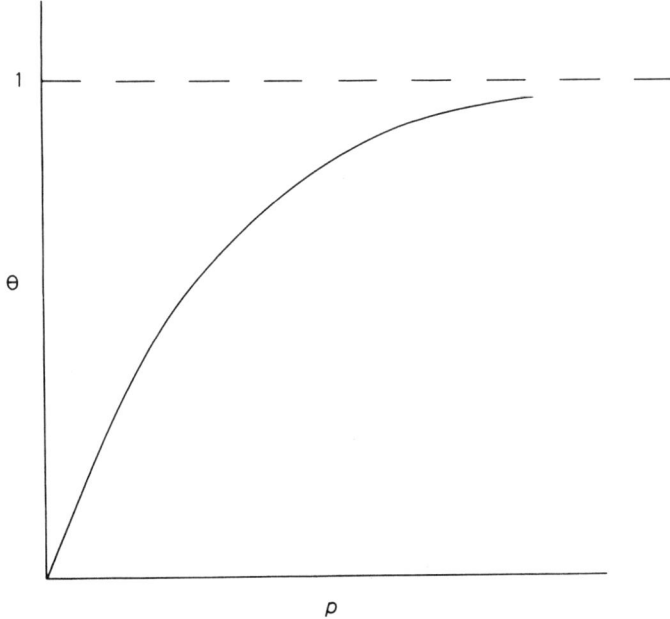

Fig. 3.1   The Langmuir adsorption isotherm.

The initial slope of the isotherm is $\sigma(2\pi MRT)^{-1/2}$, and from experimental measurement of this low-pressure part of the isotherm one can determine the value of σ, i.e. α/ν. This ratio has often been interpreted in the following way. The condensation coefficient α is equated to a steric factor S, times the probability of overcoming an activation barrier to adsorption E_a (E_a may be zero in some cases), thus:

$$\alpha = S \exp(-E_a/RT) \tag{3.4}$$

The rate of desorption from a complete monolayer is given by:

$$\nu = Nf \exp[-(E_a + Q_a)/RT] \tag{3.5}$$

where N is the number of molecules in a unit area of the monolayer, and f is their vibration frequency normal to the surface i.e. the frequency at which they attempt to surmount the energy barrier for desorption, which is composed of the heat of adsorption Q_a and the activation energy E_a as illustrated in Fig. 3.2.

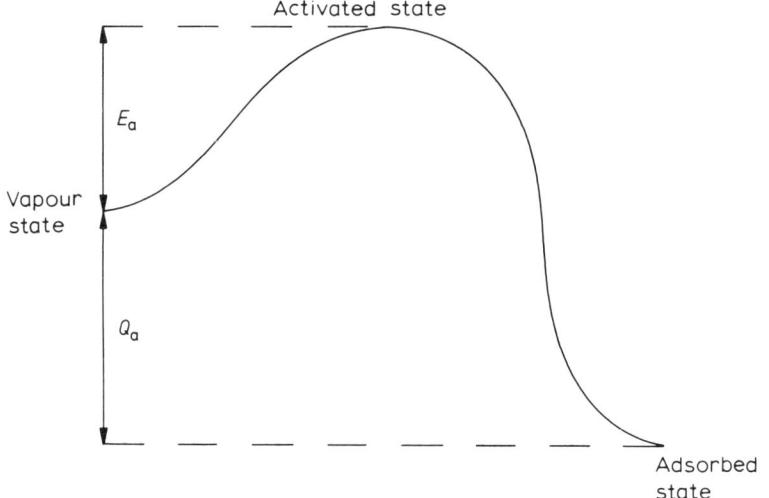

Fig. 3.2   Energy levels in activated adsorption.

Thus:

$$\frac{\alpha}{\nu} = \sigma = \frac{S e^{Q_a/RT}}{Nf} \tag{3.6}$$

From the temperature dependence of the shape of the isotherm, that is, of σ, the heat of adsorption Q_a can be found.

Langmuir's theory can be extended to deal with the simultaneous adsorption of more than one species. If the different species 1,2,3 ... have values of σ: $\sigma_1, \sigma_2, \sigma_3, \ldots$ then:

$$\theta_1 = \frac{\sigma_1 \mu_1}{1 + \sigma_1 \mu_1 + \sigma_2 \mu_2 + \ldots}$$

The occurence of terms derived from all the adsorbed species in the denominator represents the competition between the various species for the available surface sites.[86]

Langmuir extended his theory[85] to describe adsorption on a surface made up of more than one type of site, and thence to a surface composed of a continuous range of sites. However, the more types of sites one postulates, the more parameters must be determined experimentally, and the harder it is to interpret the observed isotherms.

Langmuir's simple theory is open to criticism on two main counts, namely that it ignores the possibility of building up more than one monolayer of adsorbate (though in the theory as first published he attempts to deal with this) and that it postulates a definite number of surface sites. The assumption, often made, that the adsorbed atoms or molecules do not interact, and hence that the heat of adsorption Q_a is independent of the coverage θ, can in principle be replaced by a relationship between Q_a and θ, derived from an atomistic model. In any case, the restriction of no more than a complete monolayer of adsorbate, inherent in the Langmuir theory, suggests that it is best applied to strong adsorption from rarefied gases, e.g. oxygen at low partial pressure on metal surfaces.

When molecules are adsorbed on a surface from a gas at some appreciable fraction of their saturated vapour pressure, we have to consider multilayer adsorption, for which there is ample experimental evidence. In many chemical vapour transport systems, the conditions for multilayer adsorption are achieved.

Brunauer, Emmett, and Teller[87] advanced a theory of multilayer adsorption which fits many experimentally observed adsorption isotherms, and which is usually called the BET theory for short. In their picture of multilayer adsorption, the first monolayer (complete or partial) is chemisorbed at the surface with heat of adsorption Q_1; on top of that are several more part-layers, with heats of adsorption $Q_2, Q_3 \ldots$ which in general are all different, but which are often taken as being equal to the heat of evaporation of the adsorbed species in bulk form. The argument put forward by Brunauer, Emmett and Teller is as follows. Let fractions $\theta_0, \theta_1, \theta_2, \ldots \theta_i$ of

GROWTH OF CRYSTALS FROM THE VAPOUR

the surface be covered by 0,1,2,3 ... i monolayers of adsorbate. Consider the uncovered fraction θ_0. This fraction may be increased by desorption from parts of the surface covered by one monolayer, at a rate $\nu_1 \theta_1$, and decreased by adsorption on the uncovered parts, at a rate $\mu\alpha\theta_0$. At equilibrium therefore,

$$\nu_1 \theta_1 = \mu\alpha\theta_0 \tag{3.7}$$

or using the same interpretation of ν_1 and α as in Langmuir's theory.

$$p\theta_0 = a_1 e^{-Q_1/RT} \theta_1 \tag{3.8}$$

where

$$a_1 = Nf\sqrt{(2\pi MRT)}$$

Now consider the fraction θ_1 of the surface covered by one layer. This fraction θ_1 is increased by adsorption on the bare parts of the surface and by desorption from parts covered by two layers, and is decreased by adsorption on, or desorption from, the parts covered by one layer. At equilibrium therefore,

$$\mu\alpha\theta_0 + \nu_2\theta_2 = (\mu\alpha + \nu_1)\theta_1$$

But we have already seen that $\nu_1 \theta_1 = \mu\alpha\theta_0$. Hence

$$\nu_2 \theta_2 = \mu\alpha\theta_1$$

or

$$p\theta_1 = a_2 e^{-Q_2/RT} \theta_2 \tag{3.9}$$

Clearly, for the equilibrium of parts covered by i layers:

$$p\theta_i = a_{i+1} e^{-Q_{i+1}/RT} \theta_{i+1} \tag{3.10}$$

The total volume of gas adsorbed per unit area is given by:

$$V = V_0 \sum_i i\theta_i \tag{3.11}$$

where V_0 is the volume of gas at pressure p which would adsorb to form exactly one layer on a unit area of surface.

To convert this adsorption isotherm into a useful tool, it is helpful to make the following assumptions:

(1) $Q_2 = Q_3 = \ldots = Q$, the heat of evaporation of the adsorbed species in its bulk form.
(2) $a_2 = a_3 = \ldots = a$.

These assumptions imply that the effect of the substrate or surface of the adsorbent is confined to the first layer of adsorbate, and is not felt by subsequent layers. This is obviously an oversimplification, but

THE VAPOUR CRYSTAL INTERFACE

to some extent justifiable. We know that inter-atomic forces fall off rapidly as the atoms get further apart, so that the forces between the second layer and the substrate are very much smaller than those between the first layer and the substrate. Coulombic forces fall off least rapidly ($1/r^2$), but may be effectively screened by the first monolayer of adsorbate.

With the above assumptions, we can rewrite Equations 3.8 and 3.9 thus:

$$\theta_1 = Y\theta_0 \quad \text{where} \quad Y = \frac{p}{a_1} e^{Q_1/RT}$$

$$\theta_2 = X\theta_1 \quad \text{where} \quad X = \frac{p}{a} e^{Q/RT}$$

$$= XY\theta_0$$

Hence

$$\theta_i = X^{i-1} Y\theta_0 = X^i C\theta_0 \tag{3.12}$$

where

$$C = \frac{Y}{X} = \frac{a_1}{a} e^{(Q_1 - Q)/RT} \tag{3.13}$$

We can now substitute Equation 3.12 into 3.11:

$$V = V_0 C\theta_0 \sum_i i X^i \tag{3.14}$$

The sum of all the fractional coverages is unity, so that:

$$1 = \sum_i \theta_i = \theta_0 + C\theta_0 \sum_i X^i \tag{3.15}$$

Dividing Equation 3.14 by unity, we obtain:

$$V = \frac{V_0 C \sum_i i X^i}{1 + C \sum_i X^i}$$

Performing the summations then yields:

$$V = \frac{V_0 CX}{(1-X)(1-X+CX)} \tag{3.16}$$

When the vapour pressure reaches saturation, i.e. as $p \to p_0$, $V \to \infty$ and $Y, X \to 1$, so that from the definition of X above,

$$1 = a p_0 e^{Q/RT}$$

Thus

$$X = p/p_0$$

and Equation 3.16 becomes:

$$V = \frac{V_0 Cp}{(p_0 - p)[1 + (C - 1)p/p_0]} \tag{3.17}$$

This is the BET isotherm. It contains two parameters V_0 and C which are to be determined from experimental measurements of p and V. From the temperature dependence of C, we can obtain the heat of adsorption of the first layer.

The restrictive assumptions made above have been relaxed by various workers[88–91]. The intricacies of the various modifications to the theory are beyond the scope of this book. What we are interested in is whether the adsorbed species are present as some fraction of a monolayer coverage, or whether they form a liquid-like layer several atomic-spacings thick. Since the rate at which the surface processes take place will depend on the availability of adsorbed species which can partake in reactions, we obviously expect the rate at which a crystal grows to be much affected by the nature of the adsorbed layers.

Previous work on crystal growth processes has not emphasized this aspect, and all we aim to do here is to draw the reader's attention to two theories of crystal growth that might possibly be applied to a crystal growing from the vapour but covered with several layers of a liquid-like film of adsorbate. The surface roughness theory of Jackson[92] could be applied to the crystal-adsorbate interface. We may consider such an interface to be an atomically rough one, since the components of the crystal are present in the adsorbate layers, albeit in chemical combination with a transporting agent to some extent. Jackson's theory then tells us that such a crystal can grow with a smooth surface, rather than in the form of needles or whiskers. The presence of a liquid-like layer may permit the crystal to grow by the vapour–liquid–solid (VLS) mechanism. (See Section 2.3.1 for a description of the VLS mechanism). Thus in the following schematic chemical vapour transport system:

$$M(s) + HX(g) \rightarrow MX(g) + \tfrac{1}{2} H_2(g)$$

(X being a halogen, for example) a liquid-like, multilayer film of adsorbed MX and HX may form on the crystal surface under suitable conditions of temperature and partial pressures. If the element M has more than one valency state, other halides MX_2, M_2X_3, MX_3 etc.

may be present as well in the adsorbed film Molecules of MX gas are adsorbed on the surface of the film and can react with hydrogen from the vapour. The atoms of element M thus produced diffuse through the liquid-like layer and are accommodated easily at the atomically rough crystal-adsorbate interface.

Direct evidence for the existance of multilayer films of adsorbate in crystal growth from the vapour is lacking, though it is a familiar phenomenon in growth from solution. We observe, however, that it is often possible to grow crystals of good size and quality by chemical vapour transport at temperatures where growth by simple sublimation yields poor, polycrystalline material.

To delineate the region where multilayer adsorption should occur, let us put $V = V_0$ in Equation 3.17, i.e. find the condition that sufficient gas be adsorbed to form one complete layer (though in fact it will usually adsorb in the form of an incomplete layer with parts of more layers on top). We find that:

$$Cp = (p_0 - p)[1 + (C - 1)p/p_0] \qquad (3.18)$$

or

$$C = \frac{(1 - p/p_0)^2}{(p/p_0)^2}$$

Now C is $e^{(Q_1 - Q)/RT}$, apart from the ratio of the normal vibration frequencies of atoms in the first layer and in subsequent layers, which is of the order of unity. So we find:

$$\frac{Q_1 - Q}{RT} = 2 \ln\left[\frac{p_0 - p}{p}\right] \qquad (3.19)$$

This equation is plotted in Fig. 3.3.

While adsorption of active species is clearly an essential step in crystal growth, adsorption of non-active (impurity) species on the surface of the growing crystal can modulate the growth rate markedly. In its most pronounced form, adsorption of impurities can effectively 'poison' the growth of certain faces, leading to the formation of platelets or whiskers. Less extreme is the suppression of the normally observed forms (habit) with the production of new forms, analogous to the suppression of the cube faces of NaCl grown from solution containing urea[93] and the production of crystals bound by octahedral faces. In each case, the impurity molecules or atoms are strongly adsorbed on certain crystal faces, forming a complete monolayer which acts as a barrier to adsorption of the active species. On other crystal faces, the impurity is not so strongly

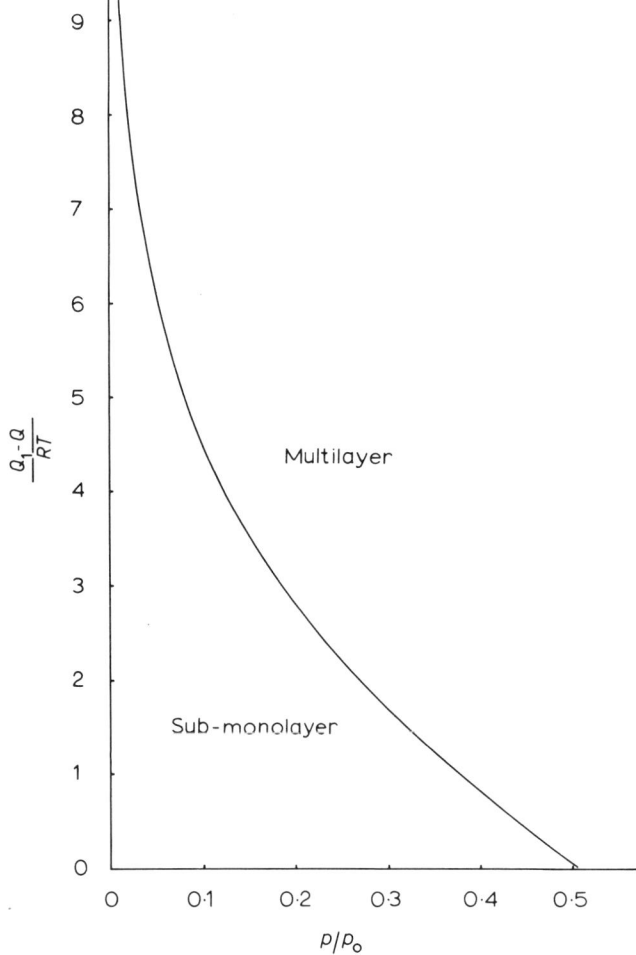

Fig. 3.3 Mono-layer and multi-layer adsorption.

adsorbed, thus allowing the active species some share of the gains in the competition for adsorption sites; alternatively, the impurity molecules require a large activation energy to become adsorbed, so that their rate of adsorption is slow and growth may continue, hindered to greater or lesser extent. *Note that an ill-chosen transporting agent may play this part in the same way as an impurity.*

Not only impurities and ill-chosen transporting agents may drastically alter the rate at which the crystal grows and the habit it assumes. In the growth of a binary or multi-component substance, the relative partial pressures of the various components are an important consideration. Here we are dealing either with the components themselves or with suitable volatile compounds. Let us consider, as the simplest case, the growth of a crystal of compound AB from a vapour of A and B molecules. If the crystal were in equilibrium with its vapour (with zero growth), there would be certain coverages of A and B (θ_A and θ_B, say,) adsorbed on the surface. At equilibrium, these coverages would be governed by an equilibrium constant relationship $K_\theta = \theta_A \theta_B$. Thus θ_A and θ_B can change (though not independently) to reflect changes in the vapour composition. The solid-vapour equilibrium is a dynamic one, and there is an 'exchange current' $J_{0,A}$ of species A adsorbing and desorbing continuously, and similarly an 'exchange current' $J_{0,B}$ of species B. These exchange currents may be large or small, their magnitude depending to a large extent on θ_A and θ_B. Clearly, if very little of the surface is covered by species B, we expect the rate of desorption of B, and hence $J_{0,B}$, to be relatively small, and similarly for species A.

When the crystal is growing from its vapour, some fraction of the molecules of A and B that are adsorbed are not desorbed again, but are incorporated as part of the crystal. The flux of these molecules is J, the growth rate of the crystal. If J is small compared with $J_{0,A}$ and $J_{0,B}$ we may still speak of an 'exchange current' of each species, which will be only slightly perturbed from the equilibrium exchange currents by the net flux J. Each atom or molecule alights on the surface and is desorbed again a great number of times before finally becoming more or less permanently incorporated, and thus has a high probability of ending up in a position of minimum potential energy, which will be its correct crystallographic position for the solid phase which is stable under the prevailing conditions of temperature, pressure and gas composition. However, if the net growth rate is driven to become comparable with the smaller of the exchange currents, say $J_{0,B}$, then B molecules will only alight on the surface a few times on average before becoming 'fixed', so that some fraction of them will remain in positions where their potential energy is a local minimum but not an absolute minimum. Thus may develop dislocations, stacking faults, misoriented grains, metastable phases (e.g. glass). This situation may be brought about either by driving the system to extreme growth rates (e.g. by quenching the vapour) or by causing one of the exchange currents to be very small. And in this latter case, several factors play a part. Firstly, since the exchange current of each species depends on the coverage of that species, we expect the exchange current to be smallest for the minority species, *ceteris paribus*. However, if adsorption of some or all species is activated, we may expect the activation energies to be functions of the coverages θ_A, θ_B .... Some range of coverages may make the sticking coefficient of one species very small, while its partial pressure is still appreciable. Thus the exchange current of that species can be very low.

We have seen the several important roles that adsorption may play in crystal growth from the vapour, and that understanding the co-adsorption of the several species involved is a large step to understanding crystal growth. Unfortunately, systematic work which is directly applicable is lacking. It is the authors' hope that once the relevance and importance of the problem is generally realized, it will receive the effort and attention it merits.

3.2 Nucleation

In the growth of a crystal, especially from a vapour, the conglomeration of atoms or molecules to form the first sub-microscopic speck or nucleus of the solid crystal is a process fraught with difficulty. The large decrease in molar entropy accompanying the process (about 120 J mol^{-1} K^{-1} for many substances) indicates that the ordered crystal is a state of very much lower thermodynamic probability than the chaotic gas. The driving force for crystallization comes from the lowering of the potential energy of the atoms or molecules when they form bonds to one another.

We may consider three separate nucleation conditons:

(a) Nucleation of a condensed phase within a gas.
(b) Nucleation of a condensed phase on a surface of a different substance (hetero-epitaxy).
(c) Nucleation of a condensed phase on a surface of the same substance (homo-epitaxy).

The first of these, homogeneous nucleation within a gas, is not of practical interest to the crystal grower. Most instances of apparent homogeneous nucleation within a gas are, in fact, more accurately described as seeded crystal growth, since minute particles of dust, etc, act as nuclei*.[94] Condensation in high-velocity gas streams emerging from convergent-divergent nozzles[95] is one of the few

---

*Wilson, in 1896, held supersaturated water vapour in his famous cloud chamber and was able to observe the trajectories of charged particles by the trail of droplets formed. More recently, rainfall has been produced by releasing silver iodide dust into clouds, thus demonstrating seeded nucleation, as well as giving a new twist to the old proverb.

instances where truly homogeneous nucleation takes place within a gas. However, the elementary theory of homogeneous nucleation brings out several of the important concepts strikingly, and we give a brief account of it below.

Consider a liquid droplet in a supersaturated vapour. The bulk free energy of the droplet is less than that of the vapour, since the vapour is supersaturated. However, the total free energy of the droplet is increased by the surface energy, which makes a relatively large contribution if the droplet is small. The Gibbs-Thompson equation[96] relates the supersaturation ratio p/p_0 to the radius r of the droplets with which it is in equilibrium:

$$\text{Excess free energy} = kT \ln p/p_0 = \frac{2\sigma v}{r} \quad (3.20)$$

where v is the volume of a molecule, and σ is the surface tension of the liquid droplet. A droplet of radius smaller than r tends to evaporate again; a droplet of radius greater than r tends to increase in size. Thus r is the critical radius, corresponding to the supersaturation p/p_0, at which a droplet or nucleus may lower its free energy whether it grows or shrinks.

Usually a spherical shape is assumed for the nucleus, and the free energy of formation is expressed as[97]:

$$\Delta G^0 = 4\pi r^2 \sigma + \frac{4}{3}\pi r^3 \Delta G_v \quad (3.21)$$

where

$$\Delta G_v = -\frac{kT}{v} \ln p/p_0 \quad (3.22)$$

the free energy change per unit volume forming the stable condensate from the vapour. A plot of Equation 3.21 is given in Fig. 3.4. It is seen that the free energy goes through a maximum at a critical radius r^*, which defines the critical nucleus.

The stationary nucleation rate is usually expressed as the product of the equilibrium concentration of critical nuclei $c(n^*)$, the rate of bombardment β of molecules on the nucleus surface $S(n^*)$, and a non-equilibrium factor Z which expresses the departure of the actual concentration and gradient of concentration of critical nuclei from the equilibrium values:

$$J = Z\beta S(n^*)c(n^*) \quad (3.23)$$

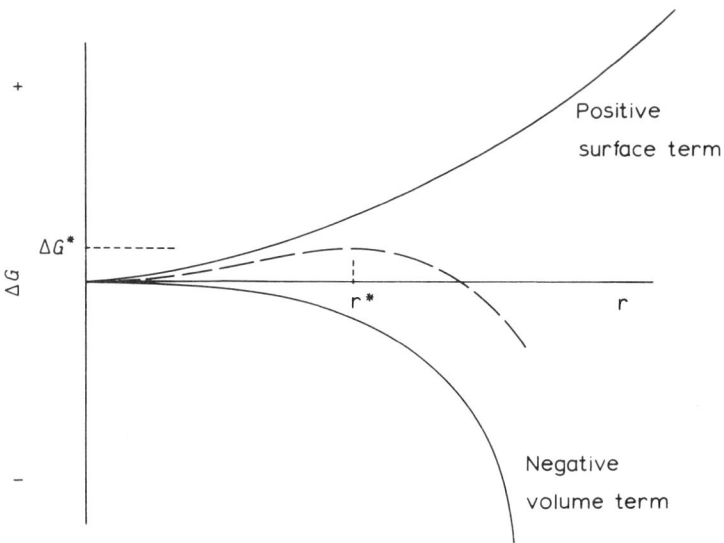

Fig. 3.4 Free energy of critical nucleus.

For homogeneous nucleation from the vapour,

$$J = Z \frac{p}{\sqrt{(2\pi MkT)}} 4\pi r^{*2} c(1)_0 \exp(-\Delta G^*/kT) \tag{3.24}$$

The factor Z has been discussed by Zeldovich[98-100] and others.

This simple theory of homogeneous nucleation brings out two essential features of every nucleation process:

(a) The vapour must be supersaturated with respect to the bulk condensate for small clusters to have any stability. The greater the supersaturation, the smaller is the cluster which stands an even chance of increasing in size.

(b) The rate of nucleation is strongly dependent on the size of the critical nucleus, that is, on the supersaturation.

Fig 3.5 illustrates this dependence, as derived from Equations (3.21)–(3.24). Over a small range of supersaturation p/p_0, the nucleation rate increases from negligible to huge. Experimental observation of this phenomenon had lead Ostwald[101] to distinguish two types of supersaturated solutions, metastable and labile. Metastable solutions, free of foreign nuclei, remain unchanged for apparently unlimited periods of time, while labile solutions crystallize in a short period.

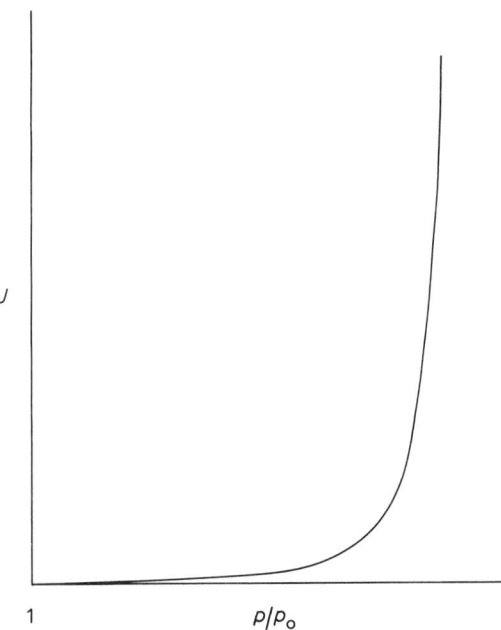

Fig. 3.5 Nucleation rate J as a function of supersaturation p/p_0.

Nucleation on a substrate, whether of the same chemical substance as the vapour, or of some other substance, presents new features. The surface is covered to greater or lesser extent by adsorbed vapour molecules. If these molecules are mobile they can come together to form two-dimensional clusters, with eventual nucleation and growth of a layer of solid from the 'two-dimensional fluid'. In many respects, this process is the two-dimensional analogue of homogeneous nucleation from the vapour. However, a large step in the exchange of entropy for enthalpy, inherent in crystallization, has already been taken in the adsorption process. Small clusters gain some degree of stability because of bonds between their component molecules or atoms and the substrate. For this reason, the critical nucleus size for a given supersaturation is very much smaller for nucleation on a substrate than for nucleation within the vapour, and at quite modest supersaturations, the critical nucleus may contain only a few atoms.[102,103] It is clearly unrealistic to use values of surface and volume free energy appropriate to bulk material to describe the stability of such small clusters. Current theories are formulated in terms of the energy of the bonds between atoms in the cluster.[104-106] A recent review paper of Frankl and

Venables[106] compares the various versions of this type of theory and lists the simplifying assumptions made in each. Without making some simplifying assumptions (for example, that there is a definite critical cluster size for given conditions of supersaturation, temperature, etc; or that clusters larger than the critical size are absolutely stable) the mathematical description of nucleation becomes abstruse.

Atomistic theories of nucleation are used to predict some or all of the following: critical nucleus size, rate of nucleation, saturation density of stable clusters, and the variation of these with temperature and supersaturation. The theoretical predictions are compared with experimental measurements of the rate of nucleation and the density of (microscopic) observable clusters, from which comparison the critical nucleus size may be estimated and something can be learnt about the energy of bonds between atoms in the clusters. The most successful results to date are probably those of Venables and Ball[107], who have studied the condensation of the inert gases on graphite. In this case it was possible to estimate the bond energy independently.

Homo-epitaxial nucleation may be treated in an atomistic way similar to the theories mentioned above. One principal difference here is that the condensing vapour is presented with a template which is the correct pattern for producing a new surface layer. Again, there is no interfacial energy between the substrate and an island of condensate formed on it. The positive surface energy term in the free energy arises from the perimeter of the island, and has a lower-power dependence on the island size than is the case for hetero-expitaxial nucleation.

These atomistic theories consider only nucleation on a pure and perfect substrate. In practice, no substrate is either pure or perfect, unless it is of microscopic dimensions. All macroscopic substrates contain some defects. We are considering now such defects as dislocations, grain boundaries, stacking faults, and the strain fields around impurity atoms. It may well be that the binding energy of adatoms and clusters of adatoms is enhanced near such defects, so that a stable island of deposit is formed much more readily. To illustrate this mechanism by an admittedly naïve example, consider a crystal formed of 'cubic' atoms, with a surface layer containing one over-sized atom (A in Fig. 3.6). Adatoms in the sites marked B have higher binding energy than adatoms elsewhere on the surface, so a new layer of crystal can spread out from impurity A. Such a mechanism may require only one of a few impurities per layer, i.e. impurity levels of 10^{-14} to 10^{-15}, which are undetectable. The

THE VAPOUR CRYSTAL INTERFACE

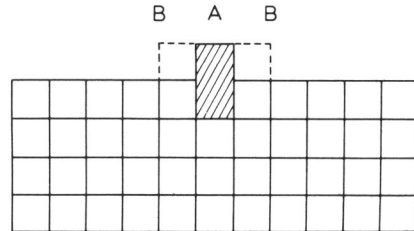

Fig. 3.6   An impurity as a nucleation site.

problem of nucleation at active centres has been discussed by Markov and Kashchiev[108] and by Stowell and co-workers.[109]

3.3   Surface roughness and surface rearrangement

The ease with which a crystal grows, that is to say, the ease with which a new layer of atoms or molecules forms on top of a previous layer, depends to a great extent on the density of adatoms on the surface and on their mobility. Obviously, when a crystal is in equilibrium with its vapour, there will be finite concentrations of adatoms and surface vacancies. These concentrations increase with temperature, and in general are larger than the concentrations of the corresponding defects (interstitial atoms and vacancies) within the bulk of the crystal. The concentration of adatoms is given by:

$$c_1 = \exp\left[\frac{-\Delta H_f}{RT} + \frac{\Delta S_f}{R}\right]$$

where ΔH_f and ΔS_f are the molar enthalpy and entropy of formation of the adatoms. The concentration of surface vacancies is given by an analogous equation. For adatoms on copper,[110] for example, $\Delta H_f = 58$ kJ mol^{-1}, $\Delta S_f = 12$ J mol^{-1} K^{-1}, so that at 1000 K, $c_1 = 5 \times 10^{-3}$. Near the melting point, then, there is a plentiful supply of mobile adatoms, and we expect the same to hold qualitatively for other substances. Of course, if the vapour pressure at the melting point is anomalously low, as for gallium, the concentration of adatoms and vacancies at the surface near the melting point is probably much lower.

Jackson[92] has discussed the growth of crystals from the melt in terms of surface roughness, and has demonstrated that the form of the interface between the crystal and the melt is governed by the parameter:

$$\alpha = \frac{L_f \xi}{kT}$$

where L_f is the heat of fusion, and $\xi \sim 1$ is the ratio of the numbers of bonds made by surface and bulk atoms. In the absence of impurities, substances for which $\alpha < 2$ grow with a smooth, planar interface. For intermediate values of α (2 – 10) the crystals grow with facetted surfaces, or pseudo-dendritically. If $\alpha > 10$ nucleation can occur ahead of the growing interface, and a polycrystalline (spherulitic) mass forms. The number of defects on the crystal surface affects the free energy of the surface to give either one or two minima, depending on the value of α (see Fig. 3.7). The fraction x of occupied surface sites at which the minima in surface energy occur are given by $e^{-\alpha}$ and $1 - e^{-\alpha}$ for $\alpha \gg 2$. Jackson's model does not carry over too happily as it stands to growth from the vapour; for example, if we put $x = 5 \times 10^{-3}$ for copper at 1000 K, we would require $\alpha = 5.3$, and we would predict that copper would deposit as facetted crystals. However, if we work out α from Jackson's equation, taking $L = 330$ kJ mol^{-1} and $\xi = 0.5$, we get $\alpha = 20$, hence $x = 2 \times 10^{-9}$, and we would predict that it would be difficult to deposit copper from the vapour as anything other than a micropolycrystalline mass. The reason for the conflict between Jackson's treatment and the experimental results lies in his way of estimating the heat of formation of adatoms, which comes out to be considerably too large for surfaces exposed to their vapour, as opposed to their melt. The main feature of Jackson's theory holds qualitatively nevertheless, namely, that if the crystal surface is sufficiently rough, growth can occur with no nucleation barrier. This situation is approached at sufficiently high temperature.

Although simple atomistic theories of nucleation often include the assumption that the surface is merely a cross-section of the bulk, there is now a large body of evidence that this is too naïve an approach. Even clean surfaces, such as may be produced by cleaving or sputtering under ultra-high vacuum conditions (10^{-10} Torr) display features which cannot be accounted for in terms of atomic arrangements characteristic of the bulk material[111-113]. It appears that the atoms on the surface have suffered some rearrangement, and it is not hard to see that it is reasonable that they should do so. Consider a crystal of an ionic compound with the NaCl structure. An ion in the bulk of the crystal is surrounded by a symmetrical array of ions, starting with six nearest neighbours of

THE VAPOUR CRYSTAL INTERFACE

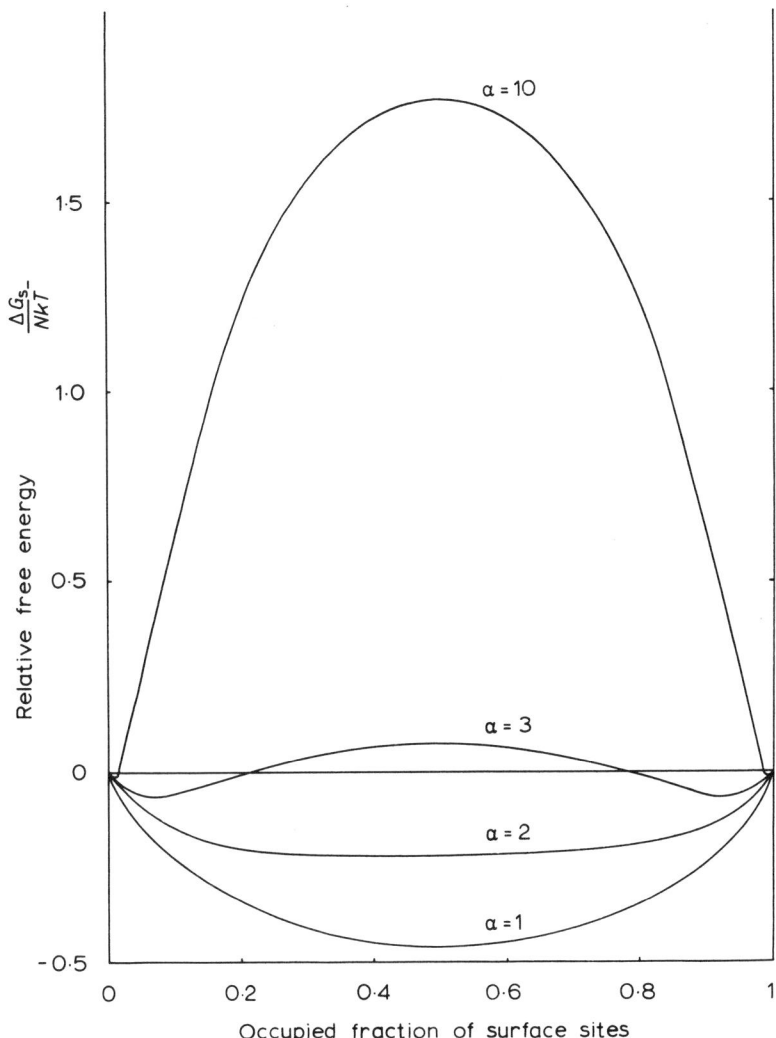

Fig. 3.7 Relative free energy versus monolayer occupation for various values of the parameter α (after Jackson[92]).

opposite charge, twelve second-nearest neighbours of the same charge, and so on. The resultant force on the central ion is zero. An ion at the surface experiences an asymmetric charge distribution, and would experience a net force if it tried to maintain its original position. The same is true to a lesser extent of ions just below the surface. The ions respond by altering their positions slightly, and by becoming polarized. There is some evidence that this rearrangement may persist to a considerable depth; this has been discussed by Weyl[114]. It is certain that, in the balance of forces leading to the equilibrium position of the ions, the resultant force from all the ions to one side of an arbitrary plane drawn through our central ion is of the order of a tenth of the bulk modulus in ionic crystals. When this force is removed, i.e. in making a surface, we expect considerable rearrangement to take place through an appreciable depth near the surface.

Such, then, is the surface which is offered to the vapour as a template for building on. In the case of homo-epitaxy, it may well be that this template is excellent for building a new surface layer on. One may imagine that, as this layer builds up, a certain amount of relaxation takes place throughout the rearranged layers at the surface, so that after the addition of a new surface layer, the depth of rearrangement is still the same, and the new surface layer has the same structure as the original surface had (Fig. 3.8).

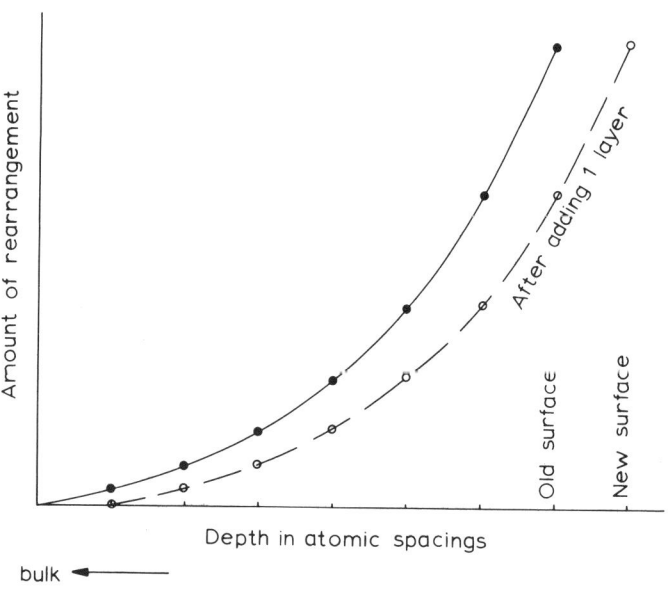

Fig. 3.8   Rearrangement of surface layers.

In this way, a perfect crystal can be built up. However, it may happen that the surface rearrangement is such that it provides a good template for a metastable phase to be developed; or again, the surface may have relieved some of its excess free energy by adsorbing a layer, or even several layers, of contamination such as water, oxygen, carbon, etc. In such a case, growth may not take place without two-dimensional nucleation (probably at defect sites) on what is essentially a substrate of a different substance, and on which the adatoms may be so strongly bound as to be immobile.

The work of Distler[115] has shed some light on growing crystals on contaminated surfaces. He has taken substrates of ionic crystals and ferro-electric crystals, and covered them with various thicknesses of amorphous materials such as amorphous ZnO and carbon. He has shown that: (1) oriented overgrowth is possible on these amorphous surfaces, (2) the orientation of the deposit is the same, even if the substrate is separated from the amorphous film and removed; (3) nucleation occurs over defect sites on the original substrate, and the information about the defect sites remains in the amorphous film even when the original substrate is removed; (4) in a similar way, the domain structure of a ferro-electric crystal persists in the overgrown material, even if the deposit is laid down after the original substrate has been removed. It thus appears that the template can be preserved through several hundred angstrom units of amorphous film on the substrate.

3.4 Surface diffusion

The adatoms on a solid surface in equilibrium with a vapour have a mean residence time τ which is the exponential function of temperature and the strength of the adsorption bond given below:

$$\tau = \tau_0 \exp(-Q/RT)$$

An adatom which does not have the necessary energy Q to surmount the potential barrier which keeps it on the surface may yet have the energy Q_m which will lift it over the smaller potential barrier separating it from an adjoining adsorption site. Thus, as well as being adsorbed and then desorbed again after a mean residence time τ, the adatoms can also migrate about the surface in a series of 'hops' from one site to an adjoining one, with a mean time τ_m between hops given by:

$$\tau_m = \tau_0 \exp(-Q_m/RT)$$

where τ_0 is again related to the vibration period of the migrating atom, in a direction normal to the surface[71].

The movement of atoms on the surface of solids is an important step in the growth of a crystal from the vapour. It is rarely possible to grow crystals of good quality under conditions of 'complete condensation', when every atom or molecule which is adsorbed stays fixed in its place without migrating or desorbing. These conditions are achieved by allowing a molecular beam from an oven containing an evaporating substance to impinge on a cold surface, for example. Material prepared in this way is often polycrystalline or glassy. The poor crystallinity is a consequence of the 'exchange current' of atoms or molecules arriving at each site and leaving again being essentially zero; every molecule stays where it lands, be it in the right place or the wrong. Surface migration, or surface diffusion, is one mechanism for achieving a larger 'exchange current'.

Surface migration plays another important part in crystal growth. We have seen, in Section 3.2, that in many cases it is necessary for a two-dimensional island or nucleus to form on the face of a crystal as the initial stage of the growth of an atomic layer. Such a nucleus may need to contain several atoms, (the number depending on the supersaturation of the vapour) to be stable in the sense of being as likely to increase in size as to decrease. The formation of such a 'critical nucleus' requires the conglomeration of several atoms, a process made vastly more facile by the mobility of the surface atoms. (It is worth pointing out that if the surface atoms are *not* mobile, they are probably so strongly attached to the surface that the 'critical nucleus' consists of one adatom.)

Finally, in those cases, to be described in the next section, where growth of the crystal can proceed without the formation of two-dimensional nuclei (by the continual reproduction of atomic ledges), it is surface diffusion which moves the adatoms alighting on the surface towards the ledges which act as sinks for them.

There is a wealth of experimental evidence for surface diffusion, apart from the observation that crystals grow. Volmer and Easterman[116], when growing platelets of mercury from mercury vapour at $-50°C$, found that the bombardment rate of gas molecules at the edges of the platelets was only 1/1000th of that required to bring about the observed rate of growth. They concluded that mercury atoms were moving over the large faces of the platelets quite freely. Alty and Clark[117] dipped cylinders of tin into mercury and observed the spreading of the amalgam up the side of the cylinder. Rates of up to a few millimetres a minute were recorded, and the rate varied widely according to the way the surface was prepared. Chariton, Semenoff and Schalinkoff[118] found that areas of a substrate surrounding seed crystals did not get covered with a crystalline deposit when the substrate was bombarded with atoms. They observed a 'depletion region' around each crystallite, a phenomenon taken into account in some theories of heterogeneous nucleation on a surface[106]. With the advent of modern experimental equipment further strides have been made in our under-

THE VAPOUR CRYSTAL INTERFACE 93

standing of the process. Recent investigations have involved the field-ion and field-emission microscopes, and even more recently, the Auger spectrometer. It is now possible to prepare and observe surface diffusion on really clean surfaces, so that the movement of atoms can be studied without much interference from surface impurities.

We can derive a theoretical expression for the diffusion coefficient for surface diffusion, using a model very similar to that which we shall use in Section 4.1 for diffusion in the solid. Fig. 3.9 represents a cross-section of the surface region. Let $s(x)$ be the line density of adatoms at position (x).

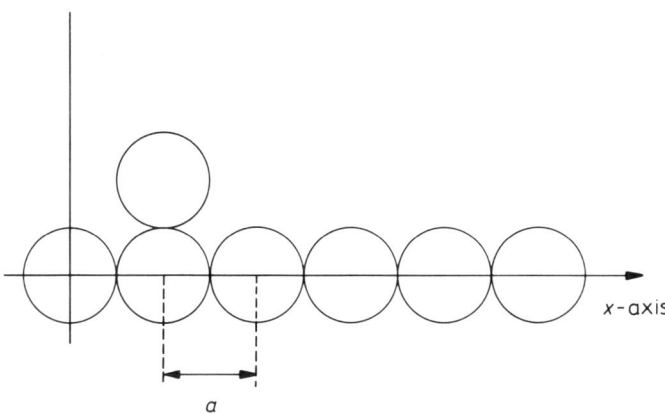

Fig. 3.9 Simple model of surface diffusion.

If the inter-atomic distance is a, then the surface density n of adatoms is $s(x)/a$. The rate at which adatoms hop in the positive-x direction is given by:

$$R_+ = \nu_m \sigma \exp(-Q_m/RT) \tag{3.25}$$

where ν_m is the vibration frequency of the adatoms, i.e. the rate at which they approach the activation barrier to migration of height Q_m. Since some adatoms will hop in the wrong direction, or sideways, we include a geometrical factor σ, which is just the probability of an adatom hopping in a given direction, if it hops at all. The flux of adatoms in the positive-x direction is thus:

$$J_+ = s(x)\nu_m \sigma \exp(-Q_m/RT) \tag{3.26}$$

If we fix our attention on the point a distance x from the origin, adatoms passing this point in the positive-x direction have come from neighbouring sites $(x - a)$ from the origin. Adatoms moving the other way have come from sites $(x + a)$ from the origin. The net flux is

$$J_{net} = [s(x - a) - s(x + a)]\nu_m \sigma \exp(-Q_m/RT)$$

$$= -2a \frac{ds}{dx} \nu_m \sigma \exp(-Q_m/RT)$$

$$= -2a^2 \nu_m \sigma \exp(-Q_m/RT) \frac{dn}{dx} \quad (3.27)$$

since $n = as(x)$. By comparison with the usual diffusion equation (Fick's Law):

$$J = -D \frac{dn}{dx} \quad (3.28)$$

we find that:

$$D = D_0 \exp(-Q_m/RT) \quad (3.29)$$

where

$$D_0 = 2a^2 \nu_m \sigma \quad (3.30)$$

If $a = 3 \times 10^{-8}$ cm, $\nu_m = 10^{13}$ s^{-1}, $2\sigma \cong 0.5$, we find that $D_0 = 0.0045$ cm^2 s^{-1}, which is typical of the values found experimentally. The value of Q_m can vary widely, so that surface diffusion coefficients span a great range. The activation energy for self-diffusion on a (100) copper surface is about 217 kJ[119], while other metals have activation energies in the range 34 kJ (for the (321) surface of silver) to 300 kJ (average over many faces on a tungsten field-emitter[119]). Barrer[120] has given values for the activation energy for migration of various hydrocarbons, halogenated hydrocarbons, SO_2, and CO_2 over carbon and silica gel. The values are in the range 8–54 kJ.

Experimental methods for measuring diffusion rates on surfaces have improved in the last few years with the introduction of Auger spectroscopy and low-energy electron diffraction studies. It is now becoming clear that much earlier work, on the rate of smoothing of scratches and grooves, for example, was carried out on contaminated surfaces. It has now been shown[121] that the presence of impurities can substantially alter the structure of the surface of metals and semiconductors. There is less detailed evidence on diffusion on insulators, as this class of materials cannot yet be studied by electron-bombardment surface techniques because the build-up of charge on the surface scatters the electron beam.

Even on a perfectly clean surface, we may still expect a wealth of different sites which complicate the theory of self-diffusion. A pure, perfect crystal of a single element has surfaces that are covered by steps with terraces between. These steps possess kinks or jogs; thus even in this simplest case there are three types of sites — terrace- ledge- and kink-sites, and these different sites hold adatoms for different lengths of time. The so-called terrace-ledge-kink model, based on the work of Kossel[122], Stranski[123] and Burton, Cabrera, and Frank[124] has been used extensively in the field of crystal growth, and Choi and Sewmon[125] have modified the theory of diffusion on surfaces to take into account the different residence times for adatoms in different sites.

Crystals of compounds present surfaces with atoms of all constituents exposed, in general. If the bonding in the crystal is appreciably hetero-polar, the surfaces present sites of alternate net positive and negative charge, which will greatly influence the motion of charged or polar molecules, and will result in different potential wells for adsorption of adatoms on different sites. In addition, the composition of the surface of a compound can change to reflect the composition of the vapour above it. Thus the (111)B surface of gallium arsenide exhibits 'gallium stabilized' and 'arsenic stabilized' surface 'phases' when examined by low-energy electron diffraction under different ratios of Ga to As_2 vapour pressures[126].

On all crystals, of elements or compounds, different surfaces results in different activation energies for migration, as we would expect from the different arrangements of saturated and dangling bonds on different faces. The intersection of crystallographic defects such as dislocations and stacking faults with the surface results in sites which again alter the residence time of adatoms. Stacking faults in particular, which will usually produce line defects on a surface, that is, lines of misplaced surface atoms, may provide paths along which migration is very rapid, or alternatively, may provide a line of traps for adatoms. The behaviour of defect sites in general we expect to be specific to the chemical species occupying them.

While most discussions of surface migration are concerned with the movements of single adatoms, which have two degrees of translational freedom and no rotational energy, some recent work by Bonzel[110] has shown that dimers (of copper atoms on copper, in the case he studied) may have enhanced mobility because of their rotational energy. Thus the sum of rotational plus potential energy may remain constant while the dimer molecule flies hundreds or thousands of atomic spacings across the surface. In this way, Bonzel has accounted for the rapid increase in diffusion coefficient for self-diffusion on copper near the melting point.

Thus we see that surface diffusion may have characteristics in common with gaseous diffusion, although the simplest theory treats it on a basis very similar to diffusion in the solid, with the dynamics governed by the probability of an atom hopping to an adjacent site. However, the picture is complicated by the presence of impurity atoms on the surface. Normally, the surface has a vastly greater affinity for impurities than the bulk of the solid does for a given partial pressure of impurity in the vapour. As a result, the simple theory of diffusion, which describes diffusion in the solid quite well, requires many modifications to cope adequately with diffusion on real surfaces. On the other hand, it is now possible to investigate processes on clean, nearly ideal surfaces using Auger spectrometry, as a result of which our understanding of surfaces is advancing rapidly.

3.5 Self sustained growth without nucleation

In Section 3.2 we considered the initial stage of the formation of a new layer of crystalline material on a surface which was assumed to be atomically flat to begin with, though containing impurities and crystallographic imperfections. To nucleate a new layer requires a certain overpotential or supersaturation, and the rate of formation of nuclei, which can be related to the rate of growth of the crystal, is very small up to a certain supersaturation, beyond which it rises rapidly. (See Fig. 3.5). According to Burton, Cabrera, and Frank[127] this critical supersaturation is around 50%, yet as those authors points out, real crystals frequently grow at negligible supersaturations (1% and even lower).

Kössel[122] and Stranski[123] put forward the idea that a monomolecular *step* on a surface can act as a sink for migrating adatoms, thus allowing the crystal to grow without the need for two-dimensional nucleation. The extra bond energy (enthalpy) available for an adatom at a step in the surface may compensate for its loss of mobility (entropy). Frenkel[128] discussed the structure of such a monomolecular step, and demonstrated that it must contain a high concentration of kinks (see Fig. 3.10).

Such a kink site is even more favourable a place for an adatom to attach itself than at a step. Burton and Cabrera[129] have shown that the concentration of kinks is even larger than Frenkel supposed. Clearly, the presence of such steps on the surface of a growing crystal would provide a means for the crystal to grow under conditions of much lower supersaturation than required for two-dimensional nucleation.

The high-index faces of crystals may be thought of as being composed of narrow terraces of low-index faces separated by

THE VAPOUR CRYSTAL INTERFACE

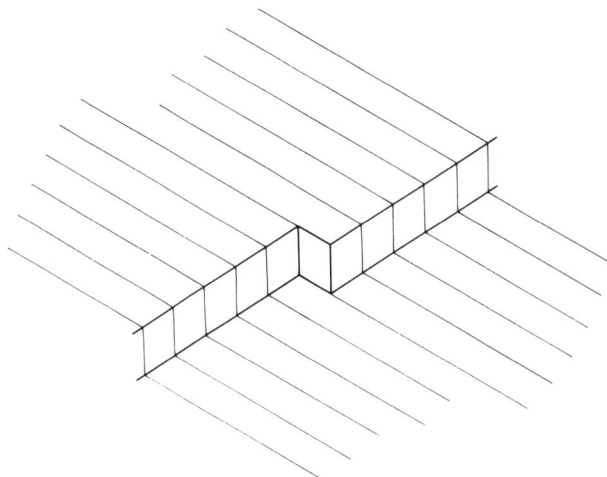

Fig. 3.10 Surface step with kink site.

monomolecular steps. However, such faces have higher surface energy than low-index faces, and are consequently not observed on real crystals. It is interesting to observe that, if an area of such a face were to form, it would grow more rapidly in a direction normal to its surface than the adjoining surfaces would, and would thus decrease in area until it disappeared.

On faces of low index, monomolecular steps would not be generated by thermodynamical fluctuations except, perhaps, very close to the melting point; therefore the steps required for growth can only be produced, on a perfect crystal surface, under a highly supersaturated environment[124]. It remains, therefore, to see how such steps may be produced, and continually reproduced, on the surface of an *imperfect* crystal, so as to give self-sustained growth, i.e. growth without two-dimensional nucleation.

In 1949, Frank[26] proposed what is probably the best-known mechanism for reproducing surface steps. He pointed out that a dislocation with a screw component would intersect the surface so as to produce a step which cannot be removed by adding atoms or molecules at the surface. Fig. 2.12(b) illustrates such a step. As the crystal grows by adding atoms or molecules along the step, the step coils up to form a spiral step, with a minimum radius of curvature given by the radius of the critical nucleas under the prevailing conditions of supersaturation, as illustrated in Fig. 3.11. Such growth spirals have frequently been observed on a great variety of crystals, both natural and synthetic[97].

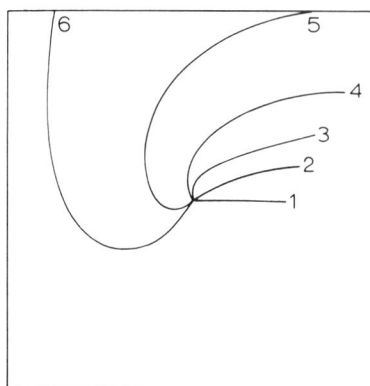 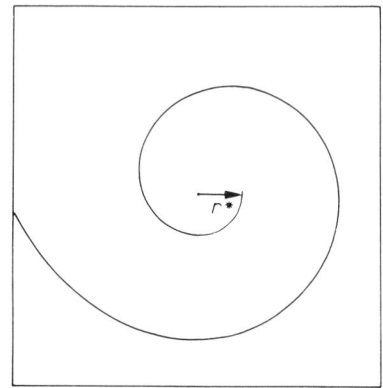

Fig. 3.11 Successive stages (1—6) and final configuration in the development of a growth spiral from a screw dislocation.

The dynamics of migration of adatoms to steps and the consequent motion of the steps has been treated comprehensively by many authors, e.g. the classic paper of Burton, Cabrera and Frank[124], and Strickland-Constable's book[130], and so we will not pursue this topic here. The theory predicts that the growth rate is proportional to the square of the supersaturation for low supersaturation, changing to a linear dependence at higher supersaturations (e.g. 10% for iodine crystals.)

A modification of the screw dislocation mechanism was proposed by Burton, Cabrera and Frank[124]. Two dislocations with screw components of opposite sign, intersecting the surface near each other, can give rise to a succession of steps in the form of closed rings, as illustrated in Fig. 3.12.

While the screw dislocation mechanism has rightly enjoyed a certain popularity, there are other ways in which surface steps may be continually reproduced. For example, twinning which results in re-entrant faces provides a line of surface sites where adatoms may attach themselves preferentially, thus initiating steps on each twin[131]. A similar mechanism may operate at re-entrant grain boundaries.

A third mechanism will be considered here which, while not being strictly self-sustained growth without two-dimensional nucleation, departs sufficiently from the simple nucleation situation to warrant inclusion in this section. It has been observed frequently that during the growth or etching of a crystal, low-index faces may break up into very small facets (micro-facets) of high index[132]. The micro-facets usually have orientations close to those of the original low-index

THE VAPOUR CRYSTAL INTERFACE

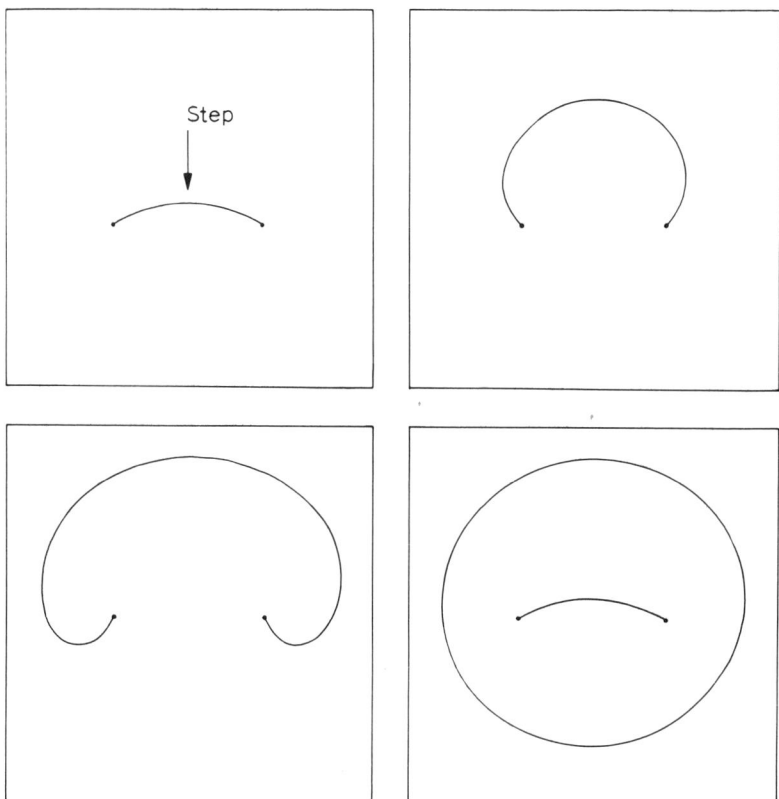

Fig. 3.12 Formation of surface steps from two dislocations with screw components of opposite sign.

planes, and may therefore be thought of as being made up of terraces parallel to the low-index plane, separated by regularly spaced monoatomic steps. This behaviour has been interpreted in terms of changes in surface free energy resulting from adsorption[133], and it would seem that in some systems, the extra positive free energy term arising from the steps is more than offset by the lowering of free energy when adsorption takes place.

Let us imagine a crystal of substance A bounded by surfaces, some of which are micro-facetted, and suppose the crystal is growing by chemical vapour transport with a transporting agent X. The micro-facetted surface has arisen as a result of the adsorption either of X or of some other species present as an impurity. Before an A-atom or molecule, which exists in the vapour mainly as a volatile compound with the transporting agent X, can be incorporated as part of the crystal, it has to adsorb on the surface and migrate to a ledge site. Both adsorption and migration will be influenced by the presence of other adsorbed species. If the other adsorbed species are strongly bound and immobile, they present a barrier to the penetration of A-atoms or molecules. On the other hand, if the other adsorbed species are only weakly bound so that they are continually adsorbing and desorbing (large exchange current and short life time τ on the surface) there is clearly plenty of opportunity for the A-atoms to penetrate, and indeed the concentration of A-atoms may be increased as a result of stabilization of A-adatoms by the other adsorbates.

If adsorption of A from the vapour and migration of A-adatoms on the surface are not seriously hindered by the presence of other species on the micro-facetted surface, we may visualize the flux of A to the atomic ledges causing these ledges to spread across the surface. It remains only to nucleate a new layer at each 'high point' where the micro-facets meet. (Fig. 3.13).

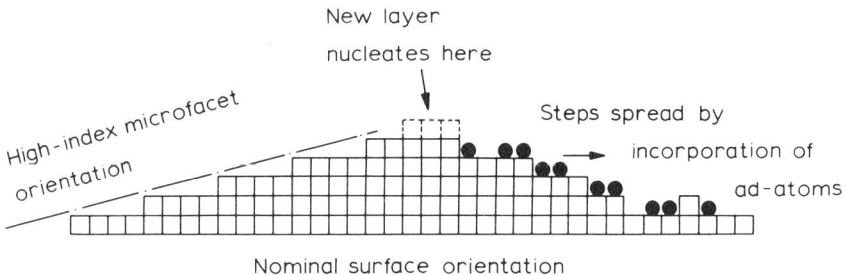

Fig. 3.13   Growth on high-index faces, with adsorption of other species.

If the adsorption of foreign atoms from the vapour lowers the free energy of atomic ledges to the extent that micro-facetting is favoured, a small island of adatoms can presumably be similarly stabilized by adsorption along its periphery. Thus the barrier to two-dimensional nucleation may be very low, and the crystal can grow at a correspondingly low supersaturation.

In conclusion, although the theories of nucleation predict rapid nucleation, and hence appreciable growth rates, only above a critical supersaturation of 20–50%, real crystals are frequently observed growing at negligible supersaturation. Clearly, some other growth mechanisms are being exploited; a few such possible mechanisms are reviewed in this section. It is obvious that the subject is far from exhausted.

CHAPTER FOUR
Vapour Transport

In this chapter we develop the concepts underlying the transport of chemical species in a vapour phase. We distinguish between molecular processes (diffusion) in which the motion involves individual molecules or atoms, and other processes (viscous flow, thermal convection, etc) in which parts of the vapour move as a whole. The basic concepts are treated in some detail, so that the transition to considering the combinations of transport processes which actually take place in real systems is not difficult. Probably most of the transport processes we shall describe are well understood, at least qualitatively, by the majority of those who read this book. The importance of these processes in crystal growth from the vapour is so great that some recapitulation of the simple theories of these processes is warranted here.

4.1 Diffusion

The earliest experiments on diffusion were performed by Priestley[134] who found that air and carbon dioxide, and air and hydrogen etc., in a cylinder became 'equally diffused' on standing for a day. The general phenomenon of diffusion, observed not only in gases and liquids but in all forms of matter, is a manifestation of the universal tendency of entropy towards a maximum. A system in which the components are initially segregated increases its entropy by the mixing of the components (decrease of ordering). We may define diffusion as the motion of each component of a system from regions where the concentration is high to regions where it is lower, the motion involving individual atoms or molecules. Thus we distinguish diffusion, which is essentially a molecular phenomenon, from convective mixing of, for example, smoke in air, or hot water in cold, where much larger packages of the components take part in the mixing process.

It is found experimentally that the flux of each component is proportional to the concentration gradient of that component, thus:

$$J_i = -D_i \nabla n_i ^* \tag{4.1}$$

a law due to Fick. The proportionality constant D_i is called the *diffusion coefficient*, and Equation 4.1 may be taken as a phenomenological definition of the diffusion coefficient. In accordance with observation, the negative sign indicates that the flux of each component is away from places where the concentration is high and towards places where it is low. Diffusion acts to level out concentration gradients. It is found that Fick's law applies not only to gases and liquids, but to all states of matter, though the diffusion coefficient is much smaller in solids.

The kinetic theory of gases provides a simple picture of diffusion in terms of the mean free path model, which we give here in a one-dimensional form. Consider a reference plane A, with molecules crossing it from the left and the right (Fig. 4.1). On average,

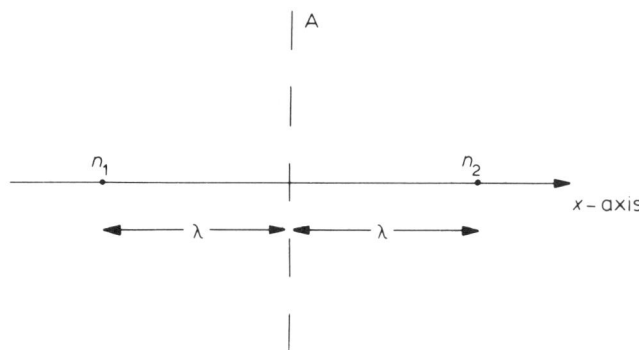

Fig. 4.1 Simple model of gaseous diffusion.

molecules crossing the plane A from the left have come from a point one mean free path, λ, away, where the concentration of those molecules is n_1. The number arriving at a unit area of A in unit time is thus $n_1 \bar{c}/4$, where $\bar{c}$ is the mean velocity of the molecules. Similarly, molecules crossing the plane A from the right have come, on average, from a point one mean free path away to the right, where the concentration is n_2, and the number of molecules crossing a unit

---

*∇ is the mathematical derivative operator $(\partial/\partial x; \partial/\partial y; \partial/\partial z)$.

area of A in unit time from the right is $n_2 \bar{c}/4$. The net flux of molecules is thus:

$$J = n_1 \bar{c}/4 - n_2 \bar{c}/4$$
$$= (n_1 - n_2)\bar{c}/4$$

If we put $(n_1 - n_2) = -2\lambda \, dn/dx$, we obtain:

$$J = -\frac{\lambda \bar{c}}{2} \frac{dn}{dx} \qquad (4.2)$$

From which it appears that:

$$D = \frac{\lambda \bar{c}}{2} \qquad (4.3)$$

In fact, in our simple analysis, we have assumed that all molecules crossing the plane A have come from a distance λ away, whereas those molecules not travelling parallel to the x-axis but at some angle θ to it will, on average, have arrived at the plane A from a point $\lambda \cos \theta$ away. Taking this factor into account alters the numerical multiplier in Equations 4.2 and 4.3 from ½ to ⅓. This simple picture does not take into account molecules of different weight (for which $\bar{c}$ and λ will be different) nor the effect of the persistence of molecular velocities after collisions. These factors and others introduce mathematical complexity into the theory of diffusion. The simple picture presented here is adequate for our purposes and we would refer readers to the excellent books by Jeans,[135] Boltzmann,[136] Chapman and Cowling,[137] and Enskog[138] for the more erudite approach.

The elementary theory we have given above brings out two important facts about diffusion. Firstly, the process of diffusion is statistical in nature. If there are more molecules of species A to the right of some imaginary reference surfaces than to the left, then because of the constant, random motion of the gas molecules, more A molecules cross the reference surface from right to left than in the reverse direction, resulting in a net current of species A. Quite clearly, it is an irreversible process, in the sense that a spontaneous separation of the different species to any appreciable degree is a very improbable event. From the standpoint of the second law of thermodynamics, diffusion results in mixing, that is, in a reduction of order and an increase in entropy. Such a process is spontaneous, while the reverse process may be carried out only by creating more entropy in the surroundings, for example, by imposing a temperature gradient and producing a certain amount of unmixing by *thermal diffusion*, described below.

Secondly, the elementary theory predicts the dependence of D on pressure and temperature to be:

$$D \propto \frac{T^{3/2}}{P}$$

which is nearly right. Experimentally it is found that a better representation is:

$$D = D_0 \frac{P_0}{P} \left(\frac{T}{T_0}\right)^n \tag{4.4}$$

where D_0 is the value of D measured at a standard temperature T_0 and pressure P_0, usually 273 K and 1 atm, and n is in the range 1.5–2.0. We will frequently use $n = 1.8$ in calculations, after Schäfer.

The elementary theory does not predict correctly the dependence of D on the composition of a gas mixture. In binary mixtures, this variation is found to be small. Tables of diffusion coefficients for binary mixtures may be found in the books by Jeans,[135] Partington,[54] Chapman and Cowling,[137] and many others. Typical values of D_0, extracted from the above works, are given in Table 4.1.

Table 4.1 *Binary gaseous diffusion coefficients*

Gases	Diffusion coefficient D_0 cm^2 s^{-1}
$H_2 - D_2$	1.13
He or H_2 – other gas	0.5 – 0.7
Two diatomic gases	0.2
Two triatomic gases	0.1

As a comparison to diffusion in gases, we will present here an elementary theory of diffusion in solids, to bring out the different magnitude of the diffusion coefficient and its more rapid change with temperature, and to emphasize the statistical nature of the process.

We consider two adjacent atomic sites in a solid, arranged along the x axis, as in Fig. 4.2. The number of solute atoms in a plane of atoms normal to the x-axis and containing the left hand site is $n_1 d$,

VAPOUR TRANSPORT

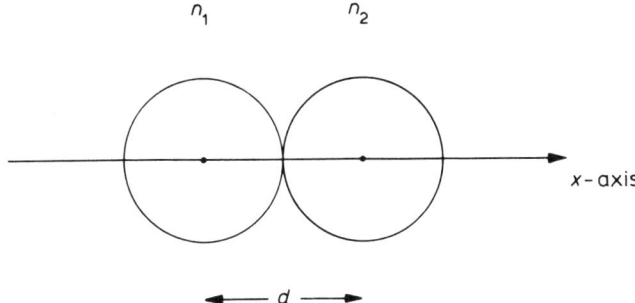

Fig. 4.2   Simple model of diffusion in solids.

where n_1 is the solute concentration and d the inter-atomic spacing along the x-axis. These atoms vibrate about their mean position with a frequency ν, and in order to hop into an adjacent site, have to have an energy of at least E_d, the height of the inter-atomic potential barrier. The probability of a given atom hopping to an adjacent site in unit time is thus $\nu e^{-E_d/kT}$. The number of atoms hopping one place to the right in unit time is thus $n_1 d\sigma\nu e^{-E_d/kT}$, where σ is the probability of the atom hopping in the right direction. The number of atoms travelling in the opposite direction is, similarly, $n_2 d\sigma\nu e^{-E_d/kT}$. The net flux is therefore given by:

$$J = \sigma d\nu(n_1 - n_2)e^{-E_d/kT}$$

If we put $n_1 - n_2 = -d(dn/dx)$ we have:

$$J = -\sigma d^2 \nu e^{-E_d/kT} \left(\frac{dn}{dx}\right) \quad (4.5)$$

an equation similar to that given by Van Liempt[139]. We may put:

$$D = \sigma d^2 \nu e^{-E_d/kT} = D_0 e^{-E_d/kT} \quad (4.6)$$

It is found experimentally that this sort of equation for the dependence of diffusion constant on temperature applies to most solids, and the order of magnitude of D_0 is correctly given by Equation 4.6; e.g. if $d^2 \simeq 10^{-15}$ cm^2, $\nu \sim 10^{13}$ s^{-1} and $\sigma \simeq 1$, we have $D_0 \simeq 0.01$ cm^2 s^{-1}. The activation energy for diffusion, E_d, is typically a few electron volts. Some examples are given in Table 4.2, to compare with gases.

So far in our discussion of diffusion, we have considered only systems which are at rest while the individual components move within the system. We now turn to consideration of the diffusion of a mixture of gases on which a macroscopic motion is imposed. The

Table 4.2 *Diffusion coefficients in solids and gases*

Temperature K	Two diatomic gases	Typical solid with $D_0 = 0.01 \ cm^2 \ s^{-1}$	
	for $D_0 = 0.2 \ cm^2 \ s^{-1}$	for $E_d = 1 \ eV$	for $E_d = 5 \ eV$
300	0.237	1.59×10^{-19}	1.00×10^{-86}
1000	2.07	9.2×10^{-8}	6.3×10^{-28}
2000		3.02×10^{-5}	2.51×10^{-15}

velocities of the molecules of each species of the gas will cover a distribution which, as a first approximation, will be the Maxwell–Boltzmann distribution, but slightly perturbed, since a macroscopic redistribution of the various species is taking place. For each species, we may define an average velocity u_i. If the gas were stationary and uniform in composition, u_i would be zero. If the gas were of a single component and moving at some macroscopic velocity v, then $u_i = v$. However, in a non-uniform gas, each species will generally have a different average velocity u_i. Under such circumstances, we can ascribe a macroscopic velocity or drift velocity to the gas as a whole in various ways. For example, we may take a weighted mean of the values of u_i, weighting each in accordance with the molar density y_i, or the mass fraction c_i, or any other property, of the separate species. Which mean velocity is chosen will depend on the problem in hand. In many chemical problems, the molar average velocity:

$$U = \sum_i y_i u_i \tag{4.7}$$

is the most useful, whereas in hydrodynamic problems, the mass average velocity:

$$u = \sum_i c_i u_i \tag{4.8}$$

is favoured, usually to some advantage. Let us generalise the discussion by taking a mean velocity **V** in terms of a general property α of the gas:

$$V = \sum_i \alpha_i u_i / \sum_i \alpha_i \tag{4.9}$$

We now define the diffusion flux by:

$$J_i = \alpha_i (u_i - V) \tag{4.10}$$

This is the flux of property α transported by diffusion by species i. If we take the summation of Equation 4.10 over all the species in

VAPOUR TRANSPORT

the gas, we obtain:

$$\sum_i J_i = \sum_i u_i \alpha_i - V \sum_i \alpha_i \equiv 0 \qquad (4.11)$$

from Equation 4.9. It follows, evidently enough, from the definition of the diffusive flux, that the sum of the diffusive fluxes over all species is zero. In consequence, it is not possible for diffusion to produce a net transport of property α, when the diffusion fluxes are defined in terms of the α-average velocity V. To take a simple illustration, imagine two reservoirs, one containing gas A, one containing gas B, connected by a tube with a valve in the middle. The valve is shut initially, and the two gases are at the same uniform temperature and pressure. Opening the valve permits mixing of the gases, and this takes place at constant pressure*, that is to say, at constant total mole concentration. In such a system, it would be natural to define diffusive fluxes in terms of the (zero) molar average velocity, Equation 4.7. It can be seen immediately that if A and B have different molecular weights, a net transport of mass takes place, so that the mass-average velocity, Equation 4.8, is non-zero. However, since it varies in both time and space, a description of diffusion in terms of it would normally be less helpful. On the other hand, if a chemical reaction between A and B took place to form further species C, D . . ., the pressure in the system would change if the atomicity of the products differed from that of the reactants, so that a change in mole concentration, hence a net molar velocity, would result. The mass-average velocity approach is then no longer at a disadvantage. This net molar velocity has profound consequences, often unappreciated, in vapour growth of crystals, which are discussed in Section 4.2.

The phenomenological law describing diffusion may be expressed in different ways, depending on how diffusion is defined. Thus we could write:

$$J_i^n = D_i^n n \nabla y_i \qquad (4.12)$$

or

$$J_i^m = D_i^m \rho \nabla c_i; \quad J_i^\alpha = D_i^\alpha \alpha \nabla \left(\frac{\alpha_i}{\alpha}\right) \qquad (4.13)$$

and so on. J_i^n is a mole flux, J_i^m is a mass flux, J_i^α is an α-flux. It may be shown[11] that $D_i^n = D_i^m$. However, it would be possible to

---

*We chose to ignore the effect of gravity, and to assume ideal gas behaviour.

define diffusion coefficients $D_i^{m\,n}$ and $D_i^{n\,m}$ by:

$$J_i^n = D_i^{nm} \rho \nabla c_i \qquad (4.14)$$

$$J_i^m = D_i^{mn} n \nabla y_i \qquad (4.15)$$

In general, these coefficients are different from each other and from D_i^m, D_i^n, being equal only if the species all have the same molecular weight.

We have dealt with the diffusion of a moving gas mixture in detail, and particularly the various diffusion constants, because later we will be considering the combined effect of various transport mechanisms, such as viscous flow, thermal convection, solutal convection, with diffusion. The convective problems necessarily introduce mass quantities into the analysis.

Thermal diffusion

If a temperature gradient is imposed on a mixture of gases, a degree of separation of species occurs. This separation is opposed and eventually limited by ordinary diffusion. The origin of this effect, known as thermal diffusion, lies in the details of momentum transfer between colliding molecules. It may be comparable in magnitude to ordinary diffusion, but no simple, satisfactory theoretical treatment has yet been advanced. Chapman and Cowling[137] have treated the effect at length in their 'Mathematical Theory of Non-uniform Gases'. We summarize the main points of the theory here.

The direction of thermal diffusion is found to depend on the intermolecular repulsive force law Taking this law to be of the form $r^{-\nu}$, then if $\nu > 5$, the lighter molecules tend to accumulate at the hot regions, if $\nu < 5$, they accumulate in the cold regions, and there is no effect if $\nu = 5$.

Including the thermal diffusion effect, the diffusion equation takes the form, for a binary gas mixture:

$$J_1 = -J_2 = -Dn \nabla y_1 - D_T n y_1 y_2 \nabla T \qquad (4.16)$$

where D_T is the thermal diffusion coefficient.

Three combinations of the ordinary and thermal diffusion coefficients have special names; these are the Soret coefficient s_T, the thermal diffusion ratio k_T, and the thermal diffusion factor α. They are defined by:

$$s_T = D_T/D, \quad k_T = D_T T y_1 y_2/D, \quad \alpha = D_T T/D$$

It is found that of the three, α varies least with composition, and is most readily measurable experimentally. Its magnitude is about 0.5

VAPOUR TRANSPORT

in mixtures consisting largely of hydrogen, and about 0.05 in mixtures of other gases, at 20°C. It generally increases as the temperature increases, and may change sign as the temperature decreases.

To get an idea of the order of magnitude of the effect, consider a mixture of 90% H_2, 10% N_2, for which $\alpha = 0.5$ at 300 K. If the mixture is sealed in a container, and a temperature gradient of 10 K cm^{-1} imposed on it, a concentration gradient will develop to counteract thermal diffusion until there is no net flow. We may then write:

$$J_1 = -Dn \left\{ \nabla y_1 + \frac{\alpha y_1 y_2}{T} \nabla T \right\} = 0$$

Hence $\nabla y_1 = 0.15\%$ cm^{-1}, i.e. 3% concentration change in 20 cm, which could be very significant in reduction of high valency metal halides (e.g. WF_6). On the other hand, if the temperature gradients are small, or if no hydrogen is present, thermal diffusion has a negligible effect. Thus we conclude that, except in cases such as the one discussed above, the practical crystal grower need not concern himself with the effects of thermal diffusion.

4.2 Vapour transport in crystal growth systems

4.2.1 Stefan's flow

The experiments of Stefan[140] in 1882 on the evaporation of liquids from pipes led him to conclude that the vapour is not transported away from the surface by diffusion alone, as had been thought by Maxwell (and many others since then). Stefan proposed, and verified experimentally, that the volume of vapour passing through a unit cross-section of pipe per unit of time was given by:

$$V = \frac{D}{P-p} \frac{dp}{dx} \tag{4.17}$$

where P is the total pressure, p the partial pressure of the vapour, and D its diffusion coefficient in the surrounding atmosphere. Such an expression may be deduced if we imagine that the whole of the gas in the pipe is moving away from the liquid surface with a velocity U, and that simultaneously the vapour and air (or other surrounding gas) are diffusing into each other. The flux may be expressed as the sum of a flow term and a diffusion term:

$$J_{vap} = \frac{U}{RT} p_{vap} - \frac{D}{RT} \frac{dp_{vap}}{dx} \tag{4.18}$$

We have used partial pressure instead of mole fraction in the flow terms (cf. Equation 4.12). This is permissible when the total pressure varies negligibly with x, and is perhaps more readily visualized. For the surrounding air:

$$J_{air} = \frac{Up_{air}}{RT} - \frac{D}{RT}\frac{dp_{air}}{dx} = 0 \qquad (4.19)$$

since no net flux of air takes place. (In fact, some flow must take place relative to the laboratory co-ordinates, to fill the space vacated by the liquid which is evaporating. This flow is small, of course, and in any case, Equation 4.19 can be made exact by using a system of co-ordinates fixed to the liquid surface.) Addition of Equations 4.18 and 4.19 gives:

$$J_{vap} = \frac{UP}{RT} - \frac{D}{RT}\frac{dP}{dx}$$

where P is the total pressure. If the velocity U is not large, we can ignore dP/dx (see Section 4.5) and express U in terms of the flux J_{vap}:

$$\frac{U}{RT} = \frac{J_{vap}}{P}$$

Substituting this expression into Equation 4.18 gives us:

$$J_{vap}\frac{RT}{P} = \frac{P}{P - p_{vap}}\frac{dp_{vap}}{dx}$$

which is the equation proposed by Stefan.

The existence of a flow velocity U is necessary here, because as we saw in the previous section, diffusion alone will not give a net flux of molecules of the evaporating liquid along the pipe. This is a point which many authors have overlooked. The origin of the flow velocity in the vapour of an evaporating substance lies in the expansion on evaporation, an effect which is equivalent to pumping.

It is clear that a flow velocity will exist in the vapour from which a crystal is growing whenever the mechanism or reaction by which the vapour turns into a crystal involves a change in the number of vapour-phase molecules, i.e. a change in volume or pressure. This is very frequently the case in crystal growth systems. Thus for the simplest case, as we have already seen:

$$\text{solid} \rightarrow \text{vapour} \rightarrow \text{solid} \qquad (4.20)$$

there is a flow velocity away from the source material, towards the

VAPOUR TRANSPORT

growing crystal. Before considering the flow velocity in transport of material by dissociative sublimation and chemical vapour transport, we will consider in detail the transport of a single component by the process represented by Equation 4.20.

4.2.2 Example: sublimation and transport of silver

Consider a capsule of silica or other refractory material, sealed at each end, and containing solid silver at one end, a small seed crystal of silver at the other end, and an amount of inert gas which is taken to include the residual gas left after evacuating the capsule and any gases evolved on heating the capsule, as well as inert gas added to the capsule purposely. The reason for adding inert gas has to do with maintaining smooth surfaces on the growing crystal, and we defer discussion of this topic until Chapter 5. The capsule is now put in a furnace, and heated so that the end containing the small seed is at a temperature $T(0)$, and the other end is at a temperature $T(l)$. Fig. 4.3 is a schematic picture of the arrangement.

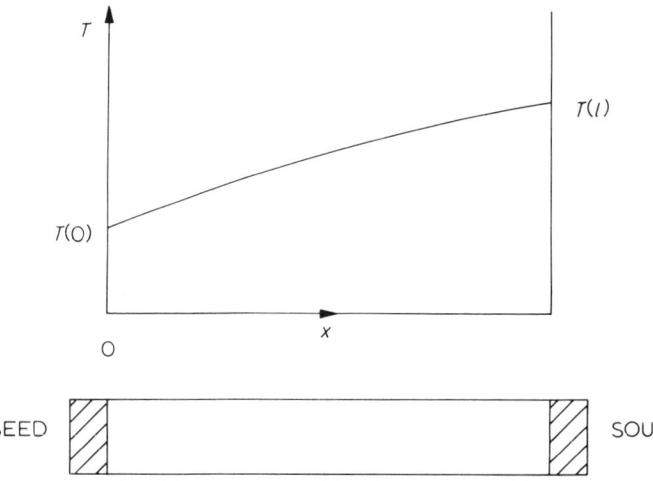

Fig. 4.3 Schematic crystal growth system and one-dimensional model.

The silver vapour is carried down the capsule by flow of the whole volume of gas at a velocity U, and also by diffusion. The inert gas is carried down the capsule at the flow velocity U, and diffuses back at such a rate as to give no net flow. The flow equations are:

$$J_{Ag} = \frac{U p_{Ag}}{RT} - \frac{D}{RT} \frac{dp_{Ag}}{dx} = J \qquad (4.21)$$

$$J_Z = \frac{Up_Z}{RT} - \frac{D}{RT}\frac{dp_Z}{dx} = 0 \qquad (4.22)$$

where the p's are the partial pressures, J is the net transport rate of silver, and Z is the inert gas. If the capsule is of the order of 1 cm or more in radius then for all reasonable transport rates the total pressure inside the capsule may be assumed not to vary along x (see Section 4.5). These equations are exactly the same as Equations 4.18 and 4.19. We can again eliminate U in terms of J by adding Equations 4.21 and 4.22:

$$J = \frac{UP}{RT}$$

Equations 4.21 and 4.22 now read:

$$J = \frac{J}{P}p_{Ag} - \frac{D}{RT}\frac{dp_{Ag}}{dx} \qquad (4.23)$$

$$0 = \frac{J}{P}p_Z - \frac{D}{RT}\frac{dp_Z}{dx} \qquad (4.24)$$

We may use Equation 4.23 to obtain an expression for J, the growth rate. Rearranging the equation, we obtain:

$$\frac{dp_{Ag}}{p_{Ag} - P} = \frac{JRT}{DP}dx \qquad (4.25)$$

This may be integrated from $x = 0$ to $x = l$, and for the moment we will assume that the variation of T with x is not great:

$$\ln\left\{\frac{p_{Ag}(l) - P}{p_{Ag}(0) - P}\right\} = \frac{JRTl}{DP} \qquad (4.26)$$

so that:

$$J = \frac{DP}{RTl}\ln\left\{\frac{p_{Ag}(l) - P}{p_{Ag}(0) - P}\right\} \qquad (4.27)$$

To evaluate J, we need to know $p_{Ag}(l)$, $p_{Ag}(0)$, and the total pressure P. We imagine that we can arrange to measure the total pressure, using a silica spiral gauge or a Bourdon spoon gauge. Alternatively, we imagine that we are able to calculate the total pressure sufficiently accurately by filling the capsule at room temperature with a known pressure of inert gas which is large compared with the pressure of residual gases in the capsule. We will

VAPOUR TRANSPORT

assume that the vapour is in equilibrium with the solid silver at the seed end of the capsule ($x = 0$) and over the source material ($x = l$), so that $p_{Ag}(0)$ and $p_{Ag}(l)$ are the saturated vapour pressures at the temperatures $T(0)$ and $T(l)$. The growth rate may now be calculated from vapour pressure data.[64] In Figs. 4.4 and 4.5 are shown the calculated growth rates as a function of the temperature difference ΔT between the source material and the growing crystal. In Fig. 4.4, the source material is at 1200 K, and in Fig. 4.5, it is at 1100 K. The inert gas pressure over the source, $p_Z(l)$, has been set at values from 10^{-8} to 10^{-4} atm in each case. We notice first of all that the growth rate increases very rapidly as ΔT increases (the growth rate scale is logarithmic), and then, when ΔT is greater than some value around 10–25 K, depending on the amount of inert gas present, the growth rate remains fairly constant. Secondly, when the amount of inert gas is small compared to the amount of silver vapour (e.g. the upper two curves of Fig. 4.4 – at 1200 K, p_{Ag} is 1.6×10^{-6} atm) variation in the inert gas pressure by an order of magnitude from 10^{-8} atm to 10^{-7} atm decreases the transport rate by less than a factor of 2. On the other hand, if the inert gas is the majority component, as it is for the lower three curves of Fig. 4.5, an order of magnitude increase in the inert gas pressure produces an order of magnitude decrease in the transport rate. These things are intuitively obvious. Let us look a bit more closely at the situation when the inert gas is the majority component. Then $P \gg p_{Ag}(l) > p_{Ag}(0)$, so that the argument of the logarithm in Equation 4.27 is near to unity. We can therefore expand it to get:

$$J \simeq \frac{DP}{RTl} \left\{ \frac{p_{Ag}(l) - P}{p_{Ag}(0) - P} - 1 \right\} \tag{4.28}$$

$$= \frac{DP}{RTl} \left\{ \frac{p_{Ag}(l) - p_{Ag}(0)}{p_{Ag}(0) - P} \right\}$$

$$\simeq \frac{-D}{RT} \left\{ \frac{p_{Ag}(l) - p_{Ag}(0)}{l} \right\} \tag{4.29}$$

which is the solution that we would have obtained had we ignored the Stefan velocity in Equation 4.21. Clearly, when the inert gas is the majority component, ignoring the Stefan velocity results in calculated rates of transport which are very close to the rates calculated when the Stefan velocity is included. This no doubt justifies the use of the 'diffusion only' approach as a useful approximation; we find it objectionable because it contains an inconsistency. If the active component diffuses towards the growing

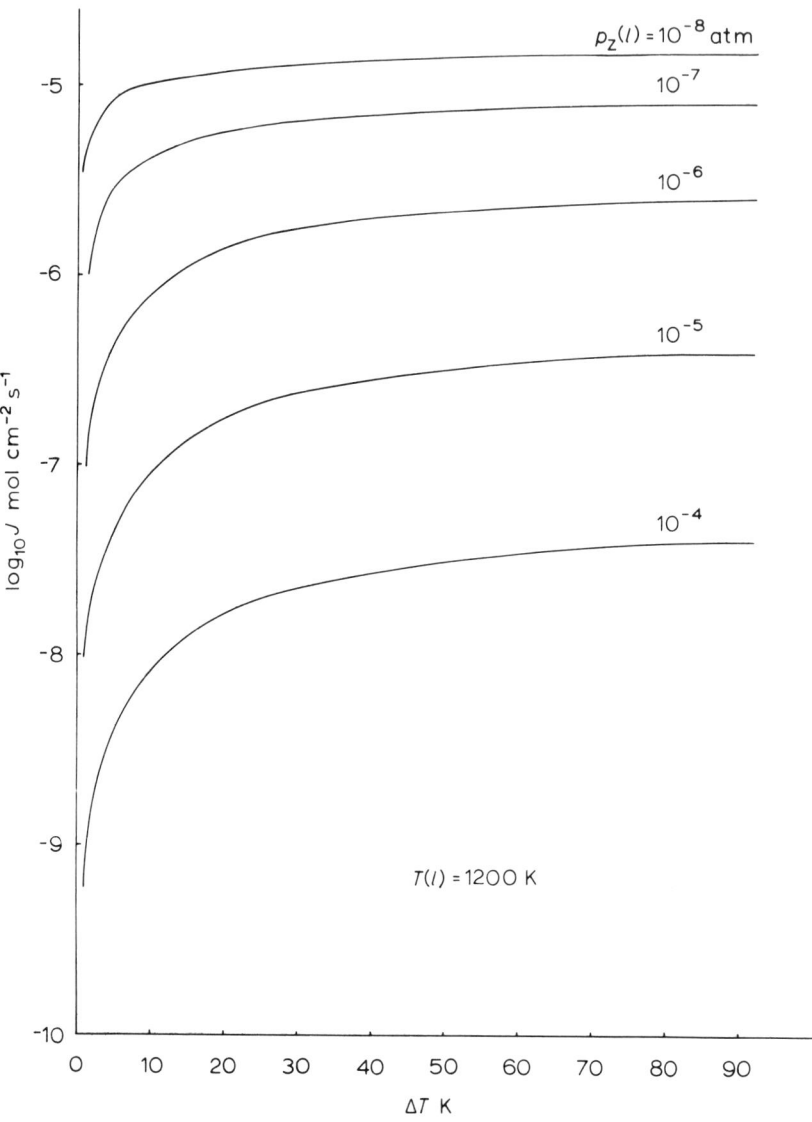

Fig. 4.4 Transport of silver. Transport rate J as a function of temperature difference between the source and the crystal, ΔT. Source temperature 1200 K.

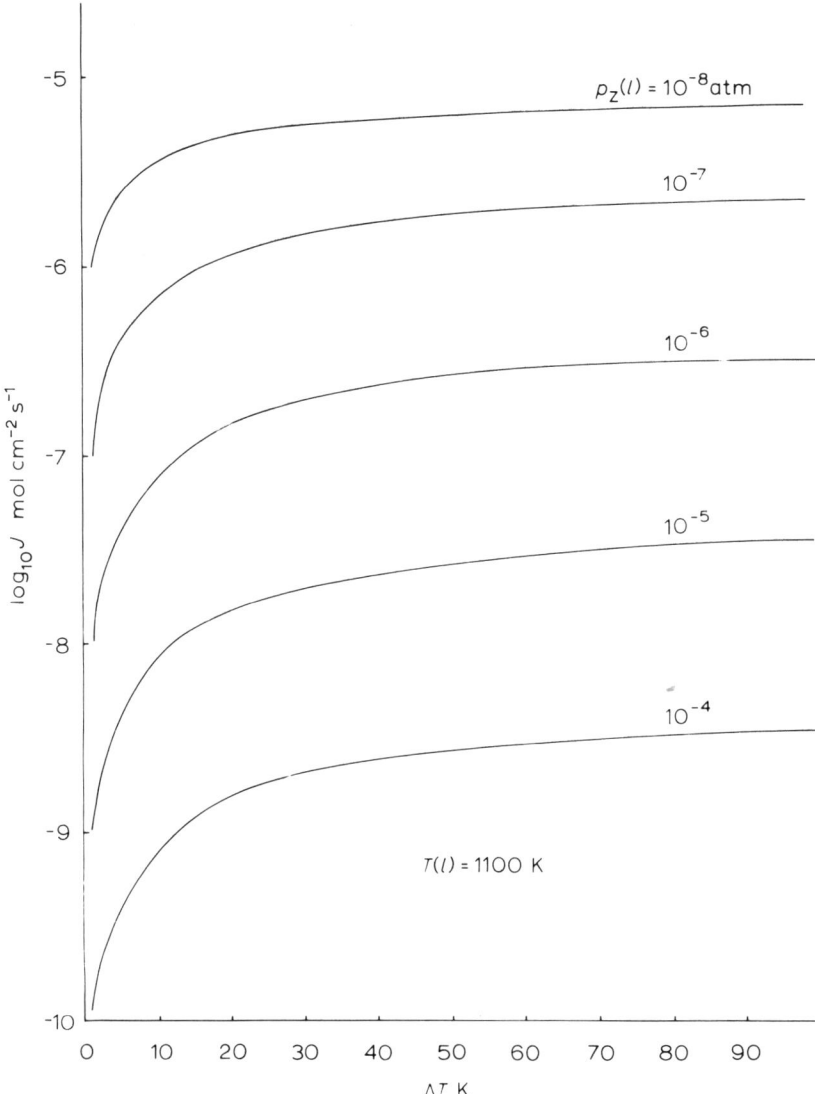

Fig. 4.5 Transport of silver. Transport rate J as a function of temperature difference between the source and the crystal, ΔT. Source temperature 1100 K.

crystal, the inert component must diffuse away at the same rate, so that its flux is not zero. This conceptual anachronism becomes more glaring when we deal with transport by dissociative sublimation (Section 4.3). If a compound AB dissociates on sublimation to A and B_2 vapour molecules, clearly both species must be transported in the same direction if crystals of AB are to be grown. The Stefan velocity, resulting from increase in molar volume when the compound sublimes, provides the mechanism for transport. The diffusion fluxes result from the difference in the vapour compositions in equilibrium with the source at temperature $T(l)$ and the growing crystal at temperature $T(0)$, and they provide for the net fluxes of the 'active' components to be in the correct ratio, and for the flux of 'inert' gas to be zero.

In Fig. 4.6 we show the calculated variation in partial pressure along the capsule for a particular case, where $T(l) = 1200$ K and

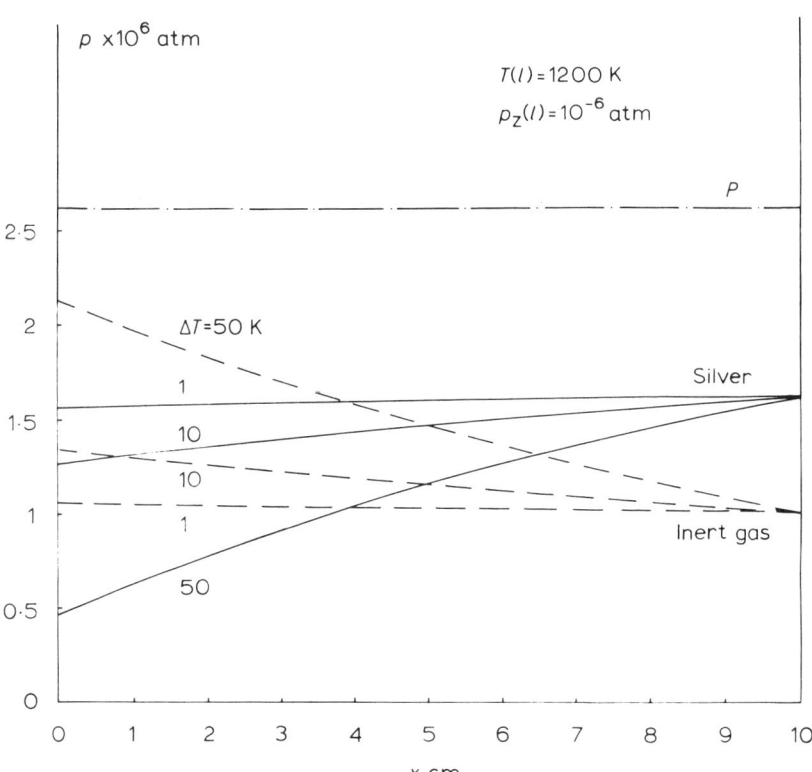

Fig. 4.6 Partial pressure profiles in the transport of silver, for ΔT of 1, 10, and 50 K. Inert gas pressure $p_Z(l) = 10^{-6}$ atm.

VAPOUR TRANSPORT

$p_Z(l)$, the inert gas pressure over the source, is 10^{-6} atm. The equilibrium vapour pressure of silver at 1200 K is about 1.6×10^{-6} atm[64] so the two components are present in roughly equal amounts. The partial pressure profiles have been plotted for three values of ΔT; 1 K, 10 K, and 50 K, and we see that there is a little curvature in the profiles, particularly for $\Delta T = 50$ K. In Fig. 4.7, similar conditions obtain except that $p_Z(l)$ is now 10^{-7} atm,

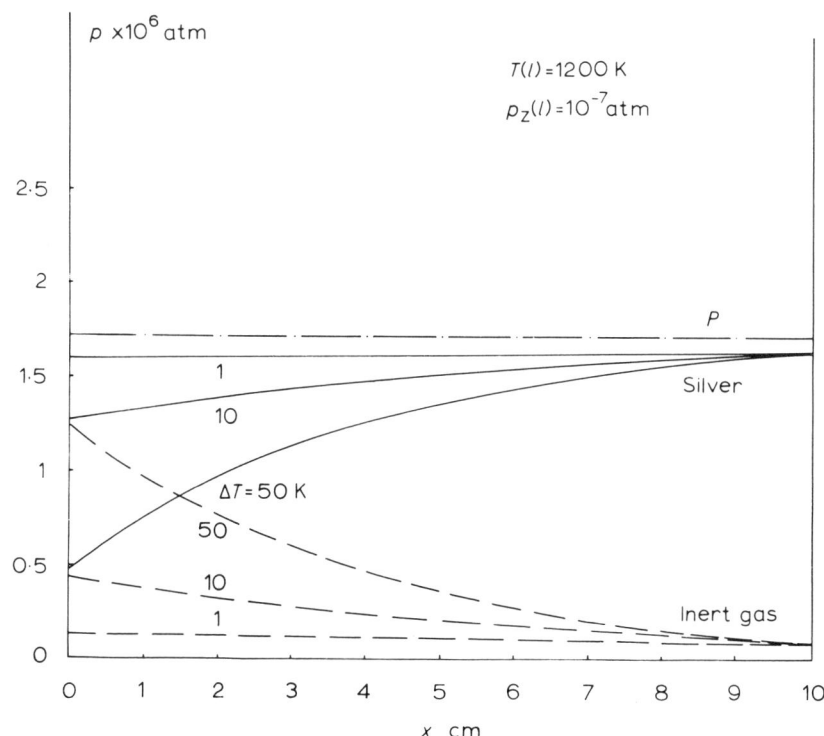

Fig. 4.7 Partial pressure profiles in the transport of silver, for ΔT of 1, 10, and 50 K. Inert gas pressure $p_Z(l) = 10^{-7}$ atm.

so that the inert gas is the minority component, perhaps representing only residual gas and impurities. The curvature on the partial pressure profiles is now quite marked. This curvature, which is caused by the Stefan velocity having a significant value, makes the partial pressure gradient steeper (i.e. larger in absolute magnitude) near the surface of the growing crystal. This fact has important repercussions when we consider the phenomenon of constitutional supercooling (or supersaturation) in the vapour, a subject first discussed by Reed and La Fleur[142] for growth of crystals by simple sublimation, and extended by Faktor et al to dissociative sublimation[141,143] and chemical vapour transport.[144] We will extend and amplify these ideas in Chapter 5.

4.2.3 Stefan's velocity in various systems

We now continue with a consideration of the flow velocity or Stefan's velocity occurring in vapour phase crystal growth systems as a consequence of a change in molar volume on evaporation or reaction. We have already seen that there is a flow of gas away from the evaporating source in the single component system just considered. The same holds good for dissociative sublimation, e.g.

$$AB \text{ (solid)} \rightarrow A \text{ (gas)} + B \text{ (gas)} \tag{4.30}$$

Transport of material and growth of crystals by this mechanism is considered in detail in Section 4.4. The chemical vapour transport system is a little more complicated, and it is perhaps best treated by considering some ideal examples. Firstly, consider the halogen transport of a binary compound, a reaction frequently used in preparing 'electronic' materials:

$$AB \text{ (solid)} + \tfrac{1}{2}X_2 \text{ (gas)} \rightarrow AX \text{ (gas)} + B \text{ (gas)} \tag{4.31}$$

where X is the halogen, component B (which may be a polymeric species $1/m$ B_m in the vapour) has a suitable vapour pressure at the temperature at which the transport is carried out, and component A is transported as a volatile monohalide. The direction of transport in a temperature gradient is determined, as in all vapour transport systems, by the sign of the enthalpy change for the transport reaction, as has been shown in Chapter 2. We consider the reaction:

$$\begin{array}{cc} \text{Reactants} & \rightarrow & \text{Products} \\ \text{\small (including some solid)} & & \text{\small (all vapour)} \end{array} \tag{4.32}$$

If this reaction is endothermic, the equilibrium is further to the right at higher temperatures, so that more products are formed in the hot region, and more reactants in the cold region. An endothermic reaction results in transport from hot to cold, and an exothermic reaction, in the other direction. Evaporation and sublimation (with or without dissociation) may be considered endothermic reactions, and it is obvious the transport there is from hot to cold. Reaction 4.31 is endothermic, so that transport is from hot to cold. The same is true of the similar system for transport using a hydrogen halide:

$$AB \text{ (solid)} + HX \text{ (gas)} \rightarrow AX \text{ (gas)} + B \text{ (gas)} + \tfrac{1}{2}H_2 \text{ (gas)} \tag{4.33}$$

and for transport via a volatile sub-oxide, using water as a transporting agent:

$$2 \text{ AB (solid)} + H_2O \text{ (gas)} \rightarrow A_2O \text{ (gas)} + 2B \text{ (gas)} + H_2 \text{ (gas)}$$

(4.34)

In all these chemical transport reactions, there are more vapour phase molecules on the right of the reaction than on the left. The exact ratio depends on the atomicity of component B in the vapour, but clearly a flow velocity exists, directed away from the source, i.e. in the same direction as the transport.

Now consider the following reactions:

$$\text{C (solid)} + O_2 \text{ (gas)} \rightarrow CO_2 \text{ (gas)} \qquad (4.35)$$

$$\text{M (solid)} + X_2 \text{ (gas)} \rightarrow MX_2 \text{ (gas)} \qquad (4.36)$$

where M is a divalent metal, and X is again a halogen. In these examples, there is no change in the number of vapour phase molecules, and consequently no flow velocity. Transport in such cases (and only in such cases) is entirely by diffusion.

Reactions such as:

$$\text{Si (solid)} + 2 Cl_2 \text{ (gas)} \rightarrow Si Cl_4 \text{ (gas)} \qquad (4.37)$$

$$\text{Si (solid)} + 4 HCl \text{ (gas)} \rightarrow Si Cl_4 \text{ (gas)} + 2H_2 \qquad (4.38)$$

involve a decrease in the number of vapour phase molecules, so that if a transport system were to be operated entirely by such a reaction, a flow velocity would exist but now opposed in direction to the net transport of silicon. In practice, compounds such as $SiCl_2$, $SiHCl_3$ SiH_2Cl_2, and SiH_3Cl would be formed as well in various proportions depending on the ratio of hydrogen to chlorine and on the temperature.[145]

In all the types of reaction considered so far, the Stefan flow velocity, where it exists, has been in the direction hot → cold, irrespective of the direction of material transport. This is because the reaction taking place at the hot end of the system has been accompanied by an increase in the number of vapour phase molecules, i.e. results in an increase in volume at constant pressure. It is interesting to see whether we may devise a transport reaction for which the Stefan flow velocity is in the direction cold to hot. We now require that $(\partial V/\partial T)_P$ should be negative for the equilibrium system solid/transporting agent/reaction products, so that a decrease in volume occurs at the hot end of the crystal growing system. We make use of one of Maxwell's relations (see Section 2.1)

$$\left(\frac{\partial V}{\partial T}\right)_P = -\left(\frac{\partial S}{\partial P}\right)_T$$

For $(\partial V/\partial T)_P$ to be negative, we require that $(\partial S/\partial P)_T$ should be positive, in other words, if the pressure of the equilibrium system solid/transporting agent/reaction products is increased while the temperature is kept constant, the reaction proceeeds in such a direction as to increase the entropy of the system. By Le Chatelier's principle, increasing the pressure drives the reaction in such a direction as to reduce the number of vapour phase molecules. We are looking, therefore, for a reaction which results in a reduction of the number of vapour phase molecules while having a positive entropy change. We would expect the vapour phase reactants in such a reaction to be diatomic molecules of low molecular weight, with little rotational entropy and even less vibrational entropy. The reaction products would be polyatomic, possibly polymeric species with high molecular weight, a large rotational entropy contribution, and an appreciable vibrational entropy contribution, which together could more than make up for the loss of translational entropy.

4.2.4 Dissociative sublimation

We next consider the growth of a crystal by dissociative sublimation,[141,143] and as an illustration, we take a binary compound AB which dissociates according to the reaction:

$$AB \text{ (solid)} \rightarrow A \text{ (gas)} + \frac{1}{m} B_m \text{ (gas)} \tag{4.39}$$

We suppose in the first instance that the only gas phase species present in the system are A atoms and B m-atomic molecules. Later we will see how the addition of a third, inert, species can drastically affect the transport rate and the surface stability.

Our growth system, then, consists of a source of material AB at a point $x = l$, and our growing crystal at $x = 0$. The temperature of the system is imposed by some sort of furnace, and will vary with x. We will suppose this variation to be moderate in the region between the crystal and the source material, so that the temperature difference, $\Delta T \equiv T(l) - T(0)$ is small compared with $T(l)$, $T(0)$. This condition we can relax later. Since sublimation is an endothermic process, we must have $T(l)$ larger than $T(0)$ if material to be transported from the source to the crystal.

The flux of each component will be the sum of a flow velocity U which acts on the gas as a whole, and a diffusion term proportional

VAPOUR TRANSPORT

to the concentration gradient of the particular component:

$$J_A = \frac{U}{RT} p_A - \frac{D}{RT} \frac{dp_A}{dx} \tag{4.40}$$

$$J_{Bm} = \frac{U}{RT} p_{Bm} - \frac{D}{RT} \frac{dp_{Bm}}{dx} \tag{4.41}$$

Addition of Equations 4.40 and 4.41 causes the diffusion terms to disappear, leaving:

$$J_A + J_{Bm} = UP/RT \tag{4.42}$$

Since we are growing solid AB, we require the net flux of the component A and B_m to be in such a ratio as to give equal fluxes of atoms A and B, that is:

$$J_A = mJ_{Bm} = J \tag{4.43}$$

where J is the growth rate of AB, and all fluxes are in molar units. Equation 4.42 now becomes:

$$\frac{UP}{RT} = J\left(1 + \frac{1}{m}\right) \equiv Js \tag{4.44}$$

where s is just a member that depends on the stoichiometry of the dissociation reaction. We can now rewrite our flow Equations 4.40 and 4.41, eliminating U:

$$J = \frac{Js}{P} p_A - \frac{D}{RT} \frac{dp_A}{dx} \tag{4.45}$$

$$J = \frac{mJs}{RT} p_{Bm} - \frac{mD}{RT} \frac{dp_{Bm}}{dx} \tag{4.46}$$

These equations may be integrated from x to l, and this is particularly simple if we choose to ignore the very small variation of P with x. As we are ignoring the relatively small variation in temperature, we will also disregard the variation in the diffusion coefficient. Integration gives:

$$p_A(x) - \frac{P}{s} = \left[p_A(l) - \frac{P}{s}\right] \exp\left\{\frac{JsRT}{DP}(x-l)\right\} \tag{4.47}$$

$$p_{Bm}(x) - \frac{P}{ms} = \left[p_{Bm}(l) - \frac{P}{ms}\right] \exp\left\{\frac{JsRT}{DP}(x-l)\right\} \tag{4.48}$$

There is a simple exponential variation in the quantities $(p_A - P/s)$ and $(p_{Bm} - P/ms)$, the strength of the exponential depending on the growth rate J. It is interesting to examine these quantities $(p_A - P/s)$ and $(p_{Bm} - P/ms)$. If the vapour were stoichiometric over the solid AB, the ratio $\alpha = p_A/p_{Bm}$ would be m, and in fact we would have:

$$p_A = \frac{mP}{1+m} = \frac{P}{s} \quad \text{and} \quad p_{Bm} = \frac{P}{1+m} = \frac{P}{ms}.$$

So these quantities that occur in Equations 4.47 and 4.48 are the departures from stoichiometry in the vapour. They vary exponentially with x, and always increase in magnitude as the distance from the source material increases. To see how this comes about, let us consider a capsule at a uniform temperature $T(l)$, containing source material at one end. The vapour in the capsule comes to equilibrium with the solid at a composition that depends on the composition of the solid and the ratio of the capsule volume to solid volume. In general, as discussed in Chapter 2, the vapour will not be stoichiometric, the ratio $\alpha, = p_A/p_{Bm}$, will not have the value m. Now imagine a seed crystal introduced at the far end of the capsule from the source, and the temperature of that end lowered to $T(0)$. Conditions are now set for the seed to grow, which it does by removing A atoms and B_m molecules from the vapour in the ratio $1:1/m$. A flow of the vapour is caused, by the reduction in molar volume on condensation at the seed and by the increase in molar volume on sublimation at the source. This flow brings up vapour to the seed in the ratio α. Clearly, if α is different from m, the vapour next to the seed becomes depleted in one component, while an excess of the other component accumulates. For example, if $\alpha > m$, an excess of A accumulates, while the vapour becomes depleted in B_m.

The change in vapour composition developing at the seed end of the capsule produces diffusion currents which carry the excess component away, and bring up more of the minority component. In the steady state, the flow of vapour and the diffusion currents of the component species combine to give net fluxes of A and B_m in the ratio m, in spite of the non-stoichiometry of the vapour.

The growth rate J of the solid may be expressed in terms of the vapour compositions over the source material and crystal by rearrangement of Equation 4.47 or 4.48:

$$J = \frac{DP}{RTsl} \ln\left\{\frac{p_A(l) - P/s}{p_A(0) - P/s}\right\} = \frac{DP}{RTsl} \ln\left\{\frac{p_{Bm}(l) - P/ms}{p_{Bm}(0) - P/ms}\right\} \quad (4.49)$$

VAPOUR TRANSPORT

The growth rate may be calculated if values for $p_A(l)$ and $p_A(0)$ are known. For the sake of some calculations, we will assume that the vapour over the source material and over the crystal is in equilibrium with the solid at the temperatures $T(l)$, $T(0)$, and we imagine that we are able, in some way, to control, or at least measure, the ratio $\alpha(l)$ of A to B_m over the source. One way in which such control is possible experimentally is described in Chapter 6. The equilibrium condition at the source end of the capsule is

$$K_p(l) = p_A(l)[p_{B_m}(l)]^{1/m} \qquad (4.50)$$

where $K_p(l)$ is the value of the equilibrium constant at the temperature $T(l)$, and we have taken the activity of AB to be unity (see Section 2.1.3). We also have

$$\alpha(l) = \frac{p_A(l)}{p_{B_m}(l)} \qquad (4.51)$$

The partial pressures $p_A(l)$ and $p_{B_m}(l)$ can be found by solving Equations 4.50 and 4.51, and the total pressure P in the capsule is simply their sum.

The equilibrium condition at the end of the capsule where the crystal is growing is:

$$K_p(0) = p_A(0)[p_{B_m}(0)]^{1/m} \qquad (4.52)$$

but we do not know the ratio $\alpha(0)$. We do know the total pressure P, however, and so the mole fractions in the vapour over the growing crystal may be calculated. The partial pressures may be made to satisfy the equilibrium Equation 4.52 by changing their ratio (i.e. $\alpha(0)$ and $\alpha(l)$ not the same) while maintaining the same total pressure. This was discussed in Section (2.1.4).

We now have all that we need to calculate the growth rate J. Since calculating the partial pressures $p_A(0)$ and $p_{B_m}(0)$ involves solving an equation of order s, it is not possible in general to write down the expression for J explicitly in terms of the quantities $T(l)$, $T(0)$ and $\alpha(l)$ which we have taken as being, as it were, the boundary conditions of the problem. The growth rate may be calculated by numerical methods without difficulty, however. The results of such calculations are reproduced graphically in Fig. 4.8 and 4.9. The compound AB in this instance is cadmium sulphide, and in the temperature range considered (1100–1400 K) sulphur is present in the vapour phase as S_2 almost entirely.

We observe that, for a given ratio $\alpha(l)$ of the components in the vapour at the source end of the capsule, the growth rate J varies very

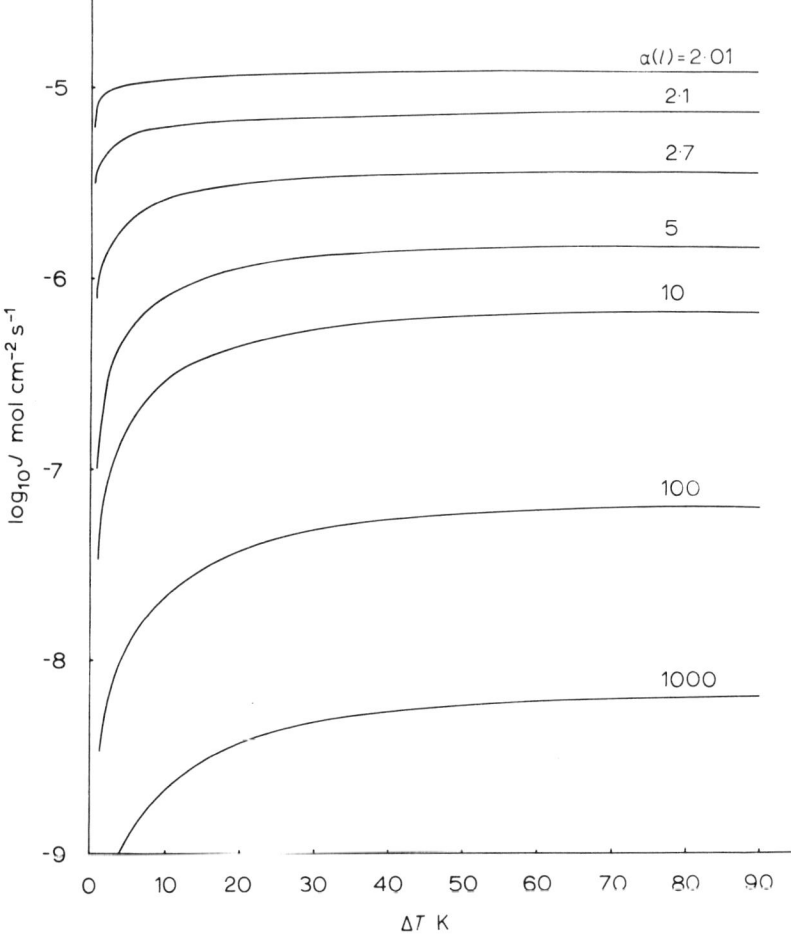

Fig. 4.8 Transport of cadmium sulphide by dissociative sublimation. Transport rate J as a function of temperature difference between the source and the crystal, ΔT, with a cadmium-rich vapour ($\alpha > 2$).

rapidly with temperature difference ΔT between the source and the crystal while ΔT is fairly small. As ΔT becomes larger, the growth rate approaches a maximum value. This is just what we would expect. The rapid rise in growth rate as ΔT increases from zero is brought about by a rapid fall in the partial pressure of the minority component over the crystal, and a corresponding rapid rise in the partial pressure of the majority component. This is illustrated in Fig. 4.10, and we see that by the time ΔT is 40 or 50 K, the minority partial pressure has become very small, while the majority partial

VAPOUR TRANSPORT

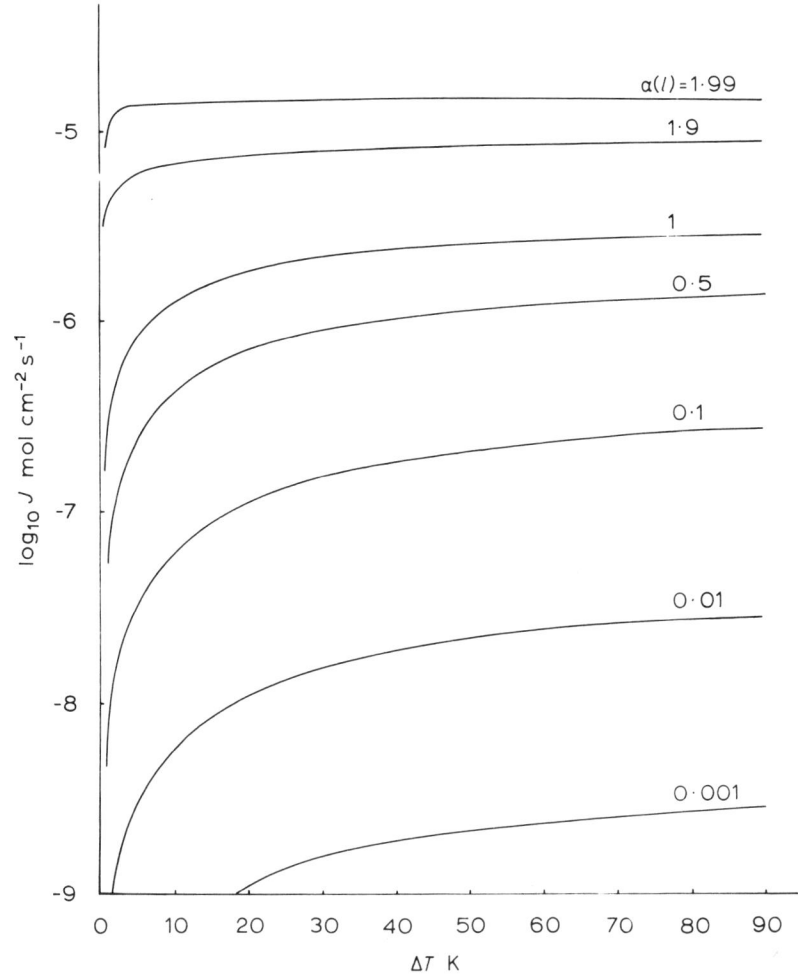

Fig. 4.9 Transport of cadmium sulphide by dissociative sublimation. Transport rate J as a function of the temperature difference between the source and the crystal ΔT, with a sulphur-rich vapour ($\alpha < 2$).

pressure is nearly the same as the total pressure. Further increase in ΔT cannot produce much change in the partial pressures, so that little change is brought about in the rate at which the excess majority component diffuses away, or the minority component arrives.

It is clear from Figs. 4.8 and 4.9 that the ratio $\alpha(l)$ of the components in the gas phase over the source material has a most dramatic effect on the rate of transport. In Fig. 4.8, values of $\alpha(l)$ ranging from 2.1 to 1000 are considered, and this variation in $\alpha(l)$

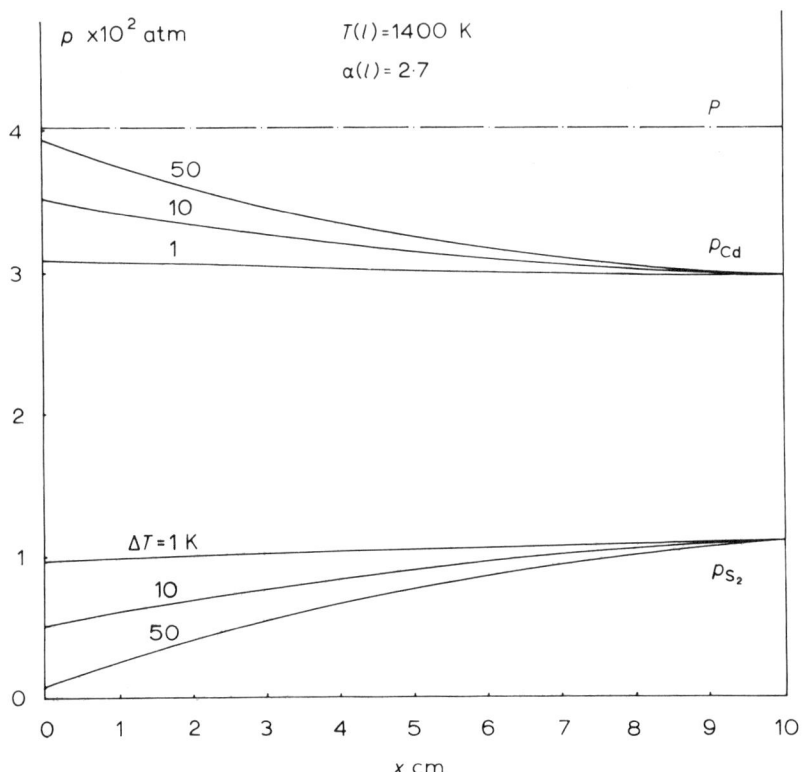

Fig. 4.10 Partial pressure profiles in the transport of cadmium sulphide, for ΔT = 1,5,10, and 50 K, and $\alpha(l)$ = 2.7.

produces a fall in growth rate of more than three orders of magnitude. It is interesting to see how much excess cadmium is required to produce a value of 1000 for $\alpha(l)$. From Equations 4.50 and 4.51 we can eliminate $p_{Bm}(l)$ to give:

$$p_A(l) = \alpha(l)^{1/3} K_p(l)^{2/3} \qquad (4.53)$$

At 1200 K, the cadmium partial pressure would be about 0.009 atm to give $\alpha(l)$ of 1000. If this partial pressure were present throughout a typical crystal growth capsule of some 30 cm^3 capacity, there would be about 0.3 mg of excess cadmium in the capsule. If this were present initially as a stoichiometric excess in 30 gm of source material, it would require a non-stoichiometry of 10 p.p.m. Alternatively, ten monolayers of oxide over 100 cm^2 of surface of the source material will result in the formation of SO$_2$ in the vapour, thus tying up the sulphur and producing a large excess of cadmium.

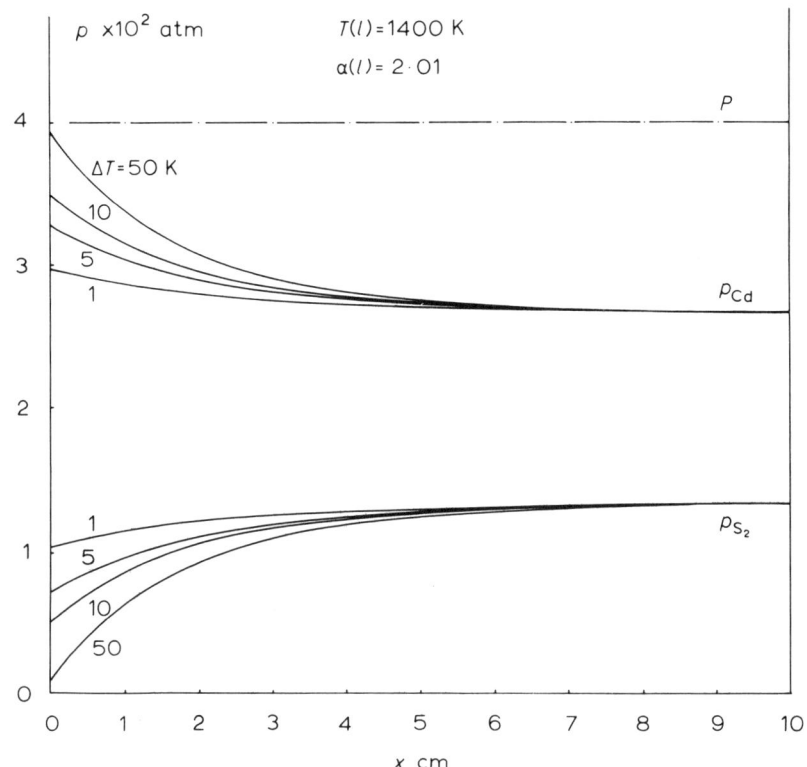

Fig. 4.11 Partial pressure profiles in the transport of cadmium sulphide, for $\Delta T = 1$, 5, and 50 K, and $\alpha(l) = 2.01$.

Many binary compounds deviate from stoichiometry by as much or more. The range of values of $\alpha(l)$ which may occur depends on the substance that is being grown. It may be desirable to produce an extreme value of $\alpha(l)$, and hence $\alpha(0)$, to give the required composition and electrical properties of the grown solid, by introducing an excess of one component into the capsule. Loading excess of one component into the capsule along with the source material is fraught with problems, since the excess material will inevitably be used up during growth, causing a change in composition of the crystal. Also, the quantities involved may be so small as to be very hard to measure out accurately. It is preferable to use some method such as the effusion hole[146] described in Chapter 6.

We have so far considered extreme values of $\alpha(l)$, where the transport rate is limited by stoichiometric excess. We will say just a little about values of $\alpha(l)$ near to the value m. As $\alpha(l) \rightarrow m$ and

$p_A(l) \to P/s$, so $p_A(0) \to P/s$ and $J \to \infty$, according to Equations 4.47 and 4.48. However, as J becomes large, the drop in total pressure between the source material and the growing crystal becomes appreciable. Let us consider the situation where $\alpha(l) = m$. There is no stoichiometric excess, so we would expect $\alpha(0) = m$: this follows also from Equation 4.47. We now have different total pressures at $x = 0$ and $x = l$, which, from Equations 4.45 and 4.46 are given by

$$P(l) = (1 + m)\left[\frac{K_p(l)}{m}\right]^{1/s} \tag{4.54}$$

$$P(0) = (1 + m)\left[\frac{K_p(0)}{m}\right]^{1/s} \tag{4.55}$$

There is now no diffusion limitation or stoichiometric excess limitation to the flow. In a capsule of centimeter dimensions, the flow would, in the ideal stoichiometric case, be limited by the rate at which the heat of sublimation could be supplied to the source or removed from the crystal. In practice, it is unlikely that this situation would be achieved, since two phenomena can contribute to upsetting it. The first we have already discussed, namely, non-stoichiometry in the vapour resulting from non-stoichiometry in the solid. The second is the accumulation of impurities and residual gas in the capsule, and outgassing of the capsule and source material, to produce an appreciable pressure of gas, some of which may not react with the vapour components A and B_m, and some of which may react to tie up part of one or both components in a less active form.

Before going on to discuss the effect of an inert gas in the capsule on the transport rate of binary compounds by dissociative sublimation, we digress for a moment to consider the 'diffusion only' approximation. Several authors[2,147], have treated vapour transport by dissociative sublimation and by chemical vapour transport on the assumption that the only transport mechanism available in a closed capsule is diffusion. This present example amply illustrates the conceptual absurdity of this approach. If sulphur diffuses towards the growing crystal with a flux J_S, there must be an equal but opposite flux of cadmium, $J_{Cd} = -J_S$, away from the growing crystal. The logical conclusion is that all the sulphur is eventually transported to one end, and all the cadmium to the other — a process which is accompanied by an *increase* in the free energy, and a *decrease* in entropy of the growth system. Clearly, one could not grow crystals of cadmium sulphide this way! Nevertheless, it turns out that the magnitudes of the fluxes of the components, calculated with this approximation, are about right if one of the components is

VAPOUR TRANSPORT

in considerable excess. If, for example, cadmium is in great excess, then from Equation 4.49 we have:

$$J_S = \frac{2DP}{3lRT} \ln\left\{\frac{p_{S_2}(l) - P/3}{p_{S_2}(0) - P/3}\right\} \tag{4.55}$$

since $m = 2, s = 3/2$. Hence:

$$J_S = \frac{2DP}{3lRT} \ln\left\{\frac{p_{S_2}(0) - P/3 + p_{S_2}(l) - p_{S_2}(0)}{p_{S_2}(0) - P/3}\right\}$$

$$\simeq \frac{2D}{RT} \frac{[p_{S_2}(l) - p_{S_2}(0)]}{l} \tag{4.56}$$

where in expanding the logarithm we have taken $p_{S_2}(0) \ll P/3$, i.e. excess of cadmium. The partial pressures will vary almost linearly with x in the highly non-stoichiometric vapour, so we may write:

$$J_S \simeq \frac{-2D}{RT} \frac{dp_{S_2}}{dx} \tag{4.57}$$

We see that our Equation 4.49 reduces to 'diffusion only' as far as the minority species is concerned, provided there is a great excess of the majority species. The corresponding equation for the majority species is:

$$J_{Cd} = \frac{2DP}{3lRT} \ln\left\{\frac{p_{Cd}(l) - 2P/3}{p_{Cd}(0) - 2P/3}\right\}$$

$$\simeq \frac{2D}{RT} \frac{[p_{Cd}(l) - p_{Cd}(0)]}{l} \tag{4.58}$$

since $p_{Cd}(0) - 2P/3 \simeq P/3$. Again assuming that the partial pressures vary linearly with x, we have:

$$J_{Cd} \simeq \frac{2D}{RT} \frac{dp_{Cd}}{dx} \tag{4.59}$$

This is *not* the diffusion flux of cadmium, since we have no minus sign; also the factor of 2 is inexplicable on the 'diffusion only' model, whereas for the flux of sulphur it was necessary since sulphur is diatomic in the gas phase in the temperature range considered. We strongly advise against the use of this approximation.

Dissociative sublimation in the presence of an inert gas

We now consider the effect of an inert third species in the crystal growth capsule, either introduced on purpose, or resulting from outgassing of the capsule and source material, formation of stable volatile SO_2 [148,149] etc. We take the species present in the capsule to be A atoms, B_m molecules and Z atoms or molecules, where Z is the inert species. The flow equations are similar to Equations 4.40 and 4.41, but we also have an equation for the zero net flow of the inert species Z:

$$J_A = \frac{U}{RT} p_A - \frac{D}{RT} \frac{dp_A}{dx} = J \qquad (4.60)$$

$$J_{Bm} = \frac{U}{RT} p_{Bm} - \frac{D}{RT} \frac{dp_{Bm}}{dx} = \frac{J}{m} \qquad (4.61)$$

$$J_Z = \frac{U}{RT} p_Z - \frac{D}{RT} \frac{dp_Z}{dx} = 0 \qquad (4.62)$$

Addition of these equations eliminates the diffusion terms, to give:

$$Js = \frac{UP}{RT} \qquad (4.63)$$

where $s = 1 + 1/m$, as before. We can again replace U in terms of J and integrate the flow equations to obtain the variation in composition of the gas along x in terms of the composition over the source:

$$[p_A(x) - P/s] = [p_A(l) - P/s] \exp\left\{\frac{JsRT(x-l)}{DP}\right\} \qquad (4.64)$$

$$[p_{Bm}(x) - P/ms] = [p_{Bm}(l) - P/ms] \exp\left\{\frac{JsRT(x-l)}{DP}\right\} \qquad (4.65)$$

$$p_Z(x) = p_Z(l) \exp\left\{\frac{JsRT(x-l)}{DP}\right\} \qquad (4.66)$$

The first two of these equations are identical to Equations 4.47 and 4.48, and we may once more express the growth rate in terms of the gas composition over the source and over the growing crystal by Equation 4.49. There is one crucial difference in the quantities that are to be substituted into this equation, however. With no inert gas present, we simply had $p_A + p_{Bm} = P$ all the way down the capsule. With the inert gas, however, $p_A + p_{Bm} = P - p_Z$, and varies exponentially with x, according to Equation 4.66. This means that in satisfying the equilibrium condition over the growing crystal,

Equation 4.52, the partial pressures $p_A(0)$ and $p_{Bm}(0)$ do not add to give the total pressure, or even to give the same sum as $p_A(l) + p_{Bm}(l)$. This has two important effects. The first is that as $\alpha(l) \to m$, the growth rate no longer increases without bound, as Equations 4.47 and 4.48 predict (or in practice, until it is limited by viscosity or heat flow). The second effect concerns the stability of small protrusions from the growing surface, and will be discussed in Chapter 5.

The calculated transport rates as functions of ΔT are shown in Fig. 4.12 for $p_Z(l) = 0.01$ atm. and in Fig. 4.13 for $p_Z(l) = 0.1$ atm. In each case, values of $\alpha(l)$ ranging from 2 to 1000 have been considered. These curves should be conpared with those in Fig. 4.8. The presence of the inert gas decreases the transport rate for all values of $\alpha(l)$, but while the decrease is very marked when $\alpha(l)$ is near 2, it is small when $\alpha(l)$ is large or small. We may think of the inert gas and the stoichiometric excess acting in a similar way in reducing the transport rate in the gas. The effect of inert gas on the growth rate is illustrated in Fig. 4.14. Here we have $\log_{10} J$ plotted as a function of $|2 - \alpha|$. A fixed value of 20 K has been used for the temperature difference, and various percentages of inert gas considered in the calculations.

The composition of the vapour as a function of x is illustrated in Fig. 4.15 and 4.16 for the capsule containing inert gas. The partial pressure of the inert gas assumes a distribution such that the flow velocity term and the diffusion term cancel out. This requires that the gradient of partial pressure should be proportional to the partial pressure at all points in the capsule.

The partial pressure ratio over the growing crystal, $\alpha(0)$, is determined by all the experimental conditions. In Fig. 4.10 and 4.11 we have values of $\alpha(l)$ of 2.01 (i.e. very nearly stoichiometric vapour over the source material) and 2.7. With a temperature difference of 1 K, $\alpha(0)$ is not too different from $\alpha(l)$, since the partial pressure profiles are fairly straight. However, if the temperature difference is increased to 50 K, $\alpha(0)$ becomes just over 50 in both cases, and if ΔT is 100 K, $\alpha(0)$ becomes 500. These large values of $\alpha(0)$ are obtained, even though the vapour over the source material is not far from stoichiometric. Of course, if $\alpha(l) < 2$, i.e. if the vapour is sulphur-rich, very small values of $\alpha(0)$ are obtained if ΔT is made large. If the vapour composition over the source is far from stoichiometric, and we have seen that values of 100 or 1000 for $\alpha(l)$ are not unreasonable, then $\alpha(0)$ may be very large, e.g. 75,000 when $\alpha(l) = 1000$, and $\Delta T = 100$ K. For sulphur-rich vapour, correspondingly large deviations from stoichiometry can occur, i.e. $\alpha(0) \simeq 10^{-5}$.

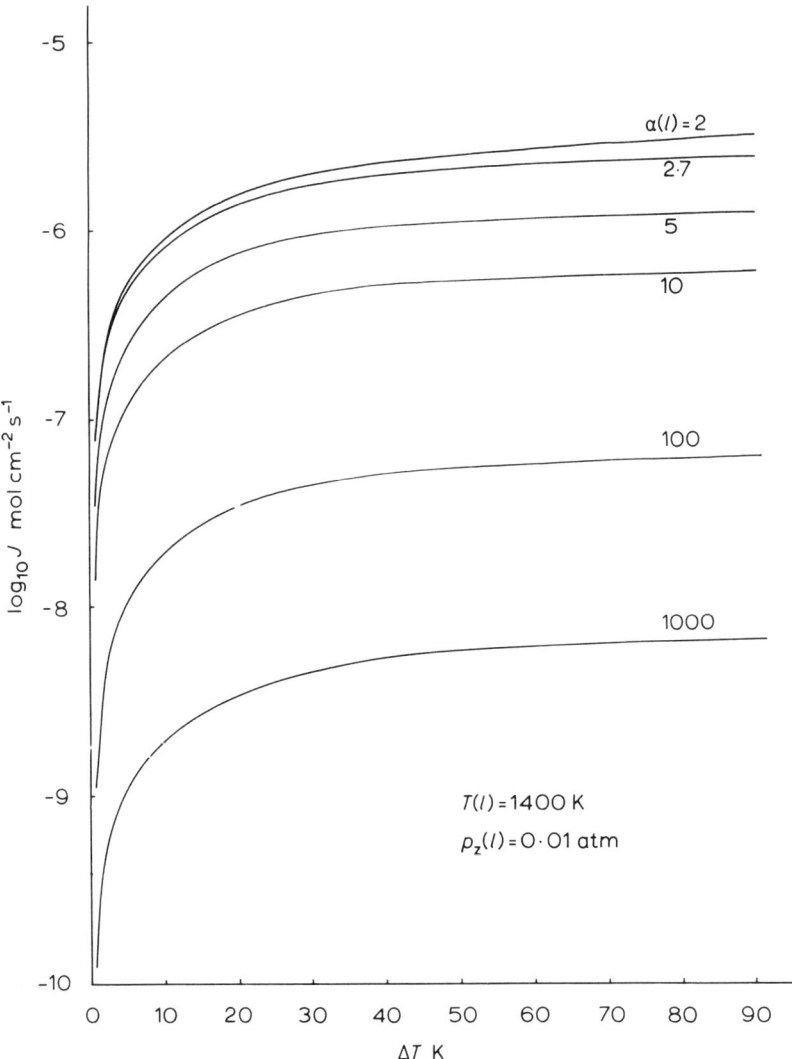

Fig. 4.12 Transport of cadmium sulphide in an inert gas. Transport rate J as a function of temperature difference between the source and the crystal, ΔT. Inert gas pressure $p_Z(l) = 0.01$ atm

These large departures from stoichiometry in the gas which may occur can have various effects on the way in which the crystal grows. An extreme value of $\alpha(0)$ in the vapour over the growing crystal may be accompanied by a great preponderance of the majority component on the surface. This will alter the heat of adsorption of the minority component, and of any impurities that may be present. If

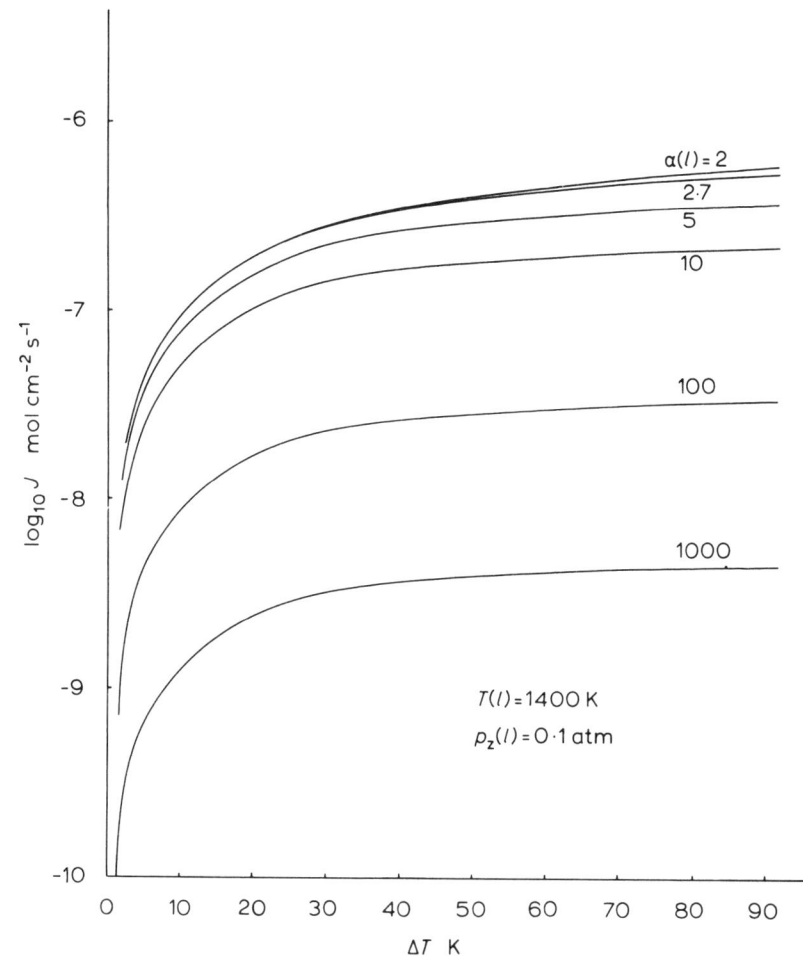

Fig. 4.13 Transport of cadmium sulphide in an inert gas. Transport rate J as a function of temperature difference between the source and the crystal T. Inert gas pressure $p_Z(l) = 0.1$ atm.

the shape of the growing crystal is sensitive to small quantities of adsorbates changing the free energy or the growth (condensation) kinetics of different faces, then the shape of the growing crystal may be altogether different when grown with extreme values of $\alpha(0)$ to when grown in vapour which is not far from stoichiometry.

The composition of the solid is, of course, determined by the composition of the vapour, via the surface composition. Adjustments in the solid composition are made by adjusting the relative numbers of point-defects on the two sub-lattices. These point defects may be

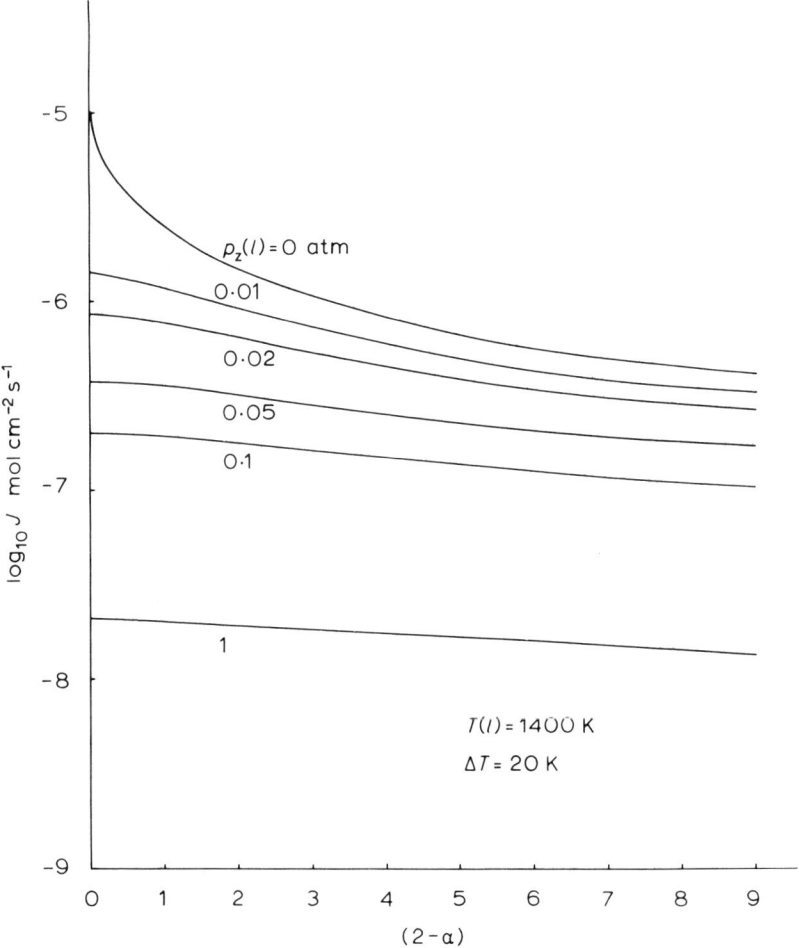

Fig. 4.14 Transport rate J of cadmium sulphide as a function of gas stoichiometry.

ionized at room temperature and contribute significantly to the intrinsic conductivity of the compound. Also, by altering the intrinsic disorder of the crystal, that is to say, by altering the vacancy concentration on one or other sub-lattice, or the concentrations of interstitial atoms of the components of the crystal, the solubility of impurities and dopant species is altered. This topic is taken up in Chapter 5. For a detailed account, see Kröger.[69]

VAPOUR TRANSPORT

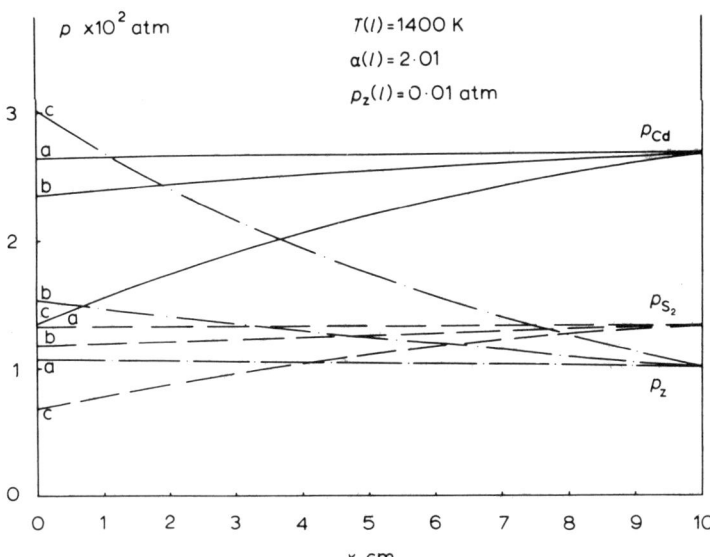

Fig. 4.15 Partial pressure profiles in the transport of cadmium sulphide in an inert gas. The letters against the curves denote the temperature difference ΔT between the source and the crystal as follows: a, 1K; b, 10K; c, 50K. Inert gas pressure $p_Z(l) = 0.01$ atm.

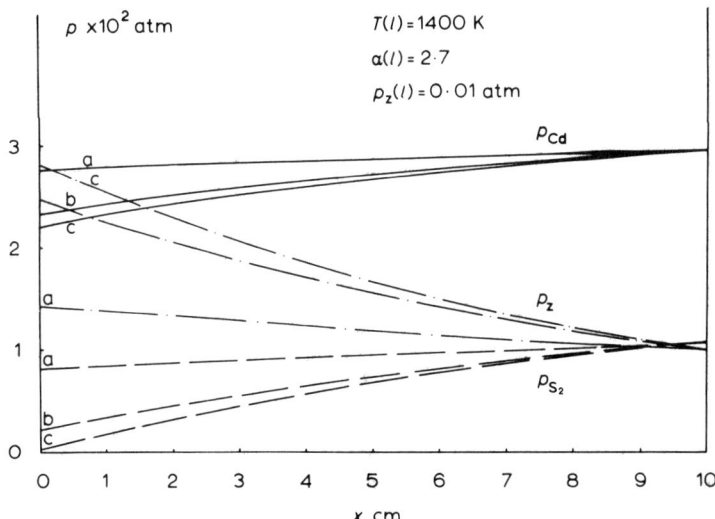

Fig. 4.16 Partial pressure profiles in the transport of cadmium sulphide in an inert gas. The letters against the curves denote the temperature difference ΔT between the source and the crystal as follows: a, 1K; b, 10K; c, 50K. Inert gas pressure $p_Z(l) = 0.1$ atm.

4.4 Chemical vapour transport

4.4.1 General introductory remarks

We do not propose to present a general formalism of vapour transport. The substances that have been transported by this powerful method range from single elements to complex substituted garnets[150]; in the one case the process may often be described by a single equation for which the equilibrium constant is known, while in the other case, many reactions contribute to the total transport rate, and frequently there is little thermochemical data available for the reactions. A general treatment embracing these two extremes would be largely redundant in simple cases, while in more complex situations it would be difficult to apply even if the equilibrium constants for all the relevant reactions were known We therefore restrict ourselves to a discussion of some simple systems, leaving the reader to apply these methods to his own system, with whatever elaboration may be necessary. We discuss one fairly complicated system of great importance to the electronic industry, namely the transport of gallium arsenide with HCl, at the end of this section.

The simplest chemical vapour transport system may be described by the reaction:

Solid + Transporting agent → Volatile product

where there is a single vapour component on each side of the reaction equation. Examples of such reactions are:

$$M(s) + X_2(g) \rightarrow MX_2(g)$$

where M is, for example, a metal with a unique valency (divalent in this example) and X is a halogen; or

$$Ni(s) + 4\ CO(g) \rightarrow Ni\ (CO)_4(g)$$

the well-known Mond process for refining nickel. We shall discuss this process in detail later as our first example. In formulating a theory of gas-phase transport using these simple reactions, we first apply the phase rule (Section 2.1.4), and find that there are two degrees of freedom in these examples. It is, perhaps, natural to choose temperature as one of the degrees of freedom, and probably total pressure as the other, as these variables are often the ones most easily controlled and measured. However, it is frequently better, when analysing the transport system, to choose a variable which represents the ratio of active to inactive species, or of transported species to transporting agent, or, as in the example of dissociative sublimation transport given in the last section, the ratio of the components themselves. This variable, which we have previously denoted by α, gives a more direct description of the chemical state of the system than does the total pressure. Thus we define:

$$\alpha = \frac{[\text{Volatile product}]}{[\text{Transport agent}]}$$

For the Mond nickel process,

$$\alpha = \frac{p_{Ni(CO)_4}}{p_{CO}}$$

For the transport of a metal as the dihalide, it so happens that α, defined in this way, is identical with the equilibrium constant and is fixed by fixing the temperature. We could therefore use α instead of temperature as the first degree of freedom, and choose some other variable such as total pressure or one of the partial pressures as the second degree of freedom.

The efficiency of the transport process may be estimated by examining the value of α at each end of the system. A large value of α at the source end indicates that the volatilization reaction has gone almost to completion (by exhausting the transport agent), while a small value of α over the growing crystal or deposit indicates that the deposition reaction has gone almost to completion. Equally obviously, if α is either very large or very small throughout the system, the vapour phase consists essentially of one component and the transport is inefficient (see Chapter 2). We require α to change by as large a fraction as possible, while having a mean value near to unity.

The parameter α is linked closely with the equilibrium constant. If the transport reaction is written:

$$A(s) + nX(g) \rightarrow AX_n(g)$$

as a general form of this simple type, we have:

$$\alpha = \frac{p_{AX_n}}{p_X}$$

$$K_p = \frac{p_{AX_n}}{p_X^n}$$

From which we can easily derive the relationships:

$$p_{AX_n} = \frac{\alpha P}{(1 + \alpha)}$$

$$p_X = \frac{P}{(1 + \alpha)}$$

where P is the total pressure, and hence:

$$K_p = \alpha \left\{ \frac{1 + \alpha}{P} \right\}^{n-1}$$

To see how α and K_p vary, let us differentiate this equation with respect to temperature, keeping the total pressure constant, as it is very nearly in most experimental systems. We obtain:

$$\frac{(1 + n\alpha)}{(1 + \alpha)} \left(\frac{\partial \ln \alpha}{\partial T} \right)_P = \frac{d \ln K_p}{dT} = \frac{\Delta H}{RT^2}$$

The quantity in the first bracket has the value n if $\alpha \gg 1$, and the value 1 if $\alpha \ll 1$, so that:

$$\left(\frac{\partial \alpha}{\partial T} \right)_P = \frac{\alpha}{n} \frac{\Delta H}{RT^2} \quad \alpha \gg 1$$

$$\left(\frac{\partial \alpha}{\partial T} \right)_P = \alpha \frac{\Delta H}{RT^2} \quad \alpha \ll 1$$

If n is a small number, like 3/2 or 1/2, etc., the second of the equations is very similar to the first, and α varies in the same manner as K_p. When n is a larger number, for example, 4, as in the transport of nickel as the carbonyl, then two distinct regions of α with T exist. The following example illustrates these points in some detail.

4.4.2 Example: Transport of nickel as nickel carbonyl

In common with some other transition metals, nickel forms a volatile carbonyl, according to the reaction:

$$\text{Ni (metal)} + 4 \text{ CO (gas)} \rightarrow \text{Ni(CO)}_4 \text{ (gas)} \tag{4.67}$$

This reaction forms the basis of the Mond process for extracting purified nickel from copper-nickel matte. The same reaction may be used for depositing layers or coatings of nickel or conceivably for growing crystals. The reaction is exothermic with a reduction in

VAPOUR TRANSPORT

entropy in the direction left → right, the equilibrium constant[6] being given by:

$$\frac{p_{Ni(CO)_4}}{p_{CO}^4} = K_p = \exp\left(\frac{36\,400}{RT} - \frac{100.3}{R}\right)$$

Transport takes place from the cold end to the hot, whereas the Stefan velocity in this system is in the opposite direction because the reaction which volatilizes the nickel causes a reduction in the number of gas phase molecules.

In the industrial process, carbon monoxide is passed over the crude nickel at about 50°C (320 K) and is decomposed at about 180°C (450 K). These temperatures are chosen to ensure a reasonable production of nickel carbonyl at the source region while remaining below the vapour pressure (nickel carbonyl boils at 42°C), and efficient decomposition at the higher temperature. The process is carried out at about atmospheric pressure, and the gas is recycled.

We discuss this example partly because it provides a simple illustration of chemical vapour transport while being a process of industrial importance, and partly because the stoichiometry of the reaction leads to a large Stefan velocity and rapid change of nickel carbonyl partial pressure with total pressure.

We can write the flow equations for this process in the usual way:

$$J_{nickel} = J_{Ni(CO)_4} = \frac{U}{RT} p_{Ni(CO)_4} - \frac{D}{RT}\frac{d}{dx} p_{Ni(CO)_4} = J \quad (4.68)$$

$$J_{carbon\,monox.} = J_{CO} + 4J_{Ni(CO)_4}$$

$$= \frac{U}{RT}(p_{CO} + 4p_{Ni(CO)_4}) - \frac{D}{RT}\frac{d}{dx}(p_{CO} + 4p_{Ni(CO)_4})$$

$$= 0 \quad (4.69)$$

If the transport takes place in a closed capsule or tube, then these equations apply throughout the vapour, unless transport by convection or mixing of the vapour by turbulence also occur. In the industrial process, the gas stream is circulated through towers in which the volatilization and deposition reactions occur, and the transport equations then apply to the diffusion boundary layer over the crude nickel matte and over the purified deposit.

The diffusion terms can be eliminated from Equations 4.68 and 4.69, as usual. We multiply the first equation by three and subtract it from the second, and obtain the relationship between J and U:

$$\frac{3J}{P} = \frac{-U}{RT} \tag{4.70}$$

We substitute this Equation into Equation 4.68 and rearrange terms to obtain the differential equation in $p_{Ni(CO)_4}$.

$$\frac{dp_{Ni(CO)_4}}{p_{Ni(CO)_4} + P/3} = \frac{-3JRT}{DP} dx \tag{4.71}$$

This equation can be integrated immediately if we are prepared to disregard the variation of T with x for the purposes of integration, as we did in the examples on transport of silver and dissociative sublimation of cadmium sulphide. However, in those examples we were considering temperature changes of perhaps 100 K in 1200–1400 K, whereas here we are considering a temperature change of 130 K in 320 K. This change can be ignored if we require only a rough estimate of the transport rate. Let us examine Equation 4.71 more closely. The diffusion coefficient, on the right hand side, can be expressed as in Equation 4.4, so that we obtain for the right hand side of Equation 4.71:

$$\frac{-3JR \times 273^{1.8}}{D^0 T^{0.8}} dx$$

and on integrating, we obtain:

$$\ln\left\{\frac{p_{Ni(CO)_4}(x) + P/3}{p_{Ni(CO)_4}(l) + P/3}\right\} = 3 \times 273^{1.8} \frac{JR}{D^0} \int_x^l \frac{dx}{T^{0.8}} \tag{4.72}$$

In practice, it would be possible to measure T as a function of x, and perform the integration numerically. We shall assume a form for $T(x)$ which makes the integration simple, while being a reasonable representation of the sort of temperature profile that might be obtained in practice:

$$\frac{1}{T^{0.8}} = \frac{1}{T(l)^{0.8}} + gx \tag{4.73}$$

Using this artifice, we may define a mean temperature $\overline{T}$ such that

$$\int_0^l \frac{dx}{T^{0.8}} = \frac{l}{\overline{T}^{0.8}}$$

VAPOUR TRANSPORT

We find, using Equation 4.73, that

$$\frac{1}{\bar{T}^{0.8}} = \frac{1}{2}\left\{\frac{1}{T(l)^{0.8}} + \frac{1}{T(0)^{0.8}}\right\}$$

and Equation 4.71 is now, after putting $x = 0$:

$$\ln\left\{\frac{p_{Ni(CO)_4}(0) + P/3}{p_{Ni(CO)_4}(l) + P/3}\right\} = 3 \times 273^{1.8} \frac{JRl}{D^0 \bar{T}^{0.8}} \quad (4.74)$$

If $T(0) = 450$ K and $T(l) = 320$ K, then $\bar{T} = 375$ K compared with the arithmetic mean value of 385 K.

From Equation 4.74 we can calculate the transport rate J if we know the partial pressure of nickel carbonyl at either end of the system. Assuming that equilibrium is achieved (or very nearly achieved, see Chapter 5), we can calculate the partial pressures quite easily. Since we are not very interested in carbon and oxygen as separate species (the dissociation constant[6] of CO is about 10^{-21} at 350 K) we may consider the components of the system to be Ni and CO, so that there are two degrees of freedom. If we specify the temperature T and total pressure P at each point in the system where there is to be equilibrium between solid and vapour (i.e. over the source and over the deposited nickel) then the system is fully determinate and the partial pressures are the solutions of the simultaneous equations:

$$K_p = \frac{p_{Ni(CO)_4}}{p_{CO}^4} \quad (4.75)$$

$$p_{Ni(CO)_4} + p_{CO} = P \quad (4.76)$$

Let us look first of all at the calculated transport rate J as a function of temperature difference ΔT, when the temperature of the source material is 320 K and the total pressure is 1 atm, (uppermost curve in Fig. 4.17). The transport rate rises rapidly at first, then when ΔT is between about 10 and 50 K, $\log_{10} J$ is roughly proportional to ΔT. Finally, the transport rate levels off at about 5×10^{-7} mol cm^{-2} s^{-1} (we have taken $l = 10$ cm in these calculations; we see from Equation 4.74 that $J \propto 1/l$). If the source temperature is raised by 50 K, faster transport is obtained for $\Delta T < 35$ K, but the curve levels off now at 2×10^{-7} mol cm^{-2} s^{-1}. A further increase in source temperature, to 420 K, reduces the transport rate by more than two orders of magnitude. We cannot reduce the source temperature below 320 K unless the total pressure is also reduced, since Ni(CO)$_4$ has a vapour pressure of about 1 atm at 320 K.

To assist the discussion of the curves of Fig. 4.17, we introduce a parameter α, as we did in the discussion of dissociative sublimation of cadmium sulphide:

$$\alpha = \frac{p_{Ni(CO)_4}}{p_{CO}} \qquad (4.77)$$

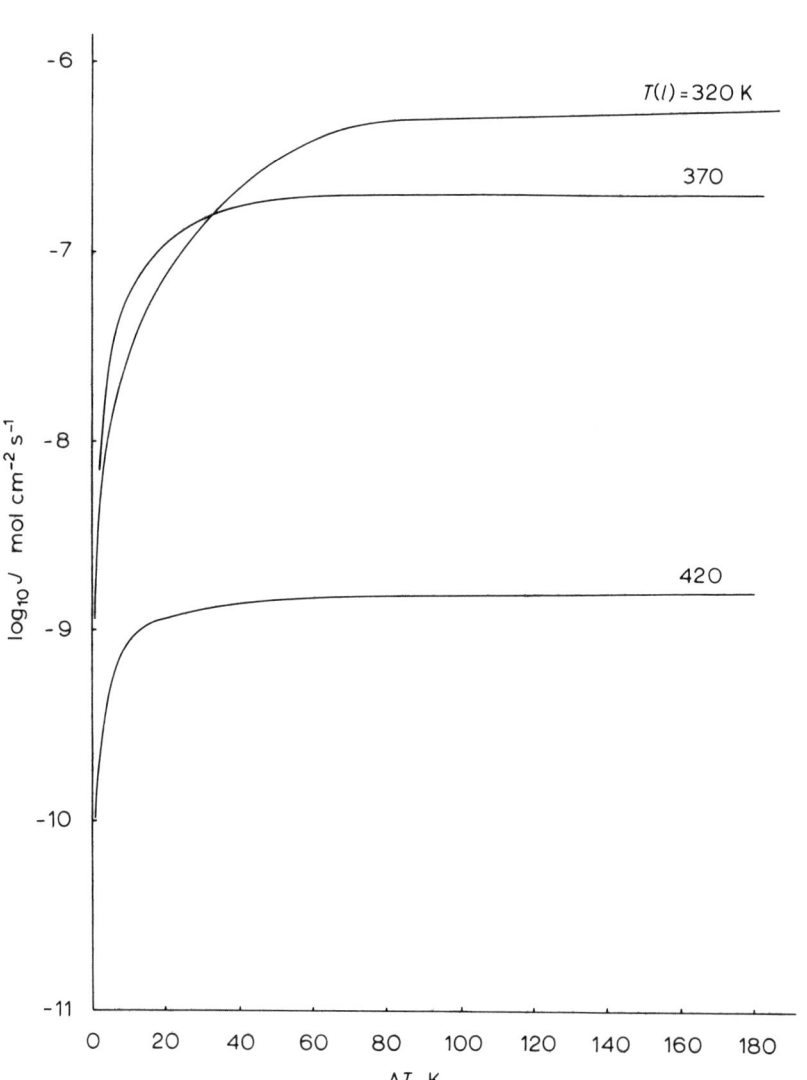

Fig. 4.17 Transport of nickel. Transport rate J as a function of temperature difference ΔT, for source temperatures of 320 K, 370 K and 420 K. Total pressure 1 atm.

VAPOUR TRANSPORT

so that α is simply the ratio of transported species to transporting agent. For the maximum transport efficiency, that is, for the maximum rate at a given temperature difference, we want α to be large over the source so that plenty of nickel is volatilized as the carbonyl, and small at the deposition end, so that most of the carbonyl is decomposed.

From Equations 4.75 – 4.77 we obtain:

$$P = (1 + \alpha) \left(\frac{\alpha}{K_p}\right)^{1/3} \tag{4.78}$$

At a given total pressure, α is a function of temperature only, as required by the phase rule. We have plotted log α as a function of temperature in Fig. 4.18, for three values of total pressure. We notice that these curves have two slopes, one of about -0.25 K^{-1} when $\alpha > 1$, and another of about -0.50 K^{-1} when $\alpha < 0.1$. We can see why this is by differentiating Equation 4.78 with respect to T:

$$\left(\frac{\partial \ln \alpha}{\partial T}\right)_P = \left\{\frac{1+\alpha}{1+4\alpha}\right\} \frac{d \ln K}{dT} = -\left\{\frac{1+\alpha}{1+4\alpha}\right\} \frac{\Delta H}{RT^2} \tag{4.79}$$

When $\alpha > 1$, the factor in curly brackets is about 1/4, whereas when $\alpha \ll 1$, this factor is about 1. The ratio of the slopes in Fig. 4.18 is less than 1:4, however, because T^2 has increased when α is small.

Clearly, to get the maximum fractional change in α, it is desirable to operate on the steep part of the log α versus T graph. However, α is small on this part, so that little nickel is volatilized, and the maximum transport rate is not high. Thus with the source material at 370 K, α over the source is about 0.2, just at the top of the steep part of the curve, and for temperature differences of up to about 30 K, transport is quite efficient. The value of α over the deposited nickel is about 0.01 with a temperature difference of 30 K, and most of the carbonyl has decomposed. Further temperature increase only removes the last little bit of the carbonyl, and the transport rate is not increased significantly. The situation is even more pronounced when the source temperature is raised to 420 K. The value of α at the source is now only 10^{-3}, so practically no nickel is volatilized.

With the source temperature at 320 K, as in the Mond process, the value of α over the source is about 4.7, so that most of the vapour is nickel carbonyl. At the other end of the system, if the temperature is below about 370 K, α is still on the shallow slope in Fig. 4.18, so that the transport rate does not rise as abruptly with ΔT as does the curve for a source temperature of 370 K. However, the maximum

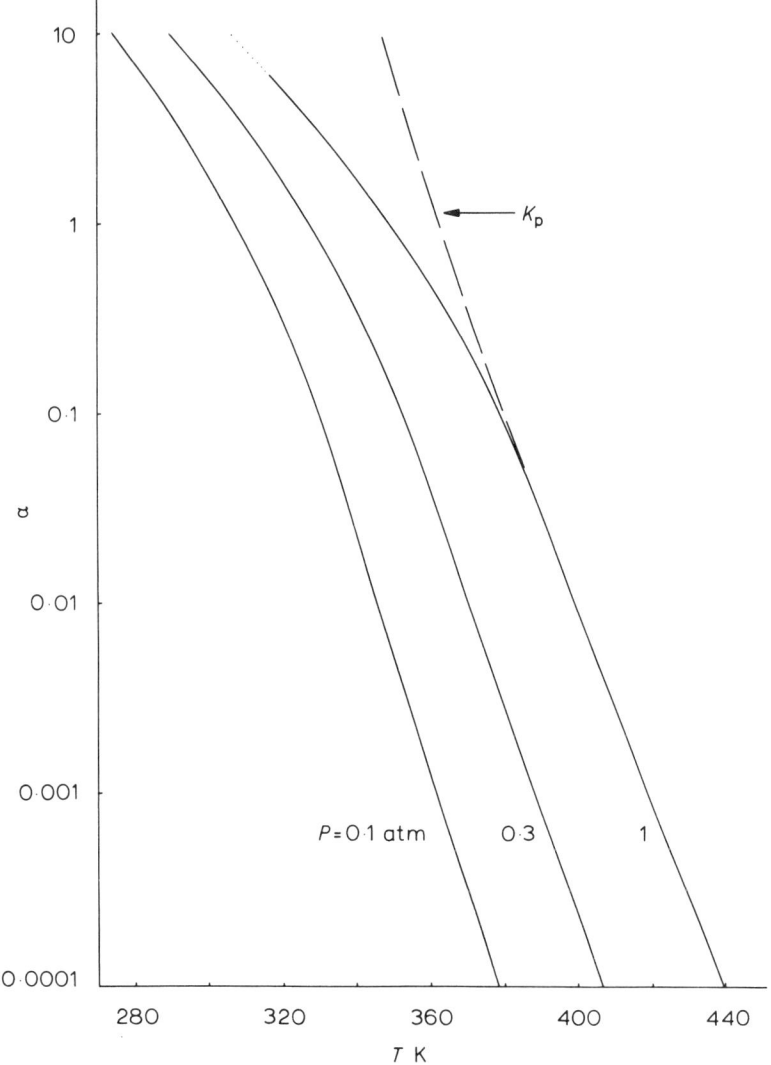

Fig. 4.18 Vapour composition in the transport of nickel as a function of temperature T, for total pressures of 1, 0.3 and 0.1 atm.

transport rate of nickel, realized at temperature differences above about 80 K, is appreciably higher. We observe that the Mond process[151] is normally operated at the temperatures which are thermodynamically most favourable for a system at atmospheric pressure. The equilibrium constant for the transport reaction, also plotted in Fig. 4.18, is near unity in the chosen temperature range.

VAPOUR TRANSPORT

If we are transporting nickel in a closed capsule, or even if the process is carried out in some more-or-less open system, we can operate at any accessible pressure, if there is an advantage in doing so. We see from Fig. 4.18 that reducing the total pressure shifts the log α versus T plot to lower temperatures. In Fig. 4.19, we have

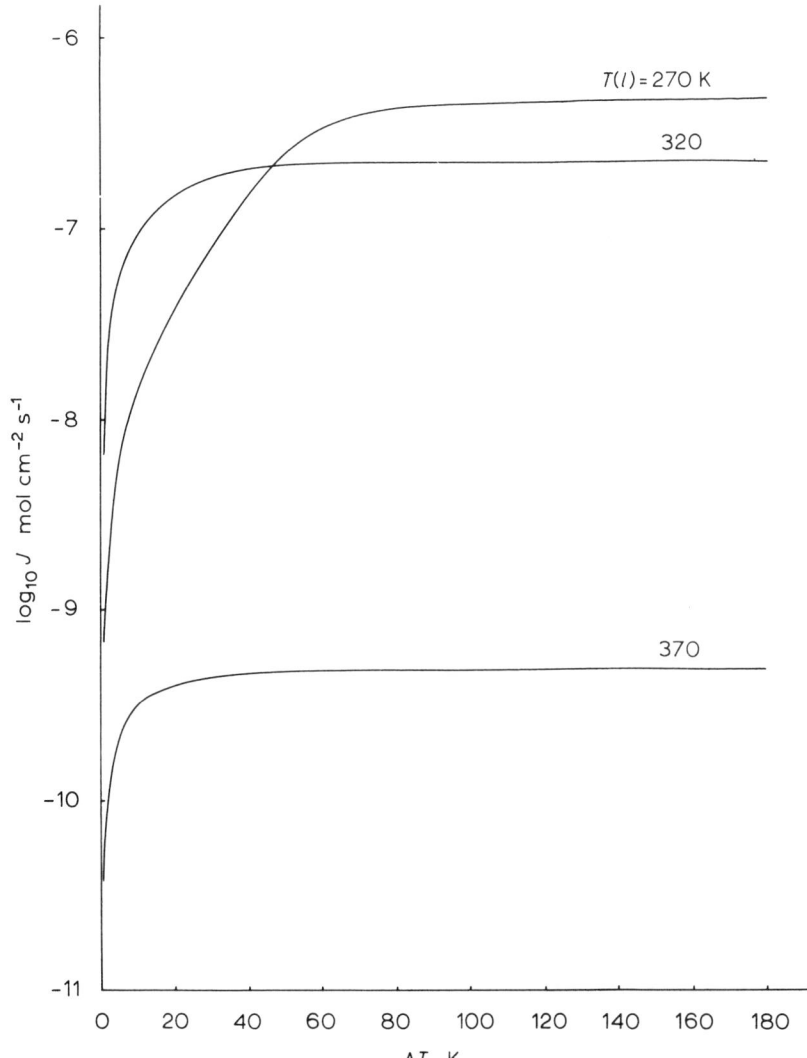

Fig. 4.19 Transport of nickel. Transport rate J as a function of temperature difference ΔT, for source temperatures of 270 K, 320 K and 370 K. Total pressure 0.1 atm.

plotted the calculated transport rate as a function of ΔT, assuming a total pressure of 0.1 atm The curves look very similar to those in Fig. 4.17; notice however that the corresponding temperatures are 50 K lower, so that the uppermost curve is for a source temperature of only 270 K, and the maximum transport rate achieved is exactly the same as for an atmospheric pressure system with a source temperature of 320 K. Although the partial pressures of the components are reduced by an order of magnitude, the transport rate is maintained, since the Stefan velocity (Equation 4.70) and the diffusion coefficient (Equation 4.4) both increase by an order of magnitude.

We have presented this fairly detailed analysis of the transport of nickel as nickel carbonyl to demonstrate how useful the basic transport equations, coupled with the thermodynamics of the system can be in designing a chemical transport system. To emphasize the importance of the variable α, which is the ratio of transported species to transporting agent, we have plotted the transport rate $\log_{10} J$ as a function of $\log_{10} \alpha$ (l) (i.e. at the source) in Fig. 4.20. These curves, which are appropriate to the temperature differences indicated, are very nearly independent of the total pressure which is chosen. Note that for a given temperature difference ΔT, there is a maximum in the transport rate versus $\alpha(l)$ plot, which occurs at a value of $\alpha(l)$ depending on ΔT. If ΔT is small, say 20 K or less, the best value of $\alpha(l)$ is around 0.4, which as we have seen, is at the top of the steep part of the $\log \alpha$ versus T curves of Fig. 4.18. The larger the temperature difference which can be used, the more it pays off to increase $\alpha(l)$, that is, to volatilize more nickel at the source end.

As a corollary to this example, we consider what happens if we ignore the Stefan flow and assume that transport takes place by diffusion only. The transport rate of nickel is given by:

$$J_{Nickel} = \frac{-D}{RT} \frac{dp_{Ni(CO)_4}}{dx}$$

which we can express as:

$$J = \frac{-D^0 \bar{T}^{0.8}}{PR \times 273^{1.8}} \left\{ \frac{p_{Ni(CO)_4}(l) - p_{Ni(CO)_4}(0)}{l} \right\}$$

since in the steady state the partial pressure profiles are straight lines on the diffusion-only model. We note in passing that the transport rate of carbon monoxide gas must, of course, be equal and opposite to that of nickel carbonyl since we are considering a binary gas mixture at constant pressure. This illustrates the conceptual defects

VAPOUR TRANSPORT

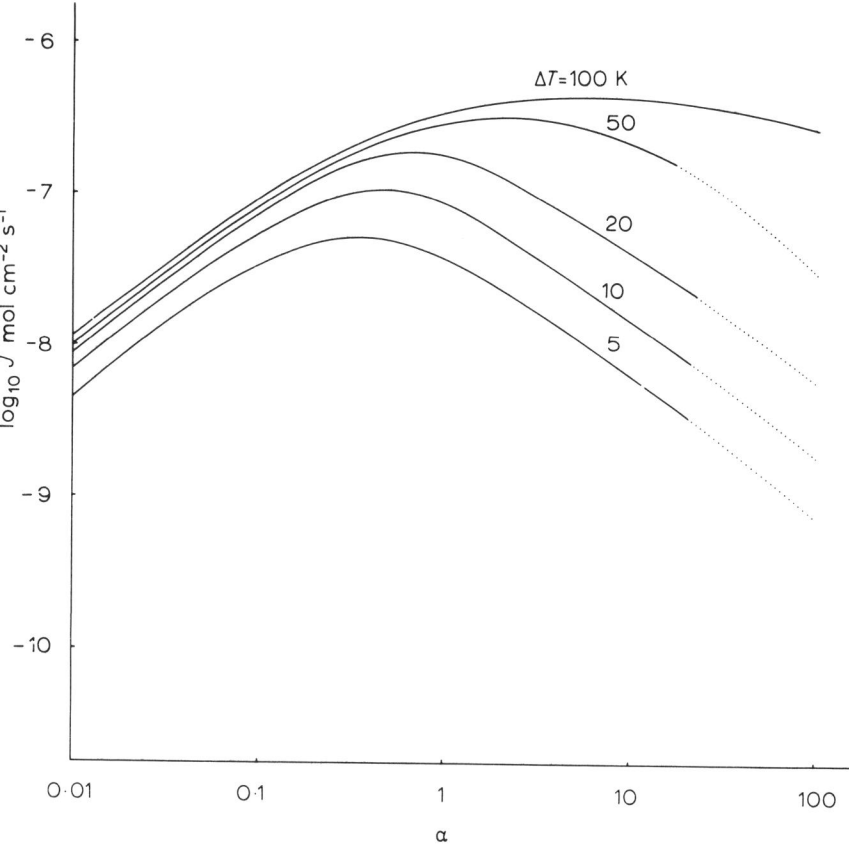

Fig. 4.20 Transport of nickel. Transport rate J as a function of vapour composition, for temperature differences ΔT of 5, 10, 20, 50 and 100 K.

of the diffusion-only model. Obviously we require the fluxes to be in the ratio 1:4, to give no net transfer of the component CO.

Fig. 4.22 shows the transport rate calculated on the diffusion-only model, and on the model including Stefan flow. The total pressure has been taken as 1 atm, and the source temperature as 320 K, as in the industrial process. The diffusion-only model predicts transport rates that are high by a factor of at least two. The Stefan velocity under these conditions is only 0.033 cm s^{-1} for large ΔT's: at lower total pressures it would be correspondingly higher, as predicted by Equation 4.70. The diffusion-only model does reproduce the shape of the $\log_{10} J$ versus ΔT curve quite well, as this shape is a reflection of the variation in α (i.e. in the equilibrium

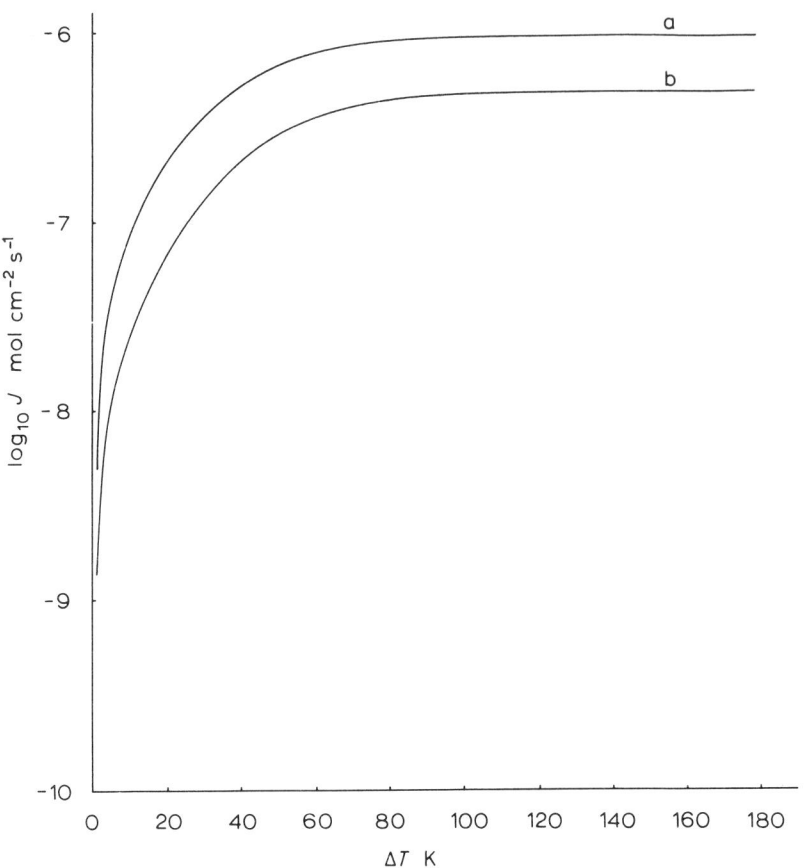

Fig. 4.21 The diffusion approximation in the transport of nickel as nickel carbonyl. Total pressure 1 atm, source temperature 320 K. Curve a: diffusion only, b: including Stefan flow.

partial pressures) with temperature. The partial pressure profiles, and particularly the partial pressure gradients over the growing crystal, are seriously in error when calculated on the diffusion-only model. The Stefan velocity increases the partial pressure gradients over the growing crystal (in magnitude, irrespective of sign) if it is directed towards the growing crystal and decreases them if it is directed away, as in this example. Thus the gradients are less than $[p(0)-p(l)]/l$ by a factor of two or more. This has interesting repercussions on the variation of the activity of the solid with x (i.e. the supersaturation or undersaturation of the vapour) which in turn may affect the

VAPOUR TRANSPORT

stability of the surface of the crystal to the development of small protrusions. We take up this topic in detail in Chapter 5.

4.4.3 Example: Transport of carbon with sulphur

The possibility of transporting carbon is of interest to the crystal grower, since carbon is often present as a contaminating species, either as elemental carbon dissolved in the source material[152] or as organic films on the surface of seed crystals and substrates, etc. The kinetics of the reaction between carbon and sulphur are known to be fast.[153]

Carbon forms two volatile compounds with sulphur according to the reactions:

$$C \text{ (solid)} + S_2 \text{ (gas)} \rightarrow CS_2 \text{ (gas)}$$

$$C \text{ (solid)} + \frac{1}{2} S_2 \text{ (gas)} \rightarrow CS \text{ (gas)}$$

The thermodynamic data for these reactions[6] give the following expressions for the equilibrium constants:

$$K_1 = \frac{p_{CS_2}}{p_{S_2}} = \exp\left\{\frac{3100}{RT} + \frac{1.73}{R}\right\}$$

$$K_2 = \frac{p_{CS}}{p_{S_2}^{1/2}} = \exp\left\{\frac{-59\,000}{RT} + \frac{22.75}{R}\right\}$$

We see that the first reaction, for the formation of the disulphide, is exothermic, resulting in transport from the cold end towards the hot, and that this reaction involves no change in the number of gas phase molecules. The second reaction, for the formation of the monosulphide, is endothermic, and results in transport from the hot end towards the cold. This reaction involves an increase in the number of vapour phase molecules, and consequently gives rise to a Stefan flow from the hot end to the cold. To evaluate this as a transport system, let us look at the equilibrium constants, K_1 and K_2. We see that K_1 changes very slowly with temperature, on account of the very small heat of reaction to form the disulphide. In fact K_1 is about 120 at 400 K, and 9 at 1200 K, only just over an order of magnitude change for a temperature difference of 800 K. On the other hand, K_2 changes very rapidly with temperature, from 5×10^{-28} at 400 K to 10^{-6} at 1200 K, on account of the large enthalpy of reaction, and is very much less than unity at readily accessible temperatures. Clearly, we do not have a useful transport system, since one equilibrium constant, though quite near to unity, is very insensitive to temperature, while the other, though sensitive to temperature, is too small to be useful.

We will pursue the analysis, however, very briefly, to give an approximate result from the transport rate. There are two degrees of freedom in this system, which we take as temperature and total pressure. We therefore have the following equations to solve for the partial pressures in equilibrium with the source and the transported carbon:

$$p_{CS_2} + p_{CS} + p_{S_2} = P$$

$$p_{CS_2} = K_1 p_{S_2}$$

$$p_{CS} = K_2 p_{S_2}^{1/2}$$

Since $K_1 \gg K_2$ at accessible temperatures, we obtain:

$$p_{S_2} = \frac{P}{K_1 + 1}, \quad p_{CS_2} = \frac{K_1 P}{K_1 + 1}, \quad p_{CS} = K_2 \left(\frac{P}{K_1 + 1}\right)^{1/2} \simeq 0$$

If we consider only sulphur and carbon disulphide in the vapour, neglecting the monosulphide for the moment, there is no Stefan flow and transport is by diffusion alone. Consequently:

$$J_{carbon} = \frac{D}{RT} \frac{dp_{CS_2}}{dx} = \frac{-D}{RT} \frac{(p_{CS_2}(l) - p_{CS_2}(0))}{l}$$

Or, since K_1 is insensitive to T,

$$J = \frac{-DP}{RT} \frac{\partial}{\partial x} \left[\frac{K_1}{1 + K_1}\right] = \frac{-DP\Delta H}{R^2 T^3 (1 + K_1)^2 l} \Delta T$$

At 1000 K, with a diffusion constant D_0 of 0.2 cm² s⁻¹, $J \simeq 6 \times 10^{-13} \Delta T$ mol cm⁻² s⁻¹, so that no appreciable transport is obtained.

At very much higher temperatures — 2000 K or more, — appreciable transport can be obtained by means of the monosulphide. If we ignore the reaction which forms the disulphide for the moment (we may think of the disulphide as 'inert gas', as it makes a very small contribution to the transport), and define:

$$\alpha = \frac{p_{CS}}{p_{S_2}}$$

Then we easily obtain the expression:

$$K_2 = \alpha \left(\frac{P}{1 + K_1 + \alpha}\right)^{1/2}$$

If we now choose $\alpha = 1$ and $T = 2000$ K, we have $K_1 \simeq 5.2$ and $K_2 \simeq 0.035$, so that the total pressure P is 0.008 atm. The partial pressures p_{CS}, p_{CS_2}, and p_{S_2} will be in the ratio $\alpha:K_1:1$, i.e. 1:5.2:1, so that an appreciable proportion of the gas is the disulphide which we are considering as inert gas. Even so, we may expect a reasonable rate of transport under these circumstances.

4.4.4 More complex chemical vapour transport systems

The transport of a binary compound by means of a halogen[154] or oxygen etc. may be described by the general equation:

$$AB \text{ (solid)} + nX \rightarrow AX_n + \frac{1}{m} B_m$$

where it is assumed thatt B_m has a sufficient vapour pressure in the temperature range in question. There are three degrees of freedom in this system. We will take the temperature as one degree of freedom, so that the equilibrium constant for the transport reaction is fixed. The remaining degrees of freedom we will take up for the moment by assigning values to two parameters α and λ, defined as follows:

$$\alpha = \frac{p_{AX_n}}{p_{B_m}}, \quad \lambda = \frac{p_X}{p_{B_m}}$$

The total pressure in the system may be calculated in terms of T (or K_p), α and λ. We find:

$$K_p = \frac{\alpha}{\lambda^n} \left(\frac{P}{1 + \alpha + \lambda} \right)^{(1/m) - n}$$

For most efficient transport (i.e. the greatest rate for a given temperature difference) it can easily be shown that, as for dissociative sublimation, we require the vapour to be stoichiometric, i.e. $\alpha = 1/m$. For a given temperature (i.e. K_p), this equation allows us to calculate the total pressure which is produced by a given value of λ, or alternatively, what value of λ will produce any required total pressure. If it is desired to operate at atmospheric pressure, we can calculate what value of λ is consistent with stoichiometric vapour.

A similar equation may be derived for the transport of a single component with, for example, a hydrogen halide or water etc:

$$A + nHX \rightarrow AX_n + \frac{n}{2} H_2$$

Again there are three degrees of freedom, and we define:

$$\alpha = \frac{p_{AX_n}}{p_{HX}} \quad \beta = \frac{p_{H_2}}{p_{HX}}$$

We take the temperature (i.e. equilibrium constant) as the third degree of freedom. We then easily obtain:

$$K_p = \alpha \beta^{n/2} \left(\frac{P}{1 + \alpha + \beta} \right)^{1 - (n/2)}$$

For efficient transport, we require α to be near unity, so that the vapour contains roughly equal quantities of the transported species AX_n and the transport agent HX. This equation now tells us what ratio of hydrogen to HX is required to give an α-value of unity at the operating pressure P.

Similar equations may be derived for other systems. If there are more components, it is neessary to define more parameters. As an example, we discuss the transport of gallium arsenide using HCl as the transport agent.[146]

4.4.5 Example: Transport of gallium arsenide

Transport of gallium arsenide using HCl as the transport agent is described entirely by a single reaction[155] in the temperature range normally used (600°C – 800°C):

$$\text{GaAs(solid)} + \text{HCl(gas)} \rightarrow \text{GaCl(gas)} + \tfrac{1}{4}\text{As}_4\text{(gas)} + \tfrac{1}{2}\text{H}_2\text{(gas)} \quad (4.80)$$

The technological importance of this process has resulted in a number of theoretical and experimental studies of this system being carried out[155,156,157]. The transport is usually carried out in a flow system, and the end product is a thin layer deposit on a suitable substrate. The dopant species necessary to produce the required electronic properties in the deposited layer are added to the gas stream in a suitable form, for example, as a volatile chloride for group II dopants, or as H_2S, etc. for group VI elements. Our discussion here will ignore the presence of the dopant species, as their concentrations are usually such as to produce only the smallest perturbation on the transport system. We will discuss the transport of gallium arsenide using a one-dimensional model, as if it were carried out in a closed capsule of length l. The results of this analysis may be carried over to the flow system if the flow rate is very low or very high. In the first case, with a very low flow rate, the system is quite similar to a capsule containing (nominally) stationary vapour, and the length l is simply the separation of the source material and the substrate. In the second case, at very high flow rates, a boundary layer forms over the source material and over the substrate, and we may imagine that all changes in the vapour composition take place within these boundary layers, so that the length l is now the combined thickness of the boundary layers. At intermediate flow rates, where most practical systems are operated, the boundary layers spread across the flow tube and also along it to a considerable extent, and the length l is not clearly defined. For a proper analysis of this situation, we clearly need a two- or three-dimensional treatment. However, we can use the one-dimensional treatment to obtain semi-quantitative results. A useful extension of this treatment for flow systems is described in Chapter 6.

In this system, there are four degrees of freedom, which we will take for the moment as the temperature T, the total pressure P, and two other parameters:

$$\alpha = \frac{p_{GaCl}}{p_{As_4}} \quad \text{and} \quad \lambda = \frac{p_{HCl}}{p_{H_2}}$$

The first of these parameters is the ratio of gallium species to arsenic species in the vapour. The vapour is stoichiometric when α has the value 4. The second parameter λ is a measure of the tendency of the gas to react with gallium to form the chloride. We may arrange for α and λ to take any values we choose by loading the capsule with suitable quantities of, for example, hydrogen, HCl, and perhaps excess gallium or arsenic. The actual calculation of the necessary quantities is fairly involved, and we shall defer discussion of it for the moment. In the open flow system the values of α and λ may be fixed by using a suitable mixture of input gases. The most common method of arranging for the input of hydrogen and HCl of sufficient purity is to pass hydrogen from a palladium diffusion purifier through liquid arsenic trichloride at around room temperature. The hydrogen picks up almost the equilibrium vapour pressure of arsenic trichloride, as well as a small amount of As_4 and HCl formed by the reaction:

$$4\,AsCl_3 + 6H_2 \rightleftharpoons As_4 + 12\,HCl \tag{4.81}$$

This reaction goes essentially to completion as the gas stream enters the hot part of the furnace, so that the gas encountering the source material consists of arsenic and HCl in the ratio 1:12 of partial pressures, fixed by the stoichiometry of the reaction, and a considerable excess of hydrogen. Over the source, reaction 4.80 takes place, and at 700–800°C the equilibrium position is well to the right, so that only some 2% of the HCl remains unreacted[158,159]. When the gas is flowed slowly over the source, this equilibrium is approached quite closely. The variation in the extent of this reaction as a function of flow rate has been studied by Ban and co-workers,[155] and on the basis of this work it seems reasonable to assume that the approach to equilibrium is limited mainly by diffusion across the flowing gas stream and not by surface reaction kinetics. For the purposes of this analysis, we will assume that equilibration is complete, which we expect to be a very good approximation for transport of gallium arsenide in a closed capsule, and also in the open flow system when the flow rate is low.

Because of the stoichiometry and position of equilibrium of reactions (4.80) and (4.81), it turns out that the value of α over the source is fixed at just below three, and cannot be altered appreciably using the gas input system described above. We can easily see how this comes about. Each mole of $AsCl_3$ which is taken up by the gas stream reacts with hydrogen to form three moles of HCl and a quarter of a mole of As_4, according to Equation 4.81. The HCl reacts with gallium arsenide, being nearly all used up, to form three moles of GaCl and a further three quarters of a mole as As_4. Thus 3 moles of GaCl form for every 1 mole of As_4, and α is pinned at 3. Because a little HCl remains unreacted, α is just below 3. Table 4.3 gives the values of α and λ as functions of the temperature of the liquid arsenic trichloride. The furnace temperature is 1100 K.

Table 4.3

$AsCl_3$ temperature °C	$\alpha = \dfrac{p_{GaCl}}{p_{As_4}}$	$\lambda = \dfrac{p_{HCl}}{p_{H_2}}$
0	2.87	0.00416
5	2.87	0.00596
10	2.86	0.00833
15	2.86	0.0115
20	2.85	0.0156
30	2.84	0.0278
60	2.83	0.122

The constancy of α over a wide range of temperatures of the arsenic trichloride has been commented on previously[160]. We wish to emphasize that although the values of α over the source material cannot be varied using this method, the value of α over the deposited layer may be varied by controlling the substrate temperature. In this way, the stoichiometry of the grown layer, and hence the solubility of dopant species, can be controlled to some small extent. The

VAPOUR TRANSPORT

transport system is buffered by a large excess of H_2, in a way comparable to the buffering of the cadmium sulphide transport system with inert gas (Section 4.3).

The transport equations may be written in a manner similar to that given in the previous examples. We will not repeat the algebra here, but will quote the main results from the literature.[146] The theoretical growth rate as a function of temperature difference ΔT between the source and the growing layer is shown in Fig. 4.22, for

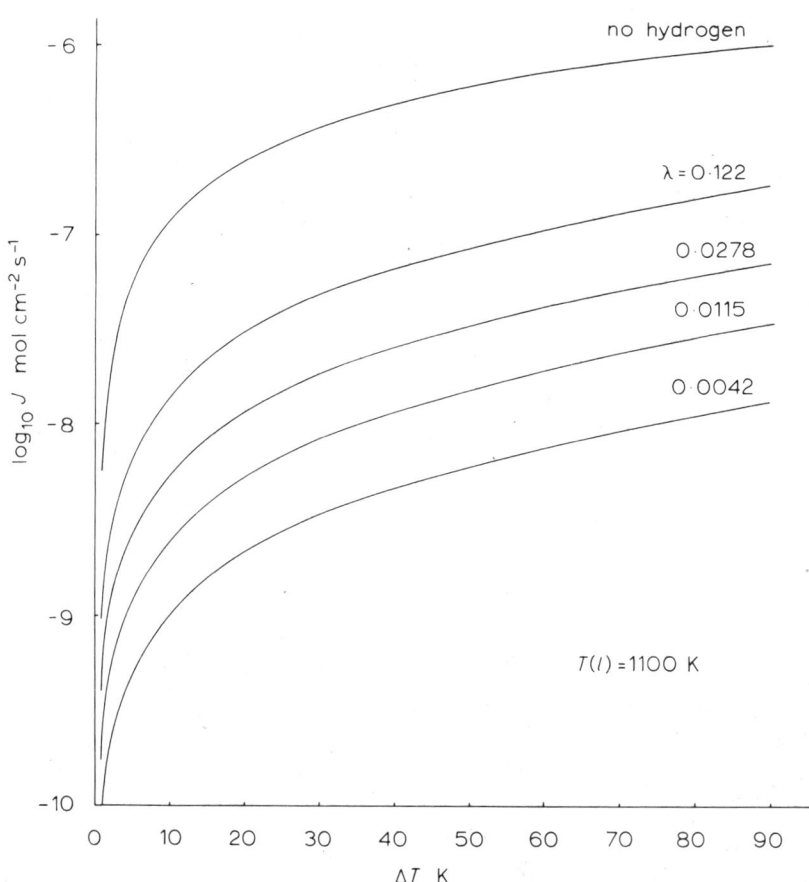

Fig. 4.22 Transport of gallium arsenide in hydrogen chloride. Transport rate J as a function of temperature difference between source and crystal, ΔT, for various values of the parameter λ. Source temperature 1100 K.

four values of λ. We observe that when ΔT is less than about 20 K, $\log_{10} J$ rises very rapidly as ΔT increases, while when ΔT is larger than 20 K, $\log_{10} J$ is roughly proportional to ΔT. If ΔT is increased very much, the growth rate eventually levels off as the vapour becomes exhausted in gallium and arsenic species over the growing layer of gallium arsenide.

By varying the value of λ, i.e. by varying the amount of arsenic trichloride picked up by the gas stream or loaded into the capsule, a considerable variation in the growth rate can be achieved. For the deposition of thin layers of gallium arsenide, it is useful to keep the transport rate fairly low: a rate of 10^{-8} mol cm^{-2}s^{-1} corresponds to a rate of growth of the layer of about 10μm s^{-1}. Since it is advantageous to operate the epitaxial deposition apparatus under conditions corresponding to the parts of the curves in Fig. 4.22 with least slope (i.e. $\Delta T = 50$ K or more) it is necessary to cool the arsenic trichloride to below room temperature or dilute the gas stream with hydrogen to suppress the rate of transport. On the other hand, if the transport system is used to grow large crystals of gallium arsenide or for purifying the source material by transport (see Chapter 5), a fast transport rate is advantageous, and this can be obtained by heating the arsenic trichloride (in the open flow system) to increase λ, or in theory by not including hydrogen in the capsule charge if the capsule system is used. The top curve in Fig. 4.22 is the calculated transport rate in a capsule containing no hydrogen. The transport reaction is then:

$$2GaAs(s) + GaCl_3(g) \rightleftharpoons 3GaCl(g) + \tfrac{1}{2}As_4(g) \qquad (4.82)$$

The value of α obtained using liquid arsenic trichloride and hydrogen as a means of producing HCl is fixed near 2.85, as we have seen. The value corresponding to a stoichiometric vapour is 4. To see how the growth rate varies with changing α, we will compare this transport system with one of our previous examples – dissociative sublimation of cadmium sulphide in the presence of an inert gas (Section 4.3). In that system the value of α corresponding to stoichiometric vapour is 2, and we find that the growth rate is fairly insensitive to α for α near 2, provided there is some inert gas present, even as little as 1% the total pressure. (Fig. 4.12). Let us consider what represents 'inert gas' in the GaAs–HCl transport system. Clearly we require the fluxes of GaCl and As_4 to be in the ratio $J : J/4$. The flux of GaCl transports Cl, and this is countered by a flux $-J$ of HCl. This in turn transports hydrogen, and is countered by a flux $J/2$ of H_2. We thus see that the combination of components

VAPOUR TRANSPORT

($2p_{H_2} + p_{HCl}$) has a zero transport rate, and may therefore be compared in its action to inert gas. Of course, part of the action of an inert gas comes from slowing the diffusion of the active components, and in assessing this effect, one considers the total pressure as the relevant parameter. However, when the active components are in nearly the stoichiometric ratio, the inert gas plays a more important part, since without it, transport would be almost entirely by Stefan flow, with diffusion playing a very small part, and would be very fast as a result, being limited eventually by, for example, heat transfer rates.

We would point out that it is impossibe to eliminate 'inert gas' in a chemical vapour transport system — that is to say, since the transport agent must be transported in the opposite direction to the active components, there must always be a diffusive contribution to transport, and there is always some combination of components with a zero transport rate. This aspect has been enlarged upon in an analysis of the GaAs–Cl_2 transport system. [144]

4.4.6 Transport systems involving several reactions

So far we have discussed only those transport systems which may be described with sufficient accuracy by a single heterogeneous reaction. However, in general other reactions take place simultaneously, for example, if the metallic component or components of a compound have more than one oxidation state, several reaction products may result, of which all may be volatile. For example, in the transport of gallium in water vapour/hydrogen mixtures, we may have:

$$2 Ga + H_2O \rightleftharpoons Ga_2O + H_2$$

$$2 Ga + 2H_2O \rightleftharpoons 2GaOH + H_2$$

The behaviour of gallium is perhaps fairly simple, though not well-documented.[161–163] If it was desired to transport a transition metal using a halide or hydrogen halide, the number of volatile compounds increases. Furthermore, if any oxygen is present in the system (e.g. outgassed water from silica ware) volatile oxides and oxychlorides will form. In such cases the analysis becomes more complex, though still possible if thermodynamic data are available for the various reactions. It may be that the formation of some of the compounds is hindered for kinetic reasons. If the kinetics are sufficiently slow at the temperature of interest, these compounds may obviously be ignored.

If several transport reactions are possible, it follows that there are possible reactions between the vapour components. In the example of gallium transported in a water/hydrogen gas mixture, we may consider the homogeneous gas-phase reaction:

$Ga_2O + H_2O \rightleftharpoons 2\,GaOH$

Depending on the reaction kinetics, this reaction may proceed effectively to equilibrium everywhere in the gas-phase, or it may proceed so slowly as to be negligible. In the latter case, no complication is introduced into our analysis, apart from the extra difficulty of calculating the equilibrium partial pressures over the source and over the growing crystal or layer (i.e. at $x = l$ and $x = 0$). In the former case, however, when the reaction goes rapidly, the composition of the vapour at any point in the capsule is determined by the temperature at that point as well as by the transport equations. Furthermore, if the gas-phase reactions involve a change in the number of molecules in the gas phase, as they will in general, the Stefan velocity U will vary along the tube or capsule, though the flux J for each element must remain constant. The extent of this variation will depend on whether the species taking part in the gas-phase reaction are present in significant concentrations, on the enthalpy of the gas-phase reaction (i.e. on the variation of the equilibrium constant with temperature) and on the stoichiometry of the reaction. It may well be possible to ignore the variation of Stefan's velocity. For example, in the GaAs HCl transport system just described, the species As_2 and $GaCl_3$ will form. Both these species may enter into homogeneous gas-phase reactions, as follows:

$As_4 \rightleftharpoons 2\,As_2$
$GaCl_3 + H_2 \rightleftharpoons GaCl + 2HCl$

Both these reactions result in a change in the number of gas-phase molecules, and hence a change of Stefan's velocity. However, under the conditions which normally obtain in practice for this system, the partial pressures of As_2 and $GaCl_3$ are very low,*[6] so that these reactions can involve only a very small fraction of the total gas. The change in Stefan's velocity can therefore be safely ignored. If the transport system were to be operated at a higher temperature, As_2 could be a significant species, and if it were operated at very low hydrogen pressure, $GaCl_3$ could be a significant species.

The solution of the transport equations when the Stefan velocity U is a function of temperature and hence of x, because of gas-phase

---

*More recent work[164] indicates that As_2 is the dominant arsenic species.

VAPOUR TRANSPORT

reactions, is beyond the scope of this book. As far as calculation of transport rates is concerned, it is probably not worthwhile. The problem is only of importance if there are two or more transport reactions giving approximately equal rates of transport. A useful estimate of the net rate may be obtained by considering each reaction separately, and taking a weighted mean of the separate transport rates according to the proportions of the different transporting species. Such an estimation procedure is probably quite good enough, considering the inherent shortcomings of the one-dimensional model we have discussed. However, when we consider the stability of the crystal-vapour interface (Chapter 5) we shall be interested in the gradients of the partial pressures over the crystal. Here gas-phase reactions may be of utmost importance.

4.5 Viscous flow of gas

The concept of viscosity is familiar to everyone, particularly as applied to liquids. It is an important phenomenon in crystal growth from the vapour since, as we saw, in most crystal growth systems, there is some movement of the gas as a whole, and this movement will be subject to viscous forces. Furthermore, in many types of apparatus, the gas is made to flow by external forces, for example, by passing a carrier gas through the apparatus from a gas cylinder. This flow of gas, like the flow produced by the change in volume on crystallization, is subject to viscous forces. These forces control the flow pattern and the velocity distribution of the gas at all points in the apparatus, including the very important region where the crystal or layer is growing. Before considering the effects of viscosity in the gas on the growth of crystals, we will examine its origin, and the laws governing it.

We will approach the subject by considering an experiment in which gas is passed down tubes of different lengths and diameters. Provision is made to measure the volume of gas flowing in a unit of time, and also the pressures at the ends of the tubes. It is found that the volume flow of the gas depends on the other parameters according to the law:

$$V \propto \frac{a^4}{l} (P_1 - P_2) \tag{4.83}$$

where P_1 and P_2 are the pressures at the inlet and outlet ends of the tube, a its radius, and l its length. The constant of proportionality is evidently characteristic of the particular gas, and turns out to be a function of temperature, but essentially independent of pressure except at very low and very high pressures. This characteristic constant is, of course, a measure of the viscosity of the gas. Such a law was derived theoretically by Poiseuille[165] (see below). The viscosity coefficient is usually defined as the shear force per unit area of an imaginary plane in the gas produced by a unit velocity gradient normal to the plane. The shear force is therefore given by:

$$\gamma = \eta \frac{dv}{dn} \qquad (4.84)$$

where v is the velocity of the gas, n the direction normal to the plane of the shear force γ, and η is the viscosity coefficient.

The origin of this shear force may be seen from a simple kinetic theory description of viscosity. Imagine a stream of gas flowing parallel to the x-axis, (Fig. 4.23) and consider the molecules crossing

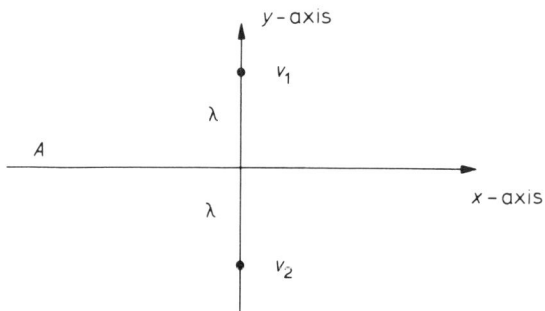

Fig. 4.23   Simple model of gas viscosity.

an imaginary reference plane A which is parallel to the gas velocity: molecules crossing this plane from above have come, on average, from a point one mean-free-path above the reference plane, and carry momentum mv_1, where m is the mass of a molecule. The number of molecules crossing a unit area of the plane in unit time is ¼ $n\bar{c}$, so that a momentum of ¼ $nm\bar{c}v_1$ is transported by these molecules across the stream per unit area in unit time. Molecules crossing the plane from below transport a momentum of ¼ $nm\bar{c}v_2$ per unit area in unit time. The net rate of transfer of momentum is thus ¼ $nm\bar{c}$ $(|v_1| - |v_2|)$, and this is equal to the effective shear force per unit area, γ. We put $(|v_1| - |v_2|)$ equal to $2\lambda\, d|v|/dy$, so that:

$$\gamma = \tfrac{1}{2} nm\bar{c}\lambda \frac{dv}{dy} \qquad (4.85)$$

Putting $nm = \rho$, the gas density, and taking account of molecules

VAPOUR TRANSPORT

travelling in all directions, we would obtain:

$$\gamma = \tfrac{1}{3}\rho\bar{c}\lambda \frac{\mathrm{d}v}{\mathrm{d}y} \tag{4.86}$$

So that

$$\eta = \tfrac{1}{3}\rho\bar{c}\lambda = \rho D \tag{4.87}$$

where the factor of ½ has become ⅓ for the reason given in the previous discussion of the diffusion coefficient D. This simple picture of viscosity is essentially correct, although we have ignored the persistance of velocities, the finite size of molecules, the forces between them, etc. We may deduce two things from this simple treatment which are supported by experimental evidence. Firstly, since the diffusion constant D is inversely proportional to the mole density, while the density ρ is directly proportional to it, we conclude that the viscosity of a gas is independent of mole density. This unexpected result was one of the early successes of the kinetic theory. Secondly, let us consider what happens to the momentum of the gas flow near a wall or other surface. Molecules impinging on the wall will be scattered in all directions, in such a way that the number of molecules leaving the wall at angles between θ and $\theta + \mathrm{d}\theta$ to a normal to the wall surface is proportional to $\sin\theta\ \mathrm{d}\theta$*. These molecules will have a normal thermal temperature distribution, so that the net momentum of the gas stream is lost. We therefore expect the flow velocity to fall to zero at the walls of a tube. It is not possible to measure the velocity of the gas stream at the surface, of course, but the velocity profile across the stream can be measured, and this profile extrapolated to the walls. It appears that the velocity is indeed zero at the walls, except for very fast flows in small capillaries.[135]

In a crystal growth system, we frequently encounter gases in which diffusion is taking place at the same time as flow of the gas as a whole. This diffusion is the result of concentration gradients brought about by, for example, a change in equilibrium concentrations between two parts of the system at different temperatures. It is interesting to consider the behaviour of a gas mixture flowing over a solid surface such as the walls of the apparatus or a substrate on which some material is to be deposited. Each component of the gas stream is travelling at a mean molecular velocity u_i specific to that

---

*For very light molecules (H_2, He) at grazing incidence ($\theta > 89°\ 56'$ at room temperature) on the best polished surfaces, a fraction is reflected specularly.[71]

component, and we may think of this velocity as being made up of a flow velocity which is some sort of average over all the components, and a diffusion velocity. For there to be a net transport of component i, u_i must be non-zero. When molecules of component i strike the apparatus walls, they are scattered randomly, and hence lose any net velocity. However, the number of molecules of component i leaving the wall is equal to the number arriving at the wall, and in consequence proportional to the mole density n_i. Now consider an imaginary reference plane perpendicular to the wall. Molecules leaving a given point on the wall will cross the plane at an angle θ if they start from a point a distance less than $\lambda \cos \theta$ from the reference plane. Clearly, more molecules cross the plane from the side where more molecules are leaving the surface, i.e. where n_i is higher, giving rise to a diffusion flux. The diffusion coefficient for this flux will not be the same as that for ordinary diffusion, since the distribution of molecules leaving the wall is a function of θ, and not equal in all directions as in the bulk of the gas. The diffusion flux at the wall causes a net momentum flow, unless the different components have the same molecular weight, since equal numbers of molecules leave the wall with a positive component parallel to it as with a negative component. We conclude further that this diffusion flux is the only flux at the wall, that is to say, that the mole average velocity U is zero at the wall. We are not taking into account diffusion of adsorbed molecules along the surface of the wall at this stage.

The temperature dependence of the viscosity coefficient is found to be more pronounced than $T^{1/2}$ which arises from the variation of $\bar{c}$, since the collision diameter σ is also a function of temperature, i.e. of $\bar{c}$. It is found empirically that a law of the form:

$$\eta(T) = \eta_0 \left\{ \frac{T}{T_0} \right\}^n \tag{4.88}$$

may be applied over a restricted range of temperature, where η_0 is the value of η at temperature T_0, and n is about 0.7 to 0.8 for many gases near $0°C$. The molecular model of Sutherland[166] gives a formula which meets with very considerable success:

$$\eta = \eta_0 \left\{ \frac{T}{273} \right\}^{3/2} \frac{(C + 273)}{(C + T)} \tag{4.89}$$

where C is Sutherland's constant for the particular gas, and η_0 is now the value of η at 273 K ($0°C$). The numerical values of C for several gases are given in Table 4.4 below.

Table 4.4 *Viscosities of gases*

Gas	Viscosity at $0°C$, μP	Sutherland's constant, K
Air	170.8	117
Argon	212	142
Arsine	145.8	(297)
CO_2	138	240
CO	166	102
Cl_2	123	350
H_2	84	72
Ne	298	56
N_2	166	104
O_2	192	125
HCl	(133)	(358)

Figures in brackets calculated from data in 'Handbook of Chemistry and Physics', 46th edition, Chemical Rubber Co., 1965. Other data from 'Tables of Physical and Chemical Constants', G. W. C. Kaye, T. H. Laby, Longmans, 1968.

When a gas flows down a tube, work is done on the gas by the external pressure, and this work is partly dissipated by viscous drag at the walls of the tube (i.e. as heat) and is partly taken up as increased kinetic energy of the gas stream. For moderate flow rates this last contribution is negligible; equating the work done by the external pressure to that done against viscous drag leads to two important results. Firstly, we find, for the volume flow:

$$V = \frac{\pi a^4 (P_1 - P_2)}{8 \eta l} \tag{4.90}$$

a law first derived by Poiseuille, and secondly, for the radial variation of the velocity:

$$v = v_0 \left(1 - \frac{r^2}{a^2}\right) \tag{4.91}$$

where v_0 is the velocity at the centre of the tube. This shows us that the velocity profile across the tube is parabolic, being zero at the wall. As far as it is possible to measure experimentally, the gas velocity does approach zero near the wall in the parabolic manner predicted.

Viscosity coefficient of a mixture of gases

The simple kinetic theory description of viscosity leads to Equation 4.87 for the viscosity coefficient of a single gas, and to an expression of the form:

$$\eta = \tfrac{1}{3}(\rho_1 \bar{c}_1 \lambda_1 + \rho_2 \bar{c}_2 \lambda_2 + \rho_3 \bar{c}_3 \lambda_3 + \ldots) \tag{4.92}$$

for a mixture of gases. For a binary mixture we would expect a coefficient of viscosity η_{12} given by:

$$\eta_{12} = \tfrac{1}{3}\rho_1 \bar{c}_1 \lambda_1 + \tfrac{1}{3}\rho_2 \bar{c}_2 \lambda_2$$

which may be put into the form: [135]

$$\eta_{12} = \frac{\eta_1}{1 + A_1 \dfrac{\rho_2}{\rho_1}} + \frac{\eta_2}{1 + A_2 \dfrac{\rho_1}{\rho_2}} \tag{4.93}$$

where A_1 and A_2 are quantities depending on the masses and collision diameters of the molecules of type 1 and 2. Such a formula leads to a maximum in η_{12} at a certain ratio of the constituents if $\eta_2 < \eta_1$ and $\eta_2 < \eta_1 A_1 A_2$. It is found that such a formula fits the observed binary viscosity coefficients with very fair accuracy.

Magnitude of the pressure difference for a given flow rate

It is useful to have in mind the order of magnitude of the pressure difference required to sustain a given flow rate, whether the flow be imposed by external forces (pumps or pressurized gas cylinders) or arises intrinsically from the Stefan flow. Equation 4.90 may be rearranged to express the flow in molar units:

$$J = \frac{1.013 \times 10^6}{8R} \cdot \frac{Pa^2}{\eta T} \cdot \frac{\Delta P}{l} \tag{4.94}$$

where P and ΔP are in atmospheres, R has the value 82.06 cm^3 atm mol^{-1} K^{-1}, η is in poise, and l in cm. For hydrogen at 1 atm, 1000 K, Equation 4.94 reduces to, approximately:

$$J = 8 \times 10^4 a^2 \frac{dP}{dx}$$

so that a flow rate of 10^{-3} mol cm^{-2} s^{-1} (26.1 cm s^{-1} mean velocity at 1 atm) in a 1 cm radius tube requires a pressure gradient of some 10^{-7} atm cm^{-1}. Such a flow rate is fairly typical of the open flow type of crystal growth apparatus. In sealed capsules, rates may be much smaller, typically 10^{-6} mol cm^{-2} s^{-1}, corresponding to roughly 0.1 cm h^{-1} growth rate. The temperature and viscosity in such a system might well be higher, however, and the total pressure much lower than atmospheric. Using data for argon at 1400 K and 10^{-3} atm, Equation 4.94 reduces to, approximately:

$$J = 17a^2 \frac{dP}{dx}$$

so that a rate of 10^{-6} mol cm^{-2} s^{-1} in a tube of radius 1 cm requires a pressure gradient of some 6×10^{-8} atm cm^{-1}. For many purposes the existance of this gradient can safely be ignored, as was done in the worked example on the transport of silver.

4.6 Laminar and turbulent flow

So far in our discussion of flowing gas in crystal growth systems, we have assumed that the velocity of the gas at any point in the system is constant in magnitude and direction, or in other words, that the flow is laminar. The various components of the gas stream have a velocity U which is the mean velocity of the gas stream (defined as in Section 4.1) and a diffusion velocity appropriate to each species. The flow velocity U is subject to the laws of viscosity, and thus varies across the tube in which the gas is flowing. We have assumed that the flow velocity is parallel to the axis of the tube throughout its length. The diffusion velocity of each component is proportional to the concentration gradient of that component. Diffusion represents the only mechanism by which a gas can become mixed in laminar flow.

The experiments of O. Reynolds[167] in 1883 demonstrated that two distinct types of flow exist: laminar flow as described already, and turbulent flow. In a fluid flowing turbulently, the velocity U is a function of time as well as of space, usually oscillating in an unsteady way about a time-average value $\bar{U}$. We can define $u' = U - \bar{U}$ as the velocity fluctuation. The velocity of the fluid also fluctuates in directions at right angles to the direction of the flow, and we define v' and w' as the velocity fluctuations at right angles to u'. The degree of turbulence, or turbulent intensity, is then defined by

$$B = \frac{\sqrt{[\frac{1}{3}(\overline{u'^2} + \overline{v'^2} + \overline{w'^2})]}}{\bar{U}}$$

That is to say, B is the r.m.s. velocity fluctuation, expressed as a fraction of the time-average velocity.

In reality, all flowing gases have a finite degree of turbulence, brought about by passage through valves, round bends in the tube, over rough surfaces, and by temperature fluctuations, etc. Some degree of mixing of the gas can always take place other than by diffusion. However, if the gas is flowing only slowly, and if the tube walls are not very rough, the degree of turbulence may be very small throughout most of the volume of the gas. This means that, for example, a carrier gas, or gas stream containing a chemical transporting agent, passing over a source of material may reach the composition which is in equilibrium with the source only in the layer of gas next to the source. The gas stream leaving the source region is therefore non-uniform in composition. If the growing crystal or crystalline layer is presented with the point of the gas stream that reached equilibrium with the source, it will grow at one rate; if it is shown the part of the gas stream which did not reach equilibrium with the source, it will grow at another quite different rate, which may even be negative. Different parts of the crystal are encountered by different parts of the gas stream, producing a variation in growth rate across the crystal. For this reason, some workers have been at pains to increase the degree of turbulence in the gas stream, at least locally over the growing crystal and immediately after the source region.

Another effect of increasing the degree of turbulence is to increase the reaction rate between the vapour and the source or growing crystal, whether the reaction be sublimation or reaction with a transporting agent. This is intuitively obvious, and is a parallel to such everyday phenomena as increasing the rate of dissolution of sugar in tea by stirring it. In the unstirred fluid, the rate of reaction is limited eventually by the rate at which the reactants and products can diffuse through the whole fluid to the solid or away from it. Stirring or mixing by turbulence results in the bulk of the fluid achieving a more or less even composition, with diffusion confined to a thin boundary layer over the solid.

In relatively slow flows (which we will define presently), laminar flow may be very closely approached. The flowing gas can flow round bends and irregularities without developing eddies and vortices. At faster speeds, the flow remains laminar except in the vicinity of protrusions and other irregularities. Above some critical velocity, which depends on the geometry of the system and the nature of the gas, the whole gas is in turbulent motion, except for a thin layer next to all solid surfaces. These three regions of flow we may call laminar, locally turbulent and turbulent.

The nature of the flow is characterized by the Reynold's number, a dimensionless parameter defined by:

$$Re_x = \frac{ux}{\nu} \qquad (4.95)$$

VAPOUR TRANSPORT

when u is the flow velocity, ν is the kinematic viscosity (η/ρ) and x is a dimension of the system. Which dimension is the relevant one depends on the geometry of the system. The transition from one type of flow to another occurs at a critical value of the Reynold's number, which also depends on the geometry of the system. According to the *similarity theorem*,[168] geometrically similar systems with the same Reynold's number have mechanically similar flows. The Reynold's number may be thought of as the ratio of two types of resistance to the gas flow the first being due to work done against forces of inertia (accelerating the gas round curved surfaces, and in eddies and vortices) and the second being work done against viscous forces. At low velocities (small Reynold's number) most of the resistance to flow is due to viscous forces, the effect of which is felt at considerable distances from solid surfaces. At high gas velocities (large Re), viscous forces are confined to a thin layer of gas near each solid surface, and most of the resistance to flow is due to inertia of the gas.

To demonstrate the application of these ideas, let us consider the flow of gas over a flat plate (Fig. 4.24).

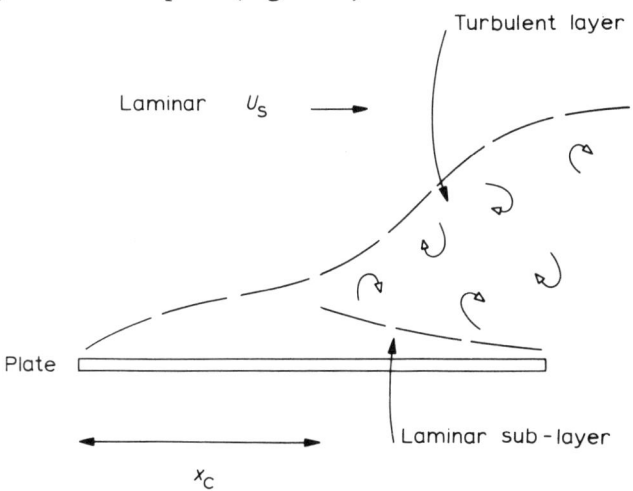

Fig. 4.24   Boundary layers over a flat plate.

The plate may in practice be a substrate on which a crystalline layer is to be grown, or the surface of a tube, or substrate holder, etc. The velocity of the free gas stream away from the plate is U_s, and we may imagine that the flow there is essentially laminar. On encountering the plate, the gas must alter its velocity locally so as to have zero velocity at the surface of the plate, and it does this by reducing the velocity to zero over a boundary layer next to the plate. The thickness δ of this boundary layer increases with distance x from the leading edge of the plate according to:

$$\delta = 4.64 \sqrt{\frac{vx}{U_s}}, \quad \text{or} \quad \frac{\delta}{x} = \frac{4.64}{\sqrt{(Re_x)}} \quad (4.96)$$

The Reynold's number is proportional to x, and so increases with distance from the front edge of the plate. At some critical Reynold's number Re_x^*, the boundary layer becomes turbulent and increases more rapidly in thickness, while a laminar sub-layer develops between it and the plate. The value of Re_x at which this transition from a laminar boundary layer to a turbulent one takes place depends on the degree of turbulence in the free stream, being 10^5 when the free stream turbulent intensity is 0.03, rising to about 4×10^6 when the free stream turbulent intensity is 10^{-3}, and thereafter remaining constant.[169]

Let us see what sort of velocity of gas corresponds to Reynold's numbers of 10^5 or 10^6. Table 4.5 gives $1/v$ s cm^{-2} for hydrogen and

Table 4.5 Reciprocals of kinematic viscosity

Temperature °C	$1/v$ s cm^{-2}	
	Hydrogen	Nitrogen
500	0.389	2.66
600	0.286	1.95
700	0.220	1.54
800	0.176	1.23
900	0.145	1.10
1000	0.122	0.853
1100	0.104	0.735
1200	0.090	0.641
1300	0.079	
1400	0.075	

nitrogen[168] (adapted from Eckert and Drake) as a function of temperature, for 1 atm pressure. To obtain the Reynold's number, we have to multiply by x and U_s, the free stream velocity, in cm and cm s^{-1} respectively. If the gas pressure is not one atmosphere, we must also multiply by the pressure in atm, since $1/v = \rho/\eta \propto P$. In a typical open-flow type of apparatus, the rate of gas flow might be a few hundred cm^3 min^{-1}, through a tube of cross-sectional area of a few cm^2, so that $U_s \cong 1$ cm s^{-1}. The distance x is not likely to be greater than 10 cm, if we are thinking of the plate as a substrate holder. Clearly the Reynold's number is in the range 1–10 in a typical apparatus. By flowing the gas faster and reducing the tube cross-sectional area, a velocity of 10^4 cm s^{-1} might be obtainable, in which case the boundary layer would become turbulent a few centimeters downstream from the front of the 'plate', according to this analysis. However, in reducing the cross-section of the gas flow (i.e. of the tube) to a few square millimeters, we have departed rather drastically from the simple model of a plate in a more of less infinite steady flow of gas. Furthermore, the restrictions would almost certainly introduce local turbulence, so that the transition to a turbulent boundary layer could occur at a lower Reynold's number.

The flow of gas in a circular pipe has been analysed in some detail[168]. The critical Reynold's number at which turbulence sets in is variously quoted at between 1000 and 5000. It is known to depend critically on how the gas enters the tube and on the roughness of the tube walls, and values as large as 500 000 have been reached[168]. The length parameter in the Reynold's number is the tube diameter. A value of 3000 is not unreasonable in many industrial situations[168]. The development of the parabolic velocity distribution (see Section 4.5) takes place over an 'entrance length' L_e, in the manner illustrated in Fig. 4.25. The velocity is

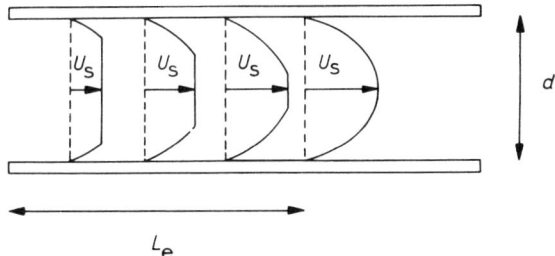

Fig. 4.25 Establishment of parabolic velocity distribution in a pipe.

approximately uniform over the central region of the tube, and falls quickly to zero in a viscous boundary layer near the wall. The thickness of the boundary layer increases with distance down the tube, until the whole width of the tube consists of 'boundary layer'. The entrance length L_e is given[168] by

$$\frac{L_e}{d} = 0.0288 Re_d \qquad (4.97)$$

If $Re_d \simeq 1$, the parabolic profile develops almost immediately. Velocities large enough to produce turbulence in the tube ($Re_d > 3000$, say) are only to be found in capillaries or at flow rates of many litres per minute, and will be rare in crystal growth apparatus.

We will now look more closely at the locally turbulent type of flow. Consider a projection from a surface over which gas is flowing. (Fig. 4.26). If the flow velocity is very small, the streamlines can

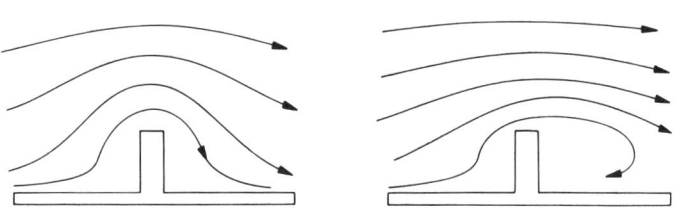

Fig. 4.26 Laminar and locally turbulent flow over a projection.

follow the shape of the projection, and no turbulence arises. At slightly higher velocity, a vortex or eddy develops behind the projection, and as the velocity increases further, a train of vortices develops. Photographic techniques have been used to study the flow of fluid around various obstacles, e.g. a flat plate of width d at right angles to the flow, or a circular cylinder of diameter d across the flow.[170] At $Re_d = 0.25$, the flow follows the outline of the obstacle. At $Re_d = 10$, vortices develop behind the obstacle, producing a turbulent region of length about $2d$, increasing to about $5d$ when $Re_d = 250$, in the case of the plate across the stream. The circular cylinder produces the well known von Karman 'trail' at a Reynold's number of about 250; this 'trail' is a series of stable vortices starting just behind the cylinder and moving along with a velocity less than that of the gas stream and persisting for a considerable distance. These somewhat idealized examples, which have been studied in detail in the hydrodynamic literature, give us an idea of the distances over which local turbulence will persist. The quantitative application of hydrodynamic principles to real systems such as crystal growth apparatus remains a major theoretical problem.

4.7 Convection

The term 'convection' is used to describe the motion of macroscopic regions or parts of a fluid, and is to be contrasted with diffusional

VAPOUR TRANSPORT

motion, in which individual atoms or molecules are concerned. Whereas diffusion is a statistical process, driven by an imaginary 'force' produced by a concentration gradient, convection is driven by the more familiar forces (usually gravity, but we can include centrifugal force, electric or magnetic forces, and so on). Some authors have used the term 'convection' to describe the movement of a fluid under a pressure gradient, and movement due to change in volume on evaporation or sublimation. We prefer to call these phenomena 'viscous flow' and 'Stefan flow', and to reserve 'convection' for describing fluid motion under gravity – we will not discuss electric or magnetic forces.

Situations in which convective motion of a gas occurs are familiar. The warm air in contact with a heated surface (e.g. a radiator) rises; that in contact with a cold surface (e.g. a window on a cold day) sinks. The forces acting on the gas may be considered as buoyancy forces arising from a change in density of the gas. The warm gas experiences an upthrust equal to the weight of unheated gas which would occupy the same volume. Since this upthrust exceeds the weight of the package of warm gas, force is available to accelerate it upwards. At a certain velocity, the viscous forces opposing the motion balance the buoyancy force.

This sort of convection – thermal convection – is perhaps the most familiar. However, density differences may be brought about in a gas by other means than by introducing temperature differences. The most significant way, from the point of view of vapour phase crystal growth, is by variations in composition as a result of chemical reaction. We have already seen in the worked examples earlier in this chapter that the composition of the vapour over a growing crystal differs from that over the source material if a temperature difference is imposed, as the crystal and the source each strive to maintain equilibrium between solid and vapour. In the examples of vapour transport given earlier in the chapter, we have seen what sort of compositional variations may arise (see Figs. 4.6 and 4.11). These will be smallest when the vapour consists largely of one component, which may be inert gas, or excess of one active component, or of the transporting agent. Thus in the transport of gallium arsenide with hydrogen/HCl gas mixtures, we expect a density difference due to composition change of about 10% in a gas which is 90% hydrogen or more, if ΔT is about 40 K. The density difference brought about by temperature change is roughly $\Delta T/T$, or about 4% in this case. The density change accompanying compositional change will be greatest when the vapour is mainly one component over the crystal, and mainly another component over the source. Such a situation arises, for example, in the transport of a single component in an inert gas, if the vapour pressure is nearly equal to the total pressure over the source, and very much smaller over the crystal. The density difference due to temperature difference may be made to cancel out this compositional density change to greater or less extent, or to augment it. The density change will be most marked if the substance being transported has a high molecular weight in the vapour, and the inert gas is light – hydrogen or helium Other examples where a large density difference may be produced by changes in composition can easily be found: dissociative sublimation with the vapour nearly stoichiometric over the source and far from stoichiometric over the growing crystal; chemical vapour transport with mainly reaction products over the source, and reactants over the growing crystal. A large enthalpy change for the transport reaction (or for sublimation) clearly enhances the effect, as this results in a large change of equilibrium constant with temperature.

To get an idea of the density changes involved, we have selected vapour compositions from the worked examples of the previous section, taking ΔT of 40 K in each case. In table 4.6 we list the compositions for various reactions.

We see that in each case, the composition of the vapour has more influence on the density than the temperature does and sometimes we may be dealing with factors of 20 or more, not just a few percent. If the geometry of the apparatus is such that the dense vapour is

Table 4.6 *Density differences in vapour transport*

Reaction	Species	Partial pressures in atm.		$\frac{\Delta \rho}{\rho}$ %
		Over Source	Over Crystal	
Sublimation of silver in helium	Ag He	1.6×10^{-6} 1.0×10^{-7}	6.15×10^{-7} 1.1×10^{-6}	84
Dissociative sublimation of CdS	Cd S_2	0.027 0.013 $\alpha = 2.01$	0.039 0.0012 $\alpha = 32$	14
Ni + 4 CO → Ni(CO)$_4$	Ni(CO)$_4$ CO	0.825 0.175	0.331 0.679	66
GaAs + HCl → GaCl + ¼ As$_4$ + ½ H$_2$	GaCl As$_4$ HCl H$_2$	0.052 0.018 0.011 0.919	0.048 0.017 0.015 0.920	5

VAPOUR TRANSPORT

above the less dense vapour, conditions are set for convection. Obviously this sort of convection, which is usually called 'solutal convection' or 'compositional convection', may be avoided, if need be, by designing the crystal growth apparatus so that the heavy gas is at the bottom and the less dense gas on top. This may not always mean that the cooler gas is at the bottom and the hotter gas on top. Alternatively, it may be preferable to arrange the apparatus with the temperature difference in a horizontal direction, as is usually done in the open flow type of apparatus, and frequently in the sealed capsule method. This geometry will not necessarily eliminate convection, because small vertical temperature gradients may well exist in the furnace, which may be sufficient to cause appreciable convective motion.

Now let us look more closely at the type of motion that can be expected. The simplest situation to visualize is that depicted in Fig. 4.27. The gas rises at the hot wall, and sinks at the cold wall.

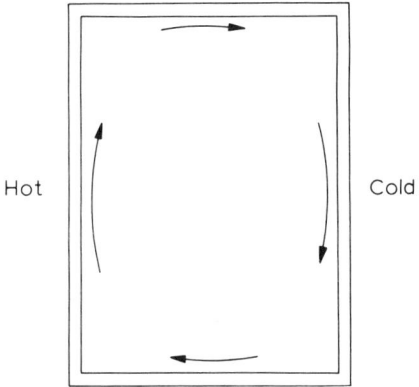

Fig. 4.27 Simple situation causing thermal convection.

The result is a circulation of gas in the container. The container shown here is rather taller than it is wide, so that the simple circulatory motion depicted could arise.

The convective motion of gas near a heated or cooled vertical wall may be described by a simple mathematical treatment.[168] It is found that the motion is characterized by two dimensionless numbers. The first is the Prandtl number, $Pr = \nu/a$, where ν is the kinematic viscosity (η/ρ) and a is the thermal diffusivity, $\kappa/C_p\rho$. This dimensionless number is a property of the gas in its particular state of temperature and pressure. The second is the Grashof number

$Gr, = gd^3\gamma\Delta T/\nu^2$, where d is some dimension of the system, γ is the volume coefficient of expansion of the gas (= $1/T$ for ideal gases), ΔT is the temperature difference across the distance d, and ν is again the kinematic viscosity. The Grashof number thus depends on the geometry and thermal distribution of the convecting system. The product of the Grashof number and the Prandtl number is called the Rayleigh number, $Ra = (gd^3/\nu a). (\Delta T/T.)$ It is found that, as a first approximation, convective motion is characterized by the Rayleigh number; for example, in the system shown above, the heat transfer between the hot wall and the cold wall is a function of the Rayleigh number. In crystal growth, we are usually more concerned with transport of matter through the vapour mixture than with transfer of heat, although the latter may sometimes be very important. Mass transfer is characterized by the Grashof number and another dimensionless number similar to the Prandtl number, the Schmidt number $Sc = \nu/D$, where D is the diffusion coefficient. The product of the Grashof number and the Schmidt number is also sometimes called the Rayleigh number; we will distinguish it with a prime:

$$Ra' = \frac{gd^3\Delta\rho}{\nu D\rho}.$$

When the temperature difference is in a vertical direction, a somewhat different situation arises (Fig. 4.28). Of course, if the upper surface is hot and the lower one cold, there is no driving force for thermal convection. The situation shown here, with the hot surface at the bottom and the cold surface at the top, was studied by Bénard[171] and analysed mathematically by Rayleigh.[172] It is found that when the Rayleigh number is low, convective motion of the gas as a whole does not occur. Packages of hot gas that leave the bottom and begin to rise lose their heat by conduction to the surrounding gas before getting very far. Similarly, packages of cold gas descending gain heat by conduction. When the Rayleigh number is above about 1700, a cellular pattern develops throughout the width of the system, with a circulatory movement of gas in each cell.

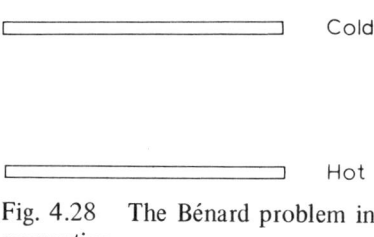

Fig. 4.28 The Bénard problem in convection.

VAPOUR TRANSPORT

The temperature difference at which thermal convection begins may readily be estimated. From the definition of the Rayleigh number, we see that

$$\frac{\Delta T_c}{T} = \frac{\nu a}{d^3} \cdot \frac{Ra_c}{g} \qquad (4.98)$$

where the subscript c denotes critical convective conditions. The quotient Ra_c/g is about 1.7 in c.g.s. units. The critical temperature difference is strongly dependent on the vertical dimensions of the system. Table 4.7 gives values of $\Delta T/T$ for hydrogen and nitrogen at 1000 K. The appreciable difference between the behaviour of hydrogen and of nitrogen is due to the anomalously high thermal conductivity and kinematic viscosity of hydrogen, so that a package of gas moving under convective forces can exchange heat more rapidly with the surrounding gas. The figures for d of 1 cm are not to be taken too literally, but they indicate that heat is transferred by conduction so rapidly as to give no cause for convection.

Table 4.7 *Critical temperature differences $\Delta T/T$ for thermal convection at 1000 K*

d (cm)	Hydrogen	Nitrogen
1	171	3.3
10	0.171	0.0033
15	0.051	0.001

Corresponding to the critical temperature difference ΔT_c for thermal convection in a gas, we may define a critical density difference for solutal convection (compositional convection) in a gas mixture. From the definition of the compositional Rayleigh number, we find

$$\frac{\Delta \rho_c}{\rho} = \frac{\nu D}{d^3} \frac{Ra'_c}{g} \qquad (4.99)$$

In Table 4.8 we give values of $\Delta \rho_c/\rho$ for a mixture consisting mainly of hydrogen ($D_0 = 0.4$ cm² s⁻¹) and a mixture of other diatomic gases, ($D_0 = 0.2$ cm² s⁻¹), again at 1000 K

Again hydrogen mixtures resist convection considerably more than mixtures of other diatomic gases, athough the difference is not so pronounced.

Table 4.8 *Critical density difference* $\Delta\rho_c/\rho$ *for solutal convection at 1000 K*

d (cm)	Hydrogen mixture	Nitrogen mixture
1	59	4.2
10	0.059	0.0042
15	0.018	0.0012

We will now see how these remarks apply to crystal growth systems. We first of all consider a sealed capsule system in a vertical furnace. Heat is applied to the vapour in the capsule by electrical heating elements or other means concentric with the capsule, and heat is lost from the furnace down the middle towards the cold ends. It is usual, therefore, for there to be a small lateral temperature gradient in the furnace, as well as the primary vertical gradient which is imposed to provide the driving force for growing the crystal. This lateral gradient can, of course, be reduced by the use of heat reflectors. If conditions are so arranged that convection sets in (if the denser gas is at the top), the lateral temperature gradient may determine the flow pattern, with the slightly hotter gas near the perimeter rising, and the cooler gas in the middle of the capsule falling. The convective velocities will be determined by a balance of the work done by buoyancy forces and work done against viscous forces.

With the combination of thermal and compositional convection, such as may arise in a crystal growth capsule, a phenomenon known as 'thermo-solutal convection'[173] may arise. When the thermal diffusivity a is considerably larger than the diffusion coefficient D, a rising package of fluid may remain almost in thermal equilibrium with its surroundings while retaining its difference in composition. Convection may then set in due to compositional gradients, even if the thermal gradient is opposed to convection (i.e. hotter fluid on top) and of sufficient magnitude to make the fluid less dense at the top in spite of the compositional gradient. This effect has been observed in liquid metals, where the thermal conductivity is large. As far as the authors have been able to ascertain, this effect has not been looked for in gaseous systems.

In horizontal furnaces, there is usually no driving force for convection except for the small temperature gradient across the furnace tube which arises because heat is fed in from the perimeter and lost down the middle to the ends. However, in processes such as deposition of silicon from $SiCl_4$, SiH_4 or similar compounds, where a relatively cool gas is decomposed on a hot substrate, vigourous convection has been observed[174], sufficient to completely mix the reactant gas and the gaseous products of reaction over most of the furnace tube. Diffusion and Stefan flow are then confined to a thin boundary layer a few millimeters thick over the substrate. The convection effects were made visible by using a smoke of titania particles.

The effect of convection on vapour transport in a sealed, vertical capsule or a similar system depends on the relative magnitude of the convective velocity and the Stefan velocity. The Stefan velocity is proportional to the transport rate, and for a given temperature difference ΔT we can make the rate large (by reducing the amount of inert gas or stoichiometric excess) or small (by the reverse processes). Clearly, if the Stefan velocity is very large compared with the convective velocity, we can ignore the effect of convection on the velocity distribution and on the variation of composition in the vapour. If the transport is very slow, on the other hand, the convection effects will dominate. If a circulation of gas is set up in the capsule, this may have the effect of shortening the distance over which partial pressure gradients are set up. The limit of this process occurs when the convective forces are so strong as to lead to turbulent convection in the capsule, so that most of the gas becomes more or less evenly mixed, with only a thin diffusion boundary layer a few millimeters thick at each end. Because the capsule has been, in effect, made shorter by the convective motion of the gas, the transport rate is greater than it would be without the convection, as the transport rate varies with the reciprocal of the length. Furthermore, the partial pressure gradients in the vapour over the growing crystal will be increased, and this affects the stability of the surface of the crystal to small protrusions. The subject of surface stability is dealt with in Chapter 5, and we will not enlarge on it here.

The phenomenon of convection has something in common with nucleated phenomena in certain systems. We have said that in the Bénard problem convection only sets in when the Rayleigh number exceeds about 1700. The convective currents take some little time to establish themselves, this time depending on the driving force for convection (i.e. on the Grashof number) and on the viscosity and other properties of the gas (i.e. on the Prandtl or Schmidt numbers). Examining the form of the Grashof number, we see that it depends on the third power of the distance over which the temperature or density difference is set up, it being very much easier to produce convection if the distance is large. Thus we may imagine a capsule in which a temperature difference is imposed, such that the Grashof number exceeds its critical value by a small amount, so that the convective motion of the gas is beginning to have a certain coherence. This produces some decrease in the temperature and composition gradients over much of the length of the capsule, confining most of the gradients to the two ends. Because of this decrease in the gradients, the Grashof number is now no longer above its critical value at any point in the system, and the convective motion decays. The growth rate of the crystal (i.e. the transport rate in the vapour) is thus an oscillatory function of time, as are the stability of the growing surface and the composition of the crystal.

The most difficult situation to analyse arises when the Stefan velocity and the convective velocity are about equal in magnitude. We can illustrate qualitatively what may happen. In Fig. 4.29 is shown the Stefan velocity, varying more or less parabolically across the capsule, and the convective velocity, assuming the gas circulates, rising up the perimeter of the capsule and sinking down the middle. The resultant velocity distribution depends on the exact magnitudes of the components, but clearly it is not difficult to produce a situation that will result in a ring of rapid growth, bordered by regions of slower growth. (Plate 1) The region of fast growth may correspond to partial pressure gradients in the vapour which make the growing surface unstable with respect to small protrusions or microfacets, so that the regions of different growth rate may be distinguished by different morphology. (There are other reasons as well, of course, why the fast-growing regions may be of different appearance from the slow-growing regions. These reasons will be touched on in Chapter 5.)

4.8 Heat flow

Apart from the economic aspect of designing efficient furnaces for growing crystals, the subject of heat flow is of interest for three reasons. Firstly, the rate at which a crystal grows may, under some circumstances, be limited by the rate at which it can lose the heat of condensation or reaction. Secondly, the stability of the growing surface of a crystal may be determined by the ease with which small protrusions can lose heat and continue to grow. Thirdly, the temperature distribution over the growing crystal or layer is one of the factors which determines whether the crystal or layer develops uniformly or not. In this section we discuss heat flow as a factor limiting the rate of growth of a crystal, and also the temperature distribution over a growing crystal. Discussion of the surface stability is postponed until Chapter 5.

The heat of condensation or reaction which must be lost as the crystal grows is, of course, considerably larger than heat of solidification of a melt. For example, the latent heat of sublimation of silver[6] is about 285 kJ mol^{-1}, compared with its latent heat of fusion[6] of 11.3 kJ mol^{-1}. The heat of sublimation and dissociation of cadmium sulphide[6] is around 326 kJ mol^{-1}, and the heat of formation of nickel carbonyl from nickel and carbon monoxide[6] is 152 kJ mol^{-1}. However, the rate at which heat is generated at a growing crystal is also proportional to the growth rate.

The crystal loses heat by conduction through its support and possibly through the gas, and by radiation. Which of these processes is the most important depends on the temperature of the crystal and its emissivity, and the thermal conductivity of the crystal, its support, and the gas. Generally, we expect radiation to play the most important part at high temperatures. To get an idea of the relative magnitude of the heat transfer rates involved, consider a crystal of surface area A and temperature T_1, enclosed in a chamber at temperature T_2, and supported on a rod of length l and cross-section B, with its further end at temperature T_3. Then the radiant heat loss is:

$$R_{radiant} = EA(T_1^4 - T_2^4)$$

where E is the emissivity (assumed independent of wave length for the moment), and the heat loss by conduction down the support is:

$$R_{cond} = \frac{KB(T_1 - T_3)}{l}$$

where K is the thermal conductivity of the rod. We have assumed that heat conduction through the crystal is fast compared to conduction down the rod. This is not necessarily the case, and in the worked example presented later, we take into account the thermal resistance of the crystal. We will put some reasonable values for the various dimensions and material properties into these equations, and see how the heat losses compare in magnitude. We take $A = 1$ cm^2, $B = 0.25$ cm^2, $l = 30$ cm, $T_3 = 300$ K (i.e. room temperature), $K = 0.015$ W cm^{-1} K^{-1} (appropriate to silica at 500°C), and equating the two heat losses, we plot the temperature difference $\Delta T = T_1 - T_2$ as a function of T_1 (see Fig. 4.30). To the right of the curve in Fig. 4.30, conditions are such that radiative heat transfer is more efficient than conduction down the support, and conversely to

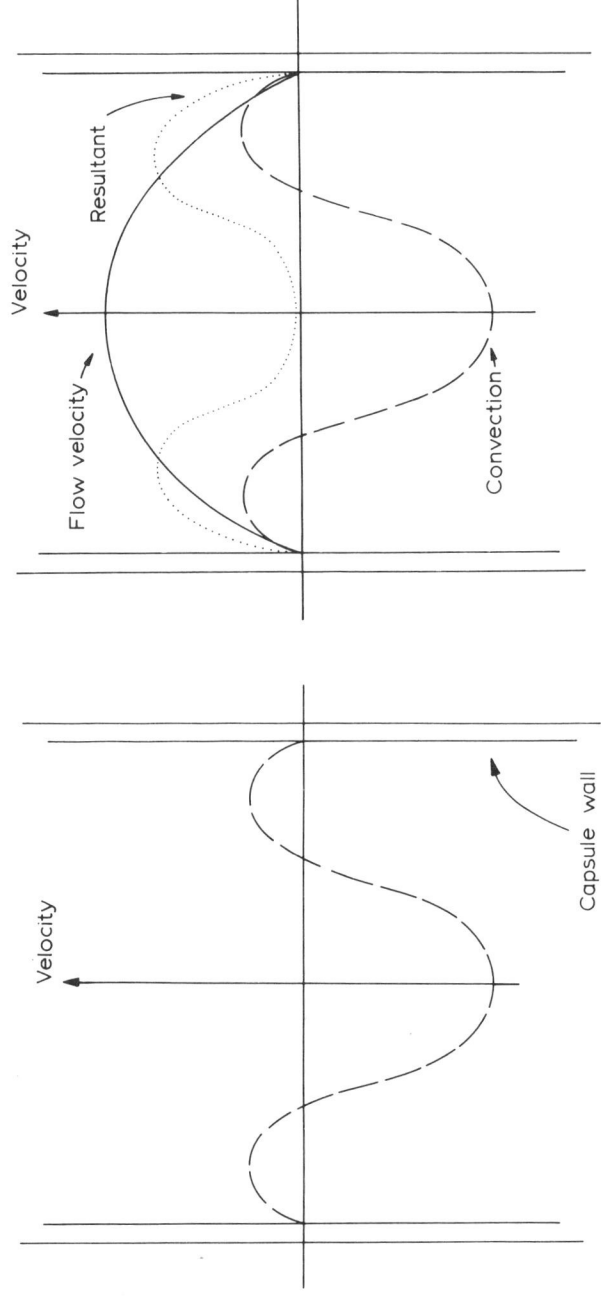

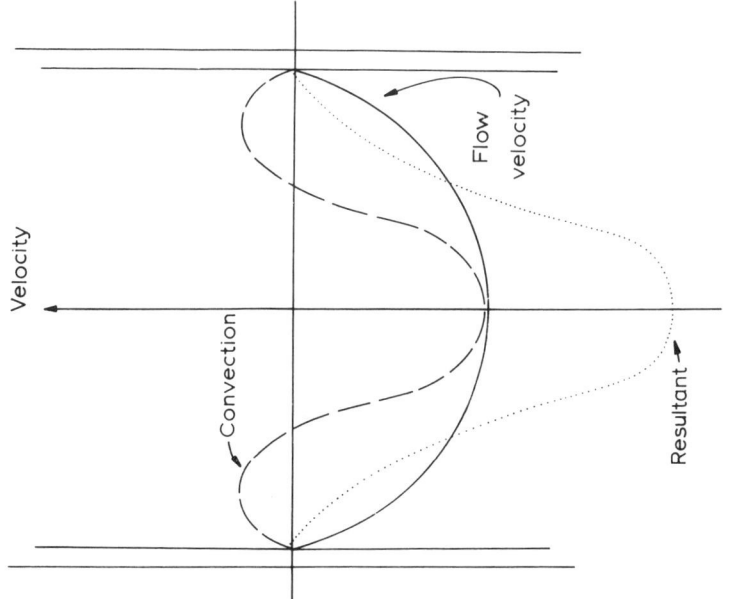

Fig. 4.29 The combined effect of convection and Stefan flow. In (a) is depicted the convective flow, in (b) an upwards Stefan flow is added, while in (c) the Stefan flow is downwards.

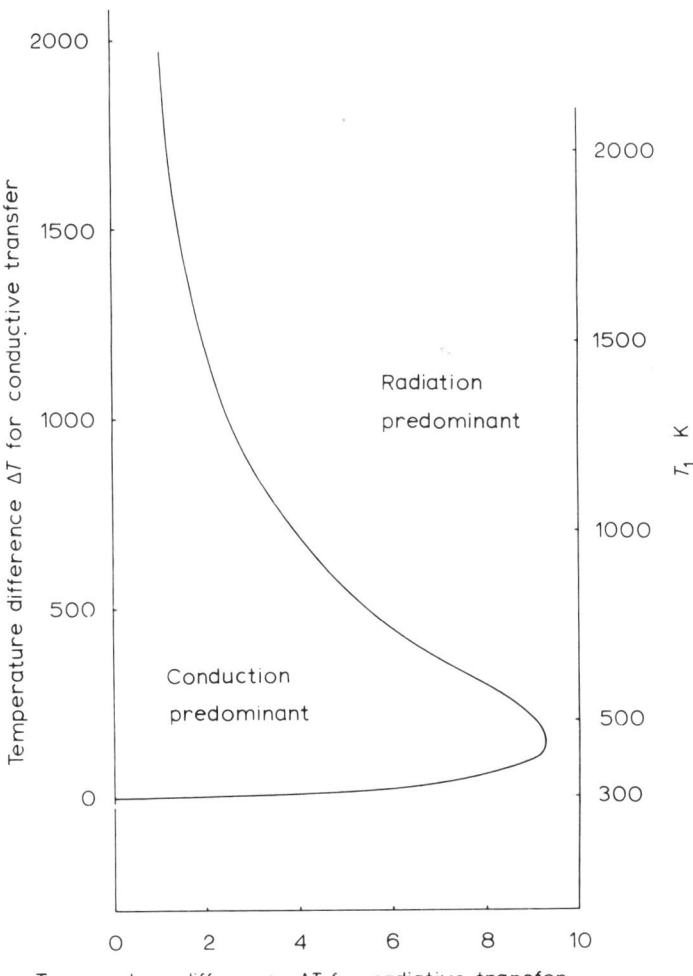

Fig. 4.30   Predominant heat transfer mechanisms for a growing crystal (see text).

the left of the curve. The low temperature horizontal portion of the curve represents the situation where the total heat transfer has become small. We have simplified the problem greatly, of course, ignoring the fact that in practice the crystal is not in an isothermal chamber, but is receiving radiative heat from hotter parts of the furnace and radiating heat to the colder parts. We should take into account uneven heat exchange between the crystal and the furnace, transfer of heat to or from the furnace gas, and many other factors. However, it is apparent that even at high temperatures, where radiation is often thought to provide extremely uniform temperatures, conduction of heat to parts of the apparatus at around room temperature can still account for a significant flow of heat. In plotting Fig. 4.30, we have taken a unit emissivity for the crystal surface. The emissivity of a real crystal may be appreciably lower, and may also be a function of wavelength, so that radiation becomes less efficient as a means of heat transfer. The emissivity of a semiconductor crystal is relatively high for short wavelength radiation, and drops off sharply for radiation of wavelengths greater than that corresponding to the bandgap.[175] The emissivity of germanium, for example, at long wavelengths ($>2\,\mu$m) is only about 5% of the short wavelength emissivity. As far as the maximum value of the emissivity as a function of wavelength is concerned, we expect this to increase with increasing temperature, as the numbers of electrons in the conduction band and holes in the valence band increase. For further details and a theoretical treatment of absorption and emission of radiation by semiconducting materials, see reference 175.

The temperature which the surface of the crystal assumes is determined by a balance of heat exchange with the surrounding parts of the furnace, supports, etc., and heat generated by condensation. When the growth rate of the crystal is extremely small, we may imagine that the crystal is in thermal equilibrium with its surroundings at a temperature T_0, which is determined by the temperature profile of the furnace, and by the radiative and conductive properties of the furnace materials and of the crystal. If the pressure in the furnace is not very low, convective heat transfer will also take place, and this is more important if the gas flows through the furnace. All these factors determine the zero growth rate equilibrium temperature T_0. Now if the crystal starts to grow, heat of condensation is generated, and the crystal assumes a new equilibrium temperature T_1. If the surroundings are not altered (as may be a first approximation) the temperature difference $T_1 - T_0$ is a direct measure of the heat transfer. If the crystal is not grown very fast, $T_1 - T_0$ may be negligibly small, and we can safely ignore heat transfer as a factor determining the rate of growth of the crystal. For example, if cadmium sulphide is grown at a rate of 10^{-7} mol cm^{-2} s^{-1} (about 0.01 cm h^{-1}), the rate of generation of heat is about 0.033 W cm^{-2}. If T_0 is 1400 K, and if all this heat is to be lost by radiation, we find T_1 = 1400.5 K, so that the temperature of the crystal is hardly raised. However, if the growth rate is increased by a factor of 30, the temperature of the crystal increases by about 15 K, and this may well be significant compared with the temperature difference between the crystal and the source material. These figures are calculated taking the emissivity as unity, and should be increased if the emissivity is less.

We now give an example of a system for growing crystals in which heat transfer is of great importance. The system uses a liquid source which is supercooled to a controlled extent, and exploits the increased saturated vapour pressure over a metastable phase as the driving force for the vapour transport.

4.8.1 Example: Isothermal growth of metanitroaniline

Metanitroaniline is an organic substance, forming crystals of class $mm2$, which has received some attention recently because of its electro-optic and electro-chromic properties[176, 177]. Crystals of this substance have been grown in the authors' laboratory, using the isothermal growth technique. The source material and a seed crystal are contained in a sealed capsule which is immersed in a thermostatic bath of silicone oil at a uniform temperature T_b, which is just below the melting point T_m of metanitroaniline, 385 K [178,179]. It is arranged that the source material is liquid, and hence supercooled, so that the vapour pressure which is in metastable equilibrium with the supercooled liquid is greater than that in equilibrium with the seed crystal. Thus transport of the vapour occurs and growth may take place. The vapour pressure curves as functions of temperature are shown in Fig. 4.31, for solid and liquid phases, using the data of Berliner and May,[178] and Ubbelohde.[179]

We can make a simple analysis of this sytem using a one-dimensional model for the transport of vapour and of heat. As the latent heat of evaporation is extracted from the surface of the liquid source, the surface takes up a temperature T_{liq} less than the temperature of the thermostatic bath T_b. Conversely, the heat of condensation liberated at the surface of the growing crystal causes it to assume a temperature T_{crys} above T_b. The heat flows may therefore be written as:

$$J = k_1 (T_{crys} - T_b)/\Delta H_{sub} \tag{4.100}$$
$$J = k_2 (T_b - T_{liq})/\Delta H_{evap} \tag{4.101}$$

where the ΔH's are the enthalpies of sublimation and evaporation, and the k's are heat transfer coefficients. For a single medium of thickness d and thermal conductivity κ, k is defined by

$$k = \frac{\kappa}{d} \tag{4.102}$$

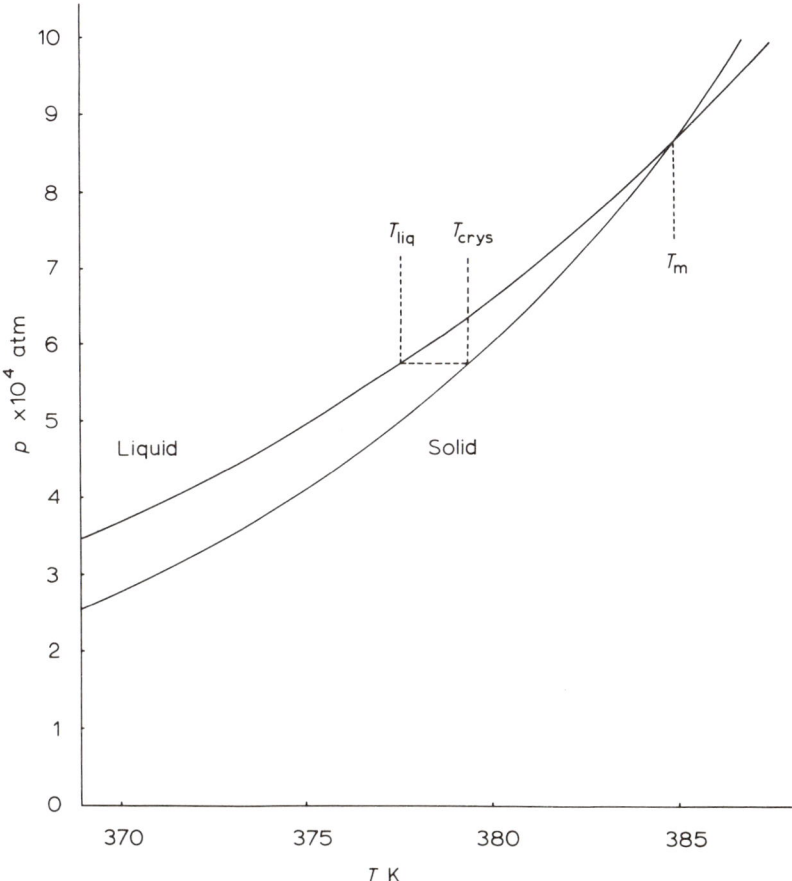

Fig. 4.31 Equilibrium vapour pressures over solid and liquid meta-nitroaniline.

For n layers of different media;

$$\frac{1}{k} = \sum_n \frac{d_i}{\kappa_i} \tag{4.103}$$

The units of heat flow in the above equations are such that the J is the same as the molar flux of material.

For the transport of material, we use the flow equations described previously. We will assume that there is a pressure $p_Z(x)$ of 'inert gas', which is taken to include residual gases and decomposition products, as well as any gas added purposely. We assume that none of this 'inert gas' dissolves appreciably in the crystal, so that the net flux of this species is zero. Our flow equations are:

for metanitroaniline:
$$J = \frac{Up}{RT} - \frac{D}{RT}\frac{dp}{dx} \quad (4.104)$$

for the 'inert' species:
$$0 = \frac{U}{RT}p_Z - \frac{D}{RT}\frac{dp_Z}{dx} \quad (4.105)$$

Addition of these equations leads to an expression for U:

$$JRT = UP$$

where $P = p + p_Z$ is the total pressure, assumed constant as in previous examples. We may substitute this expression in the transport Equations 4.104 and 4.105, and integrate along the length of the capsule. We obtain:

$$(p_{crys} - P) = (p_{liq} - P)\exp\left(\frac{JRTl}{DP}\right) \quad (4.106)$$

So that
$$J = \frac{DP}{lRT}\ln\left(\frac{P - p_{crys}}{P - p_{liq}}\right) \quad (4.107)$$

where p_{liq} and p_{crys} are the vapour pressures of metanitroaniline over the liquid source and over the crystal. If we assume that these vapour pressures are the equilibrium vapour pressures, we can express them in terms of the enthalpies and entropies of evaporation and sublimation:

$$p_{liq} = \exp\left\{\frac{-\Delta H_{evap}}{RT_{liq}} + \frac{\Delta S_{evap}}{R}\right\} \quad (4.108)$$

$$p_{crys} = \exp\left\{\frac{-\Delta H_{sub}}{RT_{crys}} + \frac{\Delta S_{sub}}{R}\right\} \quad (4.109)$$

By eliminating T_b from Equations 4.100 and 4.101 we find that J, the transfer rate for material and for heat, is proportional to the temperature difference ΔT between the surfaces of the crystal and liquid source:

$$J = \left\{\frac{T_{crys} - T_{liq}}{\frac{\Delta H_{sub}}{k_1} + \frac{\Delta H_{evap}}{k_2}}\right\} \quad (4.110)$$

VAPOUR TRANSPORT

We have plotted J versus $\Delta T, = T_{crys} - T_{liq}$, in Fig. 4.32, taking a constant value of k_2 of 0.0017 J cm^{-2} s^{-1} K^{-1}, appropriate to many organic liquids, and various values of k_1. The complete solution of the problem involves solving Equations 4.100, 4.101 and 4.107 with Equations 4.108 and 4.109 for the partial pressures, to obtain the transport rate J and the temperatures T_{liq} and T_{crys} as functions of the bath temperature T_b, which is the single experimental variable under our control, apart from the dimensions of the capsule and, possible, the amount of 'inert gas' contained in it. Fig. 4.33 shows J versus T_b, for various possible conditions, and again a fixed value of the heat transfer coefficient through the liquid. Notice that the growth rate falls to zero for $T_b = T_m$, the melting point, since there is then no difference in vapour pressure over the

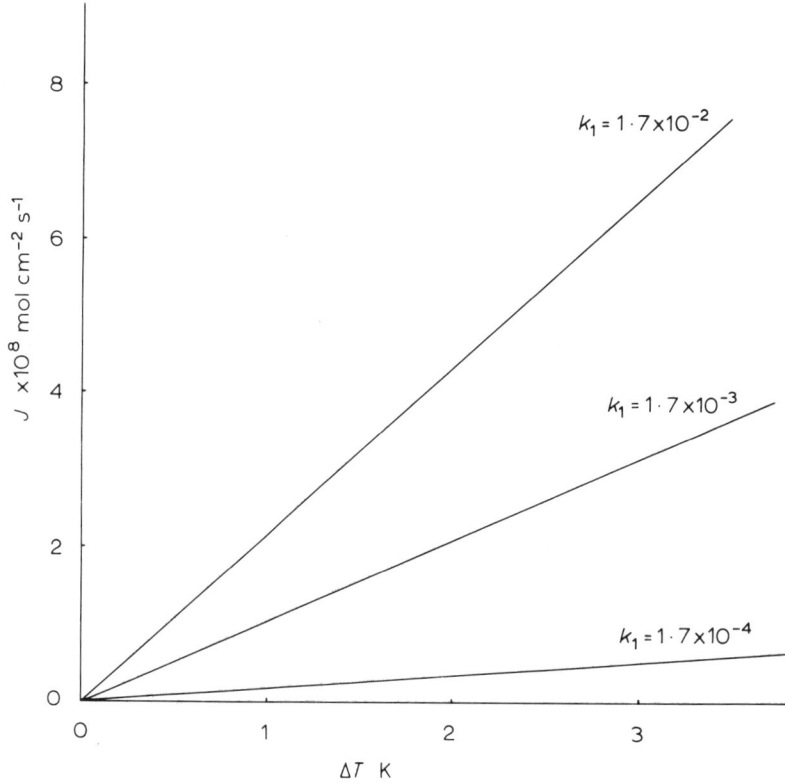

Fig. 4.32 Isothermal transport of meta-nitroaniline. Transport rate J as a function of temperature difference between the surfaces of the source and the crystal ΔT, for various heat transfer coefficients k_1. $k_2 = 0.0017$ J cm^{-2} s^{-1} K^{-1}.

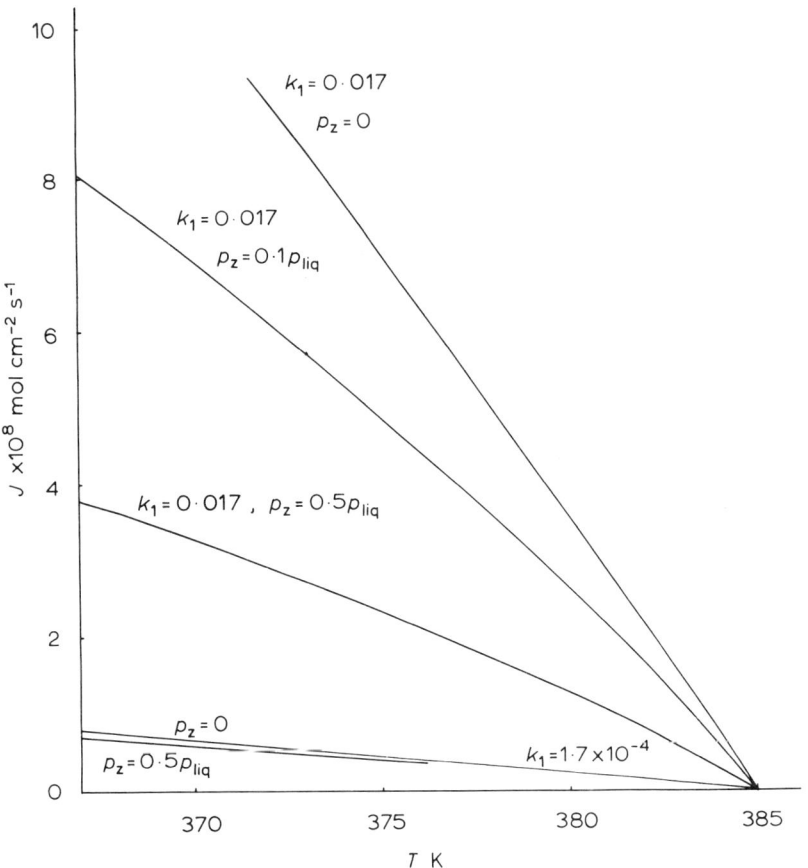

Fig. 4.33 Isothermal transport of meta-nitroaniline in an inert gas. Transport rate J as a function of temperature of the bath T_b for various inert gas pressures p_Z and heat transfer coefficients k_1. $k_2 = 0.0017$ J cm^{-2} s^{-1} K^{-1}.

liquid and over the solid. With relatively rapid heat transfer through the crystal ($k_1 = 0.017$ J cm^{-2} s^{-1} K^{-1}, corresponding to a small crystal, i.e. early in the growth) the transport rate is sensitive to the amount of 'inert gas', i.e. residual gas, decomposition products, etc. As the crystal dimensions increase, the heat transfer coefficient k_1 decreases. With $k_1 = 0.00017$ J cm^{-2} s^{-1} K^{-1} the growth rate has become small and insensitive to the amount of 'inert gas' in the capsule.

Fig. 4.34 shows the calculated temperature differences between the crystal and the thermostatic bath, and between the liquid source and the thermostatic bath. Again, the heat transfer coefficient k_2 of

VAPOUR TRANSPORT

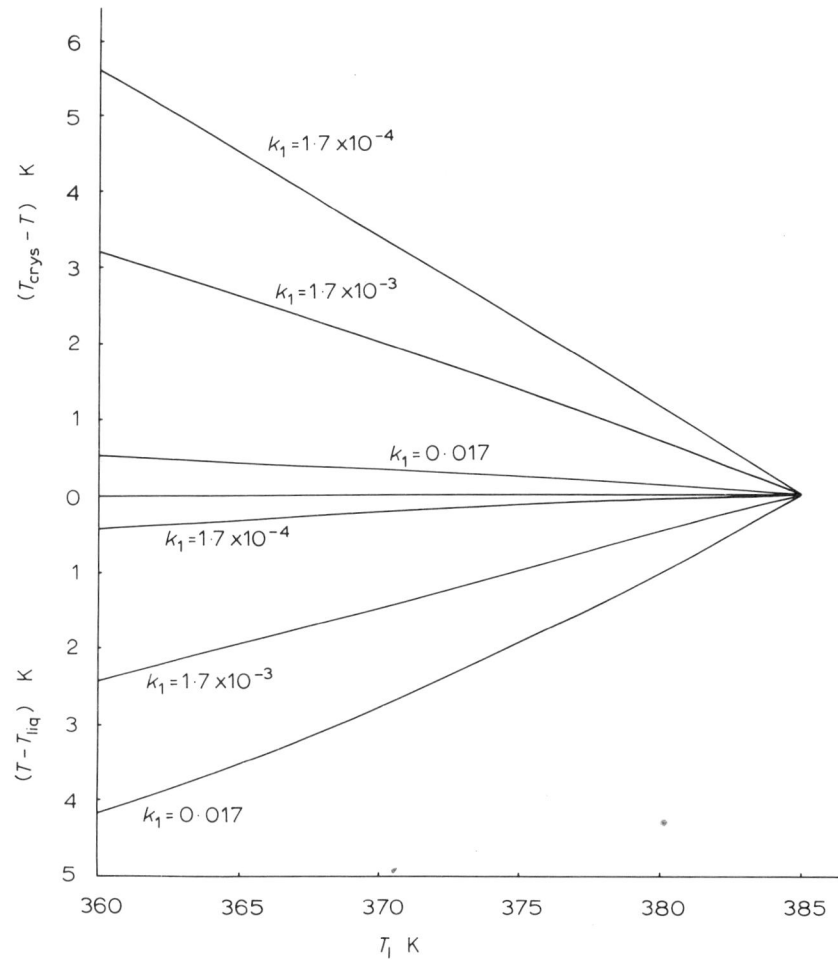

Fig. 4.34 Isothermal transport of meta-nitroaniline in an inert gas. Liquid source and crystal surface temperatures as functions of bath temperature T_b, for various values of the heat transfer co-efficient k_1. Inert gas pressure $p_Z = 0.1\, p_{liq}$, $k_2 = 0.0017$ J cm^{-2} s^{-1} K^{-1}.

the liquid source has been kept at 0.0017 J cm^{-2} s^{-1} K^{-1}, and the 'inert gas' partial pressure set at $0.1\, p_{liq}$, i.e. about 10% of the total pressure. As we would expect, when $k_1 = k_2$, $(T_{crys} - T_b)$ and $(T_b - T_{liq})$ are nearly the same, the small difference being due to the difference between ΔH_{sub} and ΔH_{evap}. If the heat transfer coefficient of the crystal, k_1, is large, the crystal is at nearly the same temperature as the thermostatic bath, and the liquid is at an appreciably lower temperature. The converse applies when $k_1 \ll k_2$. This is what we would expect.

This system has been given as an example to illustrate the importance of heat transfer as a rate-limiting step in crystal growth from the vapour, and we have chosen a situation where the temperature distribution is determined entirely by the rate of transfer of the heat of condensation away from the growing crystal, and of the heat of evaporation to the source. In more conventional systems, particularly at high temperatures, the temperature profile is imposed by suitably designed furnaces, and heats of condensation and reaction represent a small perturbation (which may be important nevertheless) on the steady state heat transfer between the furnace and the crystal.

CHAPTER FIVE
Sequential Processes in Crystal Growth

5.1 Introductory remarks

In chapter 4 we took a detailed look at the kinetics of vapour transport. This is, of course, only one process in a series of processes which go to make up crystal growth from the vapour. It is the best understood, and although the algebra may become a little intractable, the concepts involved are not too difficult. Throughout Chapter 4 we assumed that the solid phases at the growing crystal and at the source of material were in thermodynamic equilibrium with the vapour immediately adjacent to them. This assumption allowed us to derive the boundary conditions (i.e. the partial pressures), necessary to obtain numerical answers to the flow equations, from thermochemical data. It is not necessary, nor by any means always realistic, to rely on equilibrium being approached closely. Ideally, we would use for our boundary conditions the actual partial pressures over the growing crystal and over the source of material. Unfortunately, these partial pressures are not readily accessible to experimental measurement. In principle it is possible to analyse the vapour over the solid phases, or anywhere else in the system, by some means such as absorption spectroscopy, or leaking a small quantity of the gas away to a mass spectrometer. These techniques are fraught with problems, such as sampling. It is also possible to conduct a crystal growth experiment in such a way as to yield useful information about the processes which occur at the crystal-vapour interface. The first three parts of this chapter are concerned with establishing a formalism for the interplay between transport and surface processes.

We visualize crystal growth as a number of processes occurring in sequence or series. For example, consider a simple chemical vapour transport system, in which the transport reaction is:

$$M(s) + \tfrac{1}{2}X_2(g) \rightleftharpoons MX(g)$$

where X_2 is the transporting agent and MX is a volatile compound which serves as the transporting compound. We may imagine the following sequence of processes occurring:

(1) X_2 adsorbs on the M surface.
(2) X_2 dissociates on the M surface to form two adsorbed X atoms.
(3) M atoms migrate on the M surface meet the X-atoms, and react to form MX adsorbed on the surface.
(4) The adsorbed MX migrates on the surface until it is desorbed.
(5) MX is transported through the vapour, as described in Chapter 4, with X_2 moving in the opposite direction.
(6) MX is adsorbed on the surface of the crystal.
(7) MX dissociates to form single M and X atoms.
(8) The M atoms migrate to atomic steps, or coalesce into nuclei (see Chapter 3).
(9) The X atoms combine to form X_2 molecules which desorb from the crystal surface.

One could add to the list of processes in this chain; however, these processes serve to illustrate the point. In addition, each of these processes may be rivalled by other processes going on in parallel. Sometimes two or more steps may be 'short-circuited' by a single process occurring simultaneously. Thus X_2 may dissociate in the vapour, and the resulting atoms may react with the M atoms in the vapour. Transport through the vapour may take place simultaneously with surface diffusion of species MX, X_2 etc along the walls of the capsule or container. We can represent the system as a network of electrical resistors in series and parallel (Fig. 5.1).

The 'voltage' which drives the 'current' (i.e. the flux J) through this chain of resistors is the chemical potential difference or free energy difference, δG (total) usually applied as a temperature difference between the crystal and the source. As in the electrical analogue, different amounts of potential are dropped across each part of the chain, and where one part consists of several processes in parallel, different currents or fluxes flow by each process.

Continuing our analogy with an electrical circuit, we note that if one part of the chain represents a much larger resistance than the other parts, most of the potential available is dropped across this resistance and the current flowing in the chain for a given potential is governed mainly by the size of this resistance. For example, by assuming that equilibrium is closely approached between the solid

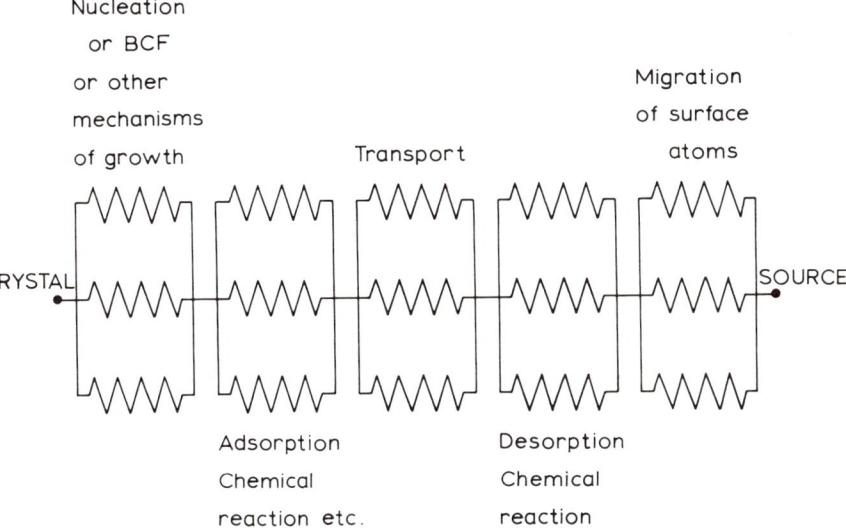

Fig. 5.1 Processes in series and in parallel.

phase and the vapour immediately adjacent to it, as we did for obtaining numerical values of the transport rate in Chapter 4, we are implying that the gas transport part of the sequence of processes presents the highest resistance. On the other hand, if the chemical reaction at the crystal surface in a chemical vapour transport system has an appreciable activation energy, that step in the sequence of processes will present the greatest resistance if the crystal growth system is operated at a low temperature. We would then see a considerable change in the partial pressure profiles in the system, with the difference $p_i(l) - p_i(0)$ between the partial pressures of each species over the source and over the crystal being very much less than would be predicted from equilibrium calculations.

Thus it is the slowest of a series of processes which dominates the picture, determining the flux for a given applied potential. However, of processes occurring in parallel, it is of course the fastest (lowest resistance) process which makes the largest contribution to the total flux. This is fairly obvious, as is the implication, that in designing an experiment to investigate one particular resistance to crystal growth, it is necessary to operate in a region where that resistance is sufficiently large compared to the resistances for other processes in sequence, yet small compared to the resistances for other parallel processes that might rival it. One must bear in mind that alterations of temperature and vapour composition, carried out to investigate the rate of a surface reaction, may also alter the vapour transport resistance significantly, as shown in Chapter 4. Again, the important surface reaction may be different in different temperature regions.

In the next sections, we demonstrate how the vapour transport theory of Chapter 4 may be coupled formally with surface reaction rate theory to yield useful information about surface processes. We then go on to discuss the effect of the partial pressures resulting from vapour transport and reaction kinetics on the stability of the growing crystal surface, and the electronic properties of the crystal.

5.2 Formalism for sequential processes in crystal growth

Although this section has an imposing title, we must start by expressing our strongly-held opinion, that anything in the way of a general formalism is next to impossible, and furthermore undesirable, because in most circumstances it would obscure more than it would explain. We limit ourselves, therefore, to considering only two (or three) processes in sequence. This is, in fact, not much of a limitation, since it lays the foundation for the consideration of more sequential processes, which the interested reader may add as the need arises.

We consider the processes of:

(1) Chemical reaction at the material source.
(2) Vapour transport.
(3) Chemical reaction at the crystal surface.

These three we reduce effectively to two, by postulating that chemical reactions proceed by the same mechanism at both source and crystal and consequently obey the same kinetic laws. It is sufficient for our purpose to take the chemical reaction at only one end of the system into consideration. We therefore ignore processes at the source end. We may imagine that the component gases are fed into the system in the required proportions from a set of taps, a situation that may often be approximated in practice[156].

We will express the rate of transport J through the vapour phase as a product of the potential for transport δG_t dropped at this step and a vapour transport conductance L_t, thus:

$$J = L_t \delta G_t \qquad (5.1)$$

Equation 5.1 thus serves as a definition of the conductance L_t. As we saw in Chapter 4, L_t is a simple function of the dimensions of the system, diffusion coefficient and total pressure, and a quite complicated function of temperature, partial pressure ratios (α, etc.)

SEQUENTIAL PROCESSES IN CRYSTAL GROWTH

amount of inert gas, and other experimental variables. We do not expect L_t to be constant over much of a range of these variables in the same way as, for example, electrical conductivity is reasonably constant over a useful range of currents. However, in those cases where the growth system is 'buffered' by a considerable amount of inert gas, or by a large excess of one component, it turns out that L_t remains fairly constant over a useful range of δG_t's, provided that the other parameters such as α, T, P, etc. are kept constant. Equation 5.1 is more of a conceptual aid or prop than a practical guide, though.

Similarly, we may express the net rate of the chemical reaction, which must obviously be exactly equal to the net rate of transport, by a similar equation[11]:

$$J = L_r A \tag{5.2}$$

where A is the affinity, which is the overpotential driving the reaction, and is defined by:

$$A = -\sum_i \nu_i \mu_i \tag{5.3}$$

Equation 5.2 is an approximation for cases where the affinity is less than RT, and is easily derived for a general reaction proceeding by a single (or a dominant) mechanism. We give a simple derivation here.

Consider the reaction:

$$\nu_a A + \nu_b B + \nu_c C + \ldots \rightleftharpoons \nu_m M + \nu_n N + \ldots$$

where A, B, etc. are species, and ν_a, ν_b, etc. are stoichiometric coefficients. We may write such a reaction as:

$$\sum_i \nu_i A_i = 0 \tag{5.4}$$

where the stoichiometric coefficients ν_i are positive for products and negative for reactants. The rate of the forward reaction is:

$$J_f = k_f [A_1]^{\alpha_1} [A_2]^{\alpha_2} [A_3]^{\alpha_3} \ldots$$
$$= k_f \prod_i [A_i]^{\alpha_i} \tag{5.4a}$$

where the square brackets denote concentrations, $\prod_i$ denotes a continued product, k_f is the kinetic rate constant (a function of temperature only), and α_i is the order of the forward reaction in species A_i. Similarly, the reverse reaction rate is:

$$J_r = k_r \prod_i [A_i]^{\beta_i} \tag{5.4b}$$

Some of the indices α_i and β_i may be zero, of course. Now, at equilibrium, $J_r = J_f$, and also the concentrations $[A_i]$ must satisfy the equilibrium constant relationship:

$$K = \prod_i [A_i]^{\nu_i} \tag{5.5}$$

Now $J_r = J_f$ if:

$$\frac{k_f}{k_r} = \prod_i [A_i]^{\beta_i - \alpha_i} \tag{5.6}$$

Since k_f/k_r is a function of temperature only, it must be a function of the equilibrium constant K. Equation 5.5 and Equation 5.6 can be reconciled only if:

$$\frac{k_f}{k_r} = K^n$$

and

$$\beta_i - \alpha_i = n\nu_i$$

Let us choose the ν_i in such a way that $n = 1$, which we can always do by multiplying all the stoichiometric coefficients by the same constant multiplier. Then we obtain the simple results that $k_f/k_r = K$, and $\beta_i - \alpha_i = \nu_i$.

Away from equilibrium the net reaction rate is:

$$\begin{aligned}
J_{net} &= J_f - J_r \\
&= k_f \prod_i [A_i]^{\alpha_i} - k_r \prod_i [A_i]^{\beta_i} \\
&= k_r \prod_i [A_i]^{\beta_i} \left\{ \frac{k_f}{k_r} \prod_i [A_i]^{\alpha_i - \beta_i} - 1 \right\} \\
&= J_r \left\{ K \prod_i [A_i]^{-\nu_i} - 1 \right\}
\end{aligned} \tag{5.7}$$

Note that since the $[A_i]$ are not now the equilibrium concentrations, $\prod_i [A_i]^{\nu_i}$ is not the equilibrium constant K, but some other quantity, which we call the concentration product (though for those species for which ν is negative it is in fact a quotient) and we denote it by K^*. Clearly, K^* is related to the affinity, as they are both quantitative measures of departure from equilibrium. We can easily find the relationship:

SEQUENTIAL PROCESSES IN CRYSTAL GROWTH

$$A = -\sum_i \nu_i \mu_i$$

$$= -\sum_i \nu_i \mu_i^0 - \sum_i \nu_i RT \ln[A_i]$$

$$= -\Delta G^0 - RT \ln K^* \qquad (5.8)$$

where ΔG^0 is the standard free energy change for the reaction. Taking exponentials of Equation 5.8 we find:

$$\exp(A/RT) = K/K^*$$

or

$$K^* = K \exp(-A/RT) \qquad (5.9)$$

Obviously, at equilibrium $A = 0$ and $K = K^*$. Returning to Equation 5.7 and substituting Equation 5.9 gives us:

$$J_{net} = J_r \left\{ \frac{K}{K^*} - 1 \right\} \qquad (5.10)$$

$$= J_r \{\exp(A/RT) - 1\}$$

Therefore

$$J_{net} \simeq \frac{J_r}{RT} \cdot A \qquad (5.10a)$$

if $A < RT$, which is the desired result. It turns out that the 'chemical conductance' L_r is J_r/RT, where J_r is the reverse reaction rate, i.e. the 'exchange current' J_0. This result fits in with the intuitive idea that a reaction with a large exchange current J_0 is driven to a given net rate by a smaller overpotential than is a more sluggish reaction with a small exchange current. The affinity or overpotential for the reaction A we may also write as δG_r, so that for the two sequential processes,

$$\delta G_{total} = \delta G_t + \delta G_r$$

Although this formalism is derived here for a chemical reaction, Equation 5.10 may be applied to the physico-chemical processes of nucleation and migration of adatoms to surface steps. In a general way, we are covering not only chemical vapour transport, but also sublimation and dissociative sublimation growth as well.

Unless the 'reaction conductance' L_r is very large compared with the vapour transport conductance L_t, the partial pressures of the active species over the growing crystal are not those which satisfy the equilibrium constant relationship. The excess of the reactant partial pressures, and the deficit of the product partial pressures, provide the driving force for the reaction to proceed. If we fix our attention on one of the reactant partial pressures, and follow it from the source or virtual source, we see it fall off in the exponential manner required by the vapour transport equations (see Fig. 5.2).

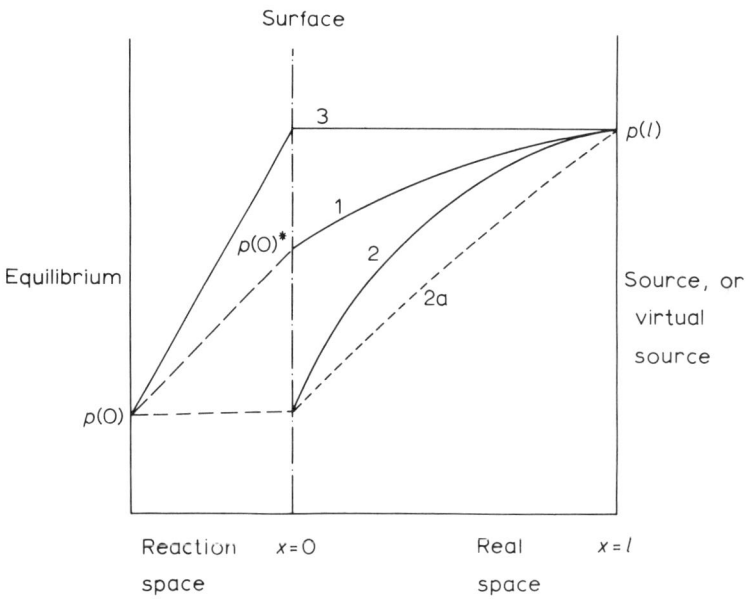

Fig. 5.2 Coupling of transport and surface processes.

Depending on the relative magnitudes of the reaction conductance L_r and the vapour transport conductance L_t, the extreme situations denoted by curves 2 and 3 may be approached. If L_t and L_r are of similar magnitude (i.e. $\delta G_r \simeq \delta G_t$), the situation denoted by curve 1 will obtain. The partial pressure $p(0)^*$ over the growing crystal departs appreciably from the equilibrium value $p(0)$. If the surface reaction were very facile, i.e. if L_r were large and δG_r small, $p(0)^*$ would be very close to $p(0)$, as represented by curve 2. This is the situation which has been assumed for convenience in most of Chapter 4, and is characteristic of experiments at high temperatures. Again, if the transport conductance is made very small, by including a high partial pressure of inert gas or producing a great excess of one component, $p(0)^*$ approaches $p(0)$. As we have seen, $p(x)$ varies almost linearly with x in this case (curve 2a).

At the other extreme, a severely hindered surface reaction (or a very small transport resistance) leads to the situation shown in curve 3. Here $p(0)^*$ is very nearly the same as $p(l)$, the partial pressure at the source. The surface reaction absorbs all the available applied potential. Such a situation could arise if the temperature were low enough to make the reaction conductance L_r very small, or alternatively if the transport conductance became enormous (e.g. stoichiometric vapour in dissociative sublimation, see Section 4.3.) In this latter case, the resistance might consist largely of a heat transfer resistance.

The region on Figure 5.2 marked 'reaction space' represents the surface of the growing crystal, and all the processes occurring on the surface take place in this reaction space. Because we have limited our consideration to a single process on the surface, only the terminal points $p(0)^*$ and $p(0)$ feature on Fig. 5.2. Had we considered more surface processes, intermediate pressures $p'(0)$, $p''(0)$ ... (or more properly, intermediate concentrations $c'(0)$, $c''(0)$ ... of adsorbed species) would feature here. The partial pressures $p(0)^*$ of the species in the vapour over the crystal surface are, of course, related to the concentrations of the corresponding adsorbed species, $c(0)^*$. If the crystal surface were in equilibrium with the vapour, $c(0)^*$ would be the equilibrium concentration $c(0)$, and would be related to the equilibrium partial pressure $p(0)$ by the appropriate adsorption isotherm, as discussed in Chapter 3. As $p(0)^*$ departs from $p(0)$, so $c(0)^*$ departs from $c(0)$. There will be a certain exchange current describing the dynamic equilibrium between $p(0)$ and $c(0)$. If the net rate of chemical reaction is small compared to this exchange current (though not necessarily small compared to the exchange current J_0 for the chemical reaction), deviations from the isotherm will be negligible. That is to say, $c(0)^*$ will be the surface concentration corresponding to $p(0)^*$ as given by the isotherm. This will be the case so long as the chemical reaction step in the sequence of surface processes has a lower 'conductance' than the adsorption step.

Now we begin to see how the transport equations developed in Chapter 4 may be applied to a practical crystal growth experiment to yield valuable information about processes occurring on the surface of a growing crystal. We may expect to obtain quantitative information of the following sort:

(1) The activation energy for the chemical reaction, $\Delta H_r^{\ddagger}$, i.e. the heat of formation of the active intermediate or active complex.
(2) The activation entropy $\Delta S_r^{\ddagger}$. Since this quantity contains the number of active sites on the surface, we may expect to be able to deduce whether the dominant path through the chemical reaction

step relies on a small number of active sites, e.g. dislocations, or whether it is distributed over all the atomic sites on the surface.

(3) The dependence of the exchange current on the partial pressures or surface concentrations, i.e. the order of the reaction in the various components. It may then be possible to postulate a particular activated complex. If nucleation on the surface is the particular surface process which determines the overall rate, it is possible, in principle at least, to determine the critical nucleus size and its free energy of formation from the above considerations.

We saw in Chapter 4 how it is possible to obtain a relationship between the transport rate or growth rate J and the partial pressures of the various species over the growing crystal and at the source. In Chapter 4 we regarded the partial pressures as boundary conditions which were to be measured experimentally or calculated from equilibrium considerations. The growth rate J was then regarded as the unknown quantity to be calculated from the transport theory. Thus the growth rates of Chapter 4 represent the maximum rates as predicted by the theory, i.e. the rates which would be observed with the given experimental conditions $T(l)$, $T(0)$, P, α, etc, if the surface processes were unhindered.

In practice it is often far easier to measure the growth rate than to measure partial pressures. In this section, therefore, we regard J as a quantity which is determined experimentally, and we apply the theory of Chaper 4 to calculate the partial pressures over the growing crystal. Thus in Chapter 4 we obtained equations of the type:

$$p_i(x) = [p_i(l) - P/s] \exp\left\{\frac{JRTs(x-l)}{DP}\right\} + P/s \qquad (5.11)$$

By putting $x = 0$ we have expressions for the partial pressures over the crystal, i.e. $p(0)^*$. Apart from the growth rate and the total pressure, which we assume the reader can measure without our advice, we have the unknowns $p_i(l)$ and D, the diffusion coefficient. We discuss each separately below.

The partial pressures $p_i(l)$ at the source may be controlled by allowing the various gaseous components to enter the system through taps from reservoirs. Or they may be controlled by various techniques such as using an effusion hole in the 'almost closed capsule' (see Chapter 6) or by having pure components as condensed phases in side-arms or reservoirs at independently controlled temperatures. Alternatively, in those experiments where no attempt is made to control the partial pressures at any point in the system, as in many closed-capsule experiments, it may still be possible to obtain

SEQUENTIAL PROCESSES IN CRYSTAL GROWTH   201

some useful analysis of the experimental data by applying chemical kinetic theory to the reaction at the source as well as to that on the crystal surface. Here we assume that the surface processes at the two ends of the vapour transport path proceed by the same mechanism, i.e. via the same activated complex. This will be a reasonable assumption if the temperatures of the source material and of the crystal are not too different. Our representation in Fig. 5.2 will now require a 'reaction space' extending to the right beyond $x = l$, in which the partial pressure drops from $p(l)$, the equilibrium value, to $p(l)^*$, the actual value over the source material.

The diffusion coefficient D, a quantity discussed in Section 4.1, was glossed over to some extent throughout Chapter 4. It is now appropriate to take a closer look at diffusion in a multicomponent mixture of gases.

According to the mathematical theory of diffusion[137, 138] the flux of species i in a non-uniform mixture of n gases is given by:

$$J_i = -D_{i1}\frac{dn_1}{dx} - D_{i2}\frac{dn_2}{dx} - \ldots - D_{ii}\frac{dn_i}{dx} - \ldots - D_{in}\frac{dn_n}{dx} \quad (5.12)$$

in one dimension. We may write this as:

$$J_i = -\sum_k D_{ik}\frac{dn_k}{dx} \quad (5.13)$$

In these formulae, D_{ik} is related to the binary diffusion coefficient of species i in species k. The n^2 coefficients are reduced in number by the following considerations: the net diffusion flux $\Sigma_i J_i$ is zero, and the n coefficients D_{ii} may be taken as zero. Even so the resulting equations are complicated, and we will therefore pursue the general case no further here. The interested reader is referred to the classic texts[137, 138].

A situation which is of frequent interest, and which is far more tractable algebraically, arises when one species is present in a great majority. The situation may then be described as diffusion of the $(n-1)$ minority species in the majority species, and this can be treated as $(n-1)$ independent but simultaneous binary diffusion systems[86].

We may justify this approach loosely as follows: (the formal justification comes out of Equation 5.12) if a concentration gradient of species A exists in the positive-x direction we would expect A molecules to flood down this gradient at high speed, were it not for collisions between the A-molecules and other molecules. At each of these collisions, momentum is exchanged, and the diffusion coefficient is a measure of the frequency of these collisions and the way in which momentum is exchanged. If species A is diffusing in the negative-x direction, and the species B in the positive-x direction, then on average, collisions between A and B molecules reduce the total momentum of the A molecules in the negative-x direction, and reduce that of the B-molecules in the positive-x direction. Note that collision between pairs of A-molecules do not alter the total momentum of species A and hence do not affect its diffusion velocity. The same applies to all collisions between pairs of molecules of the same species. In describing the interdiffusion of n species, one of which is in a great majority, as independent binary diffusion systems, we are saying that collisions between one minority species molecule and one majority species molecule are the dominant mechanism for transfer of momentum between the species, and that collisions between two dissimilar minority species molecules are sufficiently infrequent to be disregarded.

The transport equations now contain a different (pseudo-binary) diffusion coefficient for each minority species. The diffusion coefficient appears in the exponential factor which determines the shape of the partial pressures as functions of distance x, Equation 5.11. We find that for those species which diffuse rapidly, the curvature in partial pressure is reduced, while for the slowly diffusing species, the curvature is increased.

If we wish to use the transport equations to calculate the partial pressures over the growing crystal from measurements of the rate of growth and the inlet partial pressures, it is necessary to have reliable data for the diffusion coefficients. In particular, if we want to establish the order of the rate-limiting reaction in the various species which take part, it is necessary to take into account the separate rates of diffusion of the minority species, rather than to use a single 'average' diffusion coefficient[163, 180, 181].

5.3 The effect of the growth conditions on stoichiometry and electronic properties of crystals

In the earlier sections of this chapter, we have established a formalism for vapour transport and a surface reaction in sequence. There are three reasons why we have pursued this study.

(1) In order to modulate the growth rate by control of the gas transport resistance. The theory described in Chapter 4 gives us the tools for doing this.

(2) To identify the mechanism of the surface reaction. The

formalism of Section 5.2 enables us to study the dependence of surface reaction rate on partial pressures over the surface and on temperature, while allowing for changes in the transport resistance. (3) We wish to be able to calculate the actual partial pressures over the surface of the growing crystal, as these pressures determine its stoichiometry and impurity content, and hence its defect-sensitive (e.g. electronic) properties.

We start by considering the growth of a single component substance. There is clearly no possibility of non-stoichiometry. If the crystal is grown under very nearly equilibrium conditions, the concentrations of point defects will also be very near to the equilibrium values. If an appreciable departure from the equilibrium vapour pressure is required to cause growth at an acceptable rate, the resulting overpotential permits the defect concentrations to depart from their equilibrium values in directions governed by Le Chatelier's principle. Thus for a single component substance growing from a supersaturated vapour, we expect fewer vacancies and more interstitial atoms than would be present at equilibrium.

If the crystal is grown at an elevated temperature, the defect concentrations produced during growth will be determined by the thermodynamics and kinetics of defect formation at that temperature. On cooling to room temperature, the crystal will want to contain defects only in concentrations which are in equilibrium at room temperature. There will be a tendency for excess defects to diffuse out, and for those which are in deficit to be augmented by the appropriate species diffusing in. How far this tendency is realized depends on the rate of cooling compared with the rate for diffusion in the solid crystal. At room temperature, interstitial atoms or complex defects involving interstitial atoms (e.g. so-called 'crowdions', which are interstitials non-localized along close-packed directions) are thought to be mobile in f.c.c. metals such as copper and gold[182]. However, that is a somewhat exceptional situation. Vacancies are far less mobile in metals, and in general it is possible to 'freeze in' much of the excess or deficit of defects by rapid cooling to room temperature. The defect concentrations will later adjust themselves if the crystal is used in an application where it becomes heated, for example, by electrical current.

In the limit, when substances are prepared with enormous overpotentials, by rapid condensation of a jet of vapour on a cooled plate for example, it is possible to produce glassy forms or fine powder instead of crystalline material.

When binary compounds are prepared, the possibility of altering the ratio of the two components of the crystal arises. Non-stoichiometry has already been discussed in Chapter 2, and here we are concerned with the interaction of the vapour phase and the solid, and with the resulting defect-sensitive properties of the solid.

In our analysis of vapour transport by dissociative sublimation in Chapter 4, we introduced as a conceptually helpful parameter the ratio α of the partial pressures of the component substances in the vapour. The stoichiometry of the solid which is produced depends on the composition of the vapour directly over it, i.e. on $\alpha(0)$. When the vapour is stoichiometric, α has the value m if the vapour consists of A atoms and B_m molecules. We showed in Chapter 4 that when the vapour is in equilibrium with the solid, the total pressure is given by:

$$P = (1 + \alpha) \left(\frac{K}{\alpha}\right)^{1/s} \tag{5.14}$$

where

$$s = 1 + \frac{1}{m} \tag{5.15}$$

Equation 5.14 holds if the vapour is in equilibrium with the solid. We can rewrite Equation 5.14, using the 'partial pressure product' K^* instead of K, so that the equation is true whether the vapour and solid are in equilibrium or not:

$$P = (1 + \alpha) \left(\frac{K^*}{\alpha}\right)^{1/s} \tag{5.16}$$

The different 'partial pressure products' at the source and at the growing crystal are accommodated by a change in α, as was discussed in Chapter 4.

Whether the transport reaction is endothermic (as dissociative sublimation is) or exothermic, the 'partial pressure product' at the source, $K^*(l)$ must always be greater than that at the growing crystal, $K^*(0)$. This is because the total chemical potential driving the gas transport, $-RT\ln[K^*(l)/K^*(0)]$, must always be negative.

We can now see how α must alter between the source and the growing crystal to accommodate the change in K^*. After differentiating Equation 5.16 with respect to α at constant total pressure P we easily obtain the expression;

$$\left(\frac{\partial K^*}{\partial \alpha}\right)_P = \frac{K^*[1 + \alpha(s - 1)]}{\alpha(1 + \alpha)}$$

$$= \frac{K^*[m - \alpha]}{m\alpha(1 + \alpha)} \quad \text{from Equation 5.15} \tag{5.17}$$

The sign of $(\partial K^*/\partial\alpha)_P$ is determined solely by whether α is greater or less than m. Since $K^*(l) > K^*(0)$, we see that $\alpha(l) > \alpha(0)$ if $\alpha(l) < m$, and vice versa. In other words, whichever component is in excess at the source is in greater excess over the growing crystal. The reader may convince himself that the same is true when an inert gas is used in the crystal growth system.

In Section 4.3 we discussed the range of $\alpha(0)$ which might reasonably be produced in an experiment to grow cadmium sulphide in a capsule. We saw that this range is very wide: it is, in fact, limited only by the saturated vapour pressure of cadmium at one extreme and by the ability of the apparatus to contain very high sulphur pressures at the other extreme. The stoichiometry of the growing crystal is determined by the composition of the vapour from which it grows, i.e. by $\alpha(0)$. By suitable control of the experimental input conditions $\alpha(l)$, $T(l)$, and $T(0)$, any value of $\alpha(0)$ in the above range may be produced. On the other hand, if no attempt is made to control $\alpha(l)$, the composition of the vapour in the capsule will be determined by the non-stoichiometry of the source material, and may be made extreme if one component is tied up as a non-reactive oxide, for example. Furthermore, the vapour composition may change with time as an excess of one component, or an impurity, builds up.

The change of α between the source and the growing crystal for a given temperature difference ΔT is made less extreme by the presence of activational barriers to the surface reaction (since $K^*(0) > K(0)$) and by inert gas in the capsule. The effect of inert gas was discussed in Section 4.3. The effect of an activation barrier to the surface reaction is to increase the 'partial pressure product' $K^*(0)$ at the growing crystal, which results in $\alpha(0)$ being nearer to the stoichiometric value m than it would be if the vapour were in equilibrium with the crystal. In the limit, if the activation barrier is high, $K^*(0) \simeq K^*(l)$, and the vapour is of almost uniform composition. At the same time, of course, the transport rate becomes negligible.

Control over $\alpha(0)$ gives us control over the defect chemistry of the growing crystal. Since the electronic properties, which make many binary compounds scientifically and commercially important, are determined by the defects and impurities in the crystal lattice, and in particular, the ionization of the defects and impurities, an understanding of the defect chemistry is of prime importance to the crystal grower. We will not attempt to cover even the simpler aspects here, as there exist excellent textbooks on the subject already, in particular the masterly work by Kröger[69] and a very readable account of specific aspects by Van Gool[70], and by Greenwood[23].

If the gas phase contains species other than those which make up the crystal, e.g. transporting agents, dopants, or impurities, the crystal will inevitably dissolve some of the extraneous species. The topic of solubility was covered briefly in Chapter 3. As regards the defect chemistry we note the following points:

(1) In a binary compound MX, the extraneous species L may sit on an M site, an X site, or an interstitial site.
(2) The L-atoms may be associated with other defects, for example, an L-atom on an M-site may 'capture' a vacancy on an adjoining X-site if the L-atom is over-size for the M-site. The release of strain energy by association with a vacancy on an adjoining X-site reduces the chemical potential of vacancies on the X-sublattice.
(3) Impurity atoms which are ionized begin to affect the number of ionized native defects when the concentration of ionized impurities approaches the concentration of ionized native defects. Depending on the energy levels involved, this may result in impurity-controlled conductivity (i.e. doping of semiconductors) or impurity-controlled disorder.
(4) When two or more impurities are dissolved together in the host lattice, each affects the concentration, amount of ionization, and in some cases, the distribution on the M and X sites, of the other.

Again we stress that this most important topic, as well as phase stability, are merely touched upon here, with no attempt at presenting a detailed general treatment. The interested reader and crystal grower must explore particular situations of interest for himself, with reference to Kröger[69], Van Gool[70], and others. The effect of the vapour composition on the stoichiometry of the surface and the existance of distinct surface phases is a subject of great importance to the crystal grower, but which received but scant attention until the introduction in the last decade of surface analytical techniques[72].

5.4 Morphological stability

The crystal grower is frequently interested in the shape of the growing crystal. Usually, the desired shape is a large single-crystal boule, or a thin film on a substrate, and in either case, the experimenter would strive to avoid nucleation of extra crystals, or the development of platelets, whiskers, needles, and other extreme forms. Occasionally, whiskers or platelets may be the desired forms, for incoporating in fibre composites, for example. Indeed, if the growth of platelets were brought under control, much economy might result, since many single-crystal boules are sliced up to be used as substrates, with consequent wastage of expensive material. In this section, we discuss the shapes of crystals, and in particular, the stability of crystal shapes under growth conditions.

A crystal which is in internal equilibrium and in equilibrium with its surroundings, has minimized its free energy, and in particular, its surface free energy. This ideal situation is rarely achieved, if ever, since the driving force arising from differences in surface free energy is usually too small to produce a change in shape at an appreciable rate[183]. However, the ideal shape is a useful starting point for discussion, and real crystals frequently approximate this shape, if only roughly.

If the surface free energy per unit area, γ, were isotropic, the total free energy would be minimized by minimizing the surface to volume ratio, and the equilibrium shape would be a sphere. However, γ is, in general, a function of orientation, and in fact, a plot of γ versus orientation (the so-called γ-plot) has cusps in it, where the surface free energy per unit area is very much lower for particular crystal faces than for faces of slightly different orientation. These cusps determine the most stable crystal faces, which are often faces where the atoms or molecules are closely packed and the equilibrium shape of the crystal is a polyhedron bounded by such faces. The relative sizes of the different faces of the polyhedron may be determined using Wulff's theorem[184].

A crystal which is in equilibrium with its surroundings is not growing, of course. A finite rate of growth implies a departure from equilibrium, which may be large or small, and the shape of the γ-plot is altered under these circumstances. If the crystal is being grown in an atmosphere containing species which adsorb on its surface, for example, the transporting agent in a chemical vapour transport system, the shape of the crystal is again altered from the equilibrium shape[119, 132]. Thus the growing crystal may assume a shape differing from that given by Wulff's theorem, and if the perturbations of the γ-plot are large, a very different crystal habit may result.

In this section, we concern ourselves with the thermodynamic and kinetic factors which lead to stability of shape of a growing crystal, and we consider two approaches to the problem:

(1) The thermodynamic approach[142], analogous to the theories

of supercooling and constitutional supercooling first developed for solidification of molten alloys.

(2) The dynamic approach, which takes into account the kinetics of transport of heat and matter, as well as thermodynamic considerations.

The thermodynamic description of the stability of the surface of a crystal growing from the vapour was first put forward by Reed, LaFleur, and Strauss[142], who considered a single-component substance growing from a vapour which contained some inert gas as well as the vapour of the growing crystal. The partial pressure of the crystal substance in the vicinity of the crystal surface is determined by transport considerations, as discussed in Chapter 4. The partial pressure at any point may be greater or less than the saturated vapour pressure, which depends only on the temperature field imposed by the furnace. Since the temperature distribution has little effect on the rate of transport or on the partial pressure distribution (though both these depend on the temperature *difference* between the source of material and the growing crystal, as we saw in Chapter 4), we may determine whether the actual partial pressure is greater or less than the saturated vapour pressure, by imposing a suitable temperature gradient. The possible situations are illustrated in Fig. 5.3.

In Fig. 5.3a, the saturated vapour pressure $p^0(T)$ is everywhere above the actual partial pressure $p_A(x)$ of the growing crystal of substance A, a situation which has been achieved by imposing a steep temperature gradient. In Fig. 5.3b, on the other hand, a gentle temperature gradient exists, and $p^0(T)$ is below $p_A(x)$ in a region near the crystal surface, which is therefore supersaturated. If the supersaturation is great enough, new crystals may nucleate ahead of the surface of the original crystal. Notice that the supersaturation is greatest some little distance away from the surface, so that, according to this theory, nucleation should take place first at this point, and not next to the original crystal.

Suppose now that the surface of the crystal is not perfectly smooth, but has some asperities on it, which project into the supersaturated region a little way. Since the rate of growth of the crystal increases with increasing supersaturation, we expect these asperities to grow rapidly, thus amplifying the initial surface roughness. On the other hand, if the situation of Fig. 5.3a obtains, with $p^0(T)$ above $p_A(x)$ everywhere, asperities project into a region which is undersaturated, and would tend to evaporate.

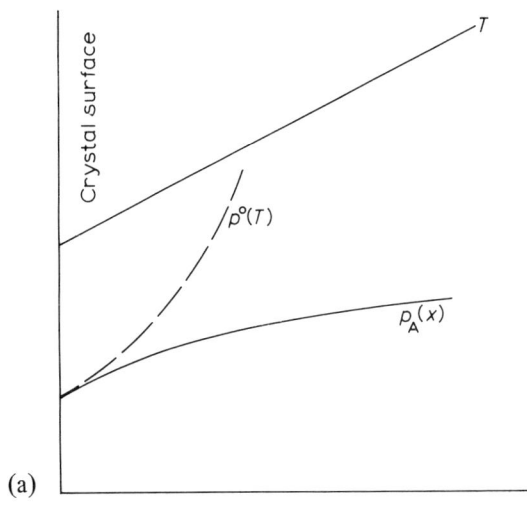

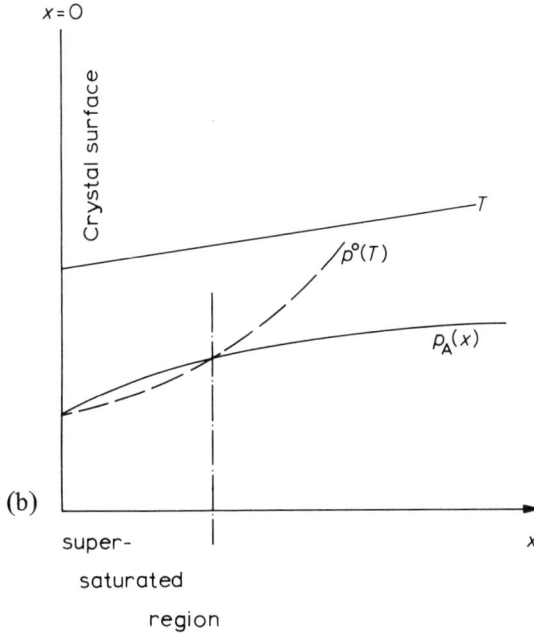

Fig. 5.3 Constitutional supercooling in the vapour.

We may write the condition for $p^0(T)$ to be above $p_A(x)$ as:

$$\frac{d}{dx}\left(\frac{p_A(x)}{p^0(T)}\right)_{x=0} < 1 \tag{5.18}$$

This is the condition for surface stability for a single component substance growing from its vapour, according to the 'consitutional supercooling in the vapour' theory. It should be realized that the quantity on the left of this inequality is a function of gas transport kinetics and of the temperature field imposed.

This simple thermodynamic theory can be assailed from various angles. To begin with, in discussing what happens to an asperity on the crystal surface, it has been assumed that the temperature of the tip of the asperity is the same as the temperature of the vapour at that point; in other words, that the asperity does not distort the temperature field. In fact, some heat will be conducted through the solid, so that the temperature field around the protrusion may be distorted. If the protrusion is growing faster than the rest of the crystal, it will have to get rid of the heat of condensation (or of reaction in a chemical vapour transport sytem), so that for this reason, it will tend to be hotter than its surroundings. In addition to affecting the temperature field in its vicinity, a fast-growing asperity will also affect the composition of the vapour, i.e. by growing, it will locally relieve the supersaturation. When this effect is translated into three dimensions, it can be seen that a fast-growing asperity can reduce the supersaturation in the vapour not only ahead of itself, but on all sides as well, so that the neighbouring, slowly-growing regions are presented with a lower over potential for growth than would exist if the asperity were not growing fast.

While the thermodynamic approach to morphological stability cannot take these effects into account, and cannot predict what will happen if the condition of Equation 5.18 is not fulfilled, it does provide a minimum criterion for the stability of the advancing interface to the development of protrusions. It may be that the interface remains stable (because of limitations on the flow of heat, for example), even when the thermodynamic condition for stability is not fulfilled.

The thermodynamic approach has been extended to binary compounds by Faktor, Heckingbottom and Garrett, and their treatment also covers multicomponent compounds [141, 143, 144, 146]. Consider the general case of a solid compound A(solid) reacting with gaseous species B, C ... to yield gaseous products M, N, ... :

SEQUENTIAL PROCESSES IN CRYSTAL GROWTH

$$A(\text{solid}) + bB(\text{gas}) + cC(\text{gas}) + \ldots \rightleftharpoons mM(\text{gas}) + nN(\text{gas}) + \ldots$$

The equilibrium constant for this reaction may be expressed in terms of the partial pressures of the gaseous components and the activity of the solid (which is unity for a pure solid):

$$K_p = \frac{p_M^m \cdot p_N^n \ldots}{a_A \cdot p_B^b \cdot p_C^c \ldots}$$

The activity a_A of solid A can thus be expressed as:

$$a_A = \frac{p_M^m \cdot p_N^n \ldots}{K_p \cdot p_B^b \cdot p_C^c \ldots}$$

If this quantity is equal to unity, solid A is in equilibrium with the gaseous species at the partial pressures $p_B, p_C \ldots p_M, p_N, \ldots$ If this quantity is greater or less than unity, the vapour is supersaturated or undersaturated with respect to solid compound A. Our condition for the interface to be stable is thus:

$$\left(\frac{\partial a_A}{\partial x}\right)_{x=0} \leqslant 0 \tag{5.19}$$

in other words, the vapour should become undersaturated with respect to solid A as one moves away from the surface of the growing crystal.

For a single component substance, this condition reduces to that of Reed, LaFleur, and Strauss,[142] equation 5.18, since then

$$K_p \equiv p^0(T) \text{ and } a_A = p_A/K_p = p_A/p^0(T)$$

For dissociative sublimation of a binary compound:

$$AB(\text{solid}) \rightleftharpoons A(\text{gas}) + \frac{1}{m} B_m(\text{gas})$$

the condition for surface stability becomes:

$$\frac{d}{dx}\left(\frac{p_A \cdot p_{B_m}^{1/m}}{K_p}\right)_{x=0} \leqslant 0 \tag{5.20}$$

The left-hand side of this inequality is a function of gas transport, which determines the gradients of $p_A(x)$ and $p_{B_m}(x)$, and also of the temperature variation along x, which determines the gradient of K_p. Using Equations 4.47 and 4.48, derived in Chapter 4, for the variation of p_A and p_{B_m} with x, and van't Hoff's equation, we can rewrite the thermodynamic condition for a stable interface as:

$$\left[\frac{1}{p_A}\left(\frac{dp_A}{dx}\right)_{x=0} + \frac{1}{mp_{Bm}}\left(\frac{dp_{Bm}}{dx}\right)_{x=0} - \frac{d}{dx}\ln K_p\right] \leq 0$$

or:

$$\left[\left(\frac{p_A(0) - P/s}{p_A(0)} + \frac{p_{Bm}(0) - P/ms}{p_{Bm}(0)}\right)\frac{JRTs}{DP} - \frac{\Delta H}{RT^2}\left(\frac{dT}{dx}\right)_{x=0}\right] \leq 0 \quad (5.21)$$

where $s = 1 + 1/m$. The first term on the left of this expression is a function of the conditions which govern the mass transport, i.e. $T(l)$, $T(0)$, and $\alpha(l)$. The second term is determined by $(dT/dx)_{x=0}$, that is, by the temperature gradient at the surface of the growing crystal.

In Section 4.3.1, we introduced the useful parameter α, which is the ratio of the partial pressures p_A/p_{Bm}. It is easy to show that the partial pressure gradients at the growing crystal are given by equations such as:

$$\left(\frac{dp_A}{dx}\right)_{x=0} = \frac{JRTs}{DP}[p_A(0) - P/s] \quad (5.22)$$

so that

$$\left(\frac{dp_A}{dx}\right)_{x=0} = \frac{P}{l}\left[\frac{\alpha(0)}{\alpha(0)+1} - \frac{m}{m+1}\right]\ln\left\{\frac{\alpha(l) - m}{\alpha(l)+1}\bigg/\frac{\alpha(0) - m}{\alpha(0)+1}\right\} \quad (5.23)$$

It is interesting that the partial pressure gradients do not depend on the diffusion coefficient D, but only on the equilibrium constants at each end of the system (i.e. $\alpha(0)$ and $\alpha(l)$), the molecularity of the sublimation reaction (i.e. on m) and on P and l. We would tend to think of diffusion working to straighten out the partial pressures gradients, so that a larger diffusion coefficient might result in less curvature in the partial pressures as functions of x. For dissociative sublimation with no inert gas present, this is not so, since if it were possible to change the diffusion coefficients (by taking a different compound, for example), while keeping the same values of $\alpha(l)$, $\alpha(0)$, m, P, and l, both J and the Stefan velocity U would be changed in proportion. The curvature in the partial pressures is brought about by a balance between Stefan flow and diffusion, and thus remains unaltered. If an inert third gas is present, changing the diffusion coefficient (by changing the inert gas, for example) while keeping the same temperature difference brings about a change in $\alpha(0)$ and hence

SEQUENTIAL PROCESSES IN CRYSTAL GROWTH

in the partial pressure gradients, as the reader may readily confirm.

The thermodynamic condition for a stable interface can be written:

$$\frac{-[m-\alpha(0)]^2}{\alpha(0)m(m+1)} \ln \left\{ \frac{\alpha(l)-m}{\alpha(l)+1} \bigg/ \frac{\alpha(0)-m}{\alpha(0)+1} \right\} \leqslant \frac{l\Delta H}{RT^2} \left(\frac{dT}{dx}\right)_{x=0} \quad (5.24)$$

The left-hand side of this expression is positive, so that the inequality can only be satisfied if $(dT/dx)_{x=0}$ is positive.

For a given temperature gradient over the crystal, $(dT/dx)_{x=0}$, $T(l)$ and $\alpha(l)$, there will be a value of $\alpha(0)$, and hence of $T(0)$ and ΔT, which makes the left and right sides of this expression equal. We denote this value of temperature difference ΔT_{crit}, and the corresponding growth rate, J_{crit}. Smaller temperature differences result in $(d\alpha/dx)_{x=0} < 0$, so that protrusions which develop on the surface are thermodynamically unstable. Large temperature differences result in protrusions being amplified, as they project into a region of higher supersaturation.

In Fig. 5.4 we reproduce the plots of calculated growth rate J versus temperature difference ΔT for cadmium sulphide at 1400 K, taking a range of values of $\alpha(l)$. This system is discussed in Section 4.3.1. We have also shown the locus of critical conditions J_{crit} and ΔT_{crit} for different values of $(dT/dx)_{x=0}$ ranging from 1 to 10 Kcm^{-1}. There are two points of interest about the curves delineating the critical conditions. If $\alpha(l)$ is near the stoichiometric value m, which is 2 in this case, ΔT_{crit} is rather small, so that it may be difficult to maintain a constant growth rate below J_{crit}. When $\alpha(l)$ is large, ΔT_{crit} approaches a constant value which depends on $(dT/dx)_{x=0}$. By expanding Equation 5.21, it can be shown that

$$\Delta T_{\text{crit}} = \frac{RT(l)T(0)}{2\Delta H} \ln \left\{ \frac{2l\Delta H}{RT^2} \frac{dT}{dx} + 1 \right\} \quad (5.25)$$

for $\alpha(l)$ very large. For cadmium sulphide at 1400 K, this becomes:

$$\Delta T_{\text{crit}} \simeq 24 \ln \left\{ 0.4 \frac{dT}{dx} + 1 \right\} \quad (5.26)$$

When the system contains a quantity of inert gas, the transport rate is markedly decreased for $\alpha(l) \simeq m$, as we saw in Section 4.3.1, and slightly decreased for $\alpha(l)$ far from m. The effect of inert gas on the critical conditions for a stable interface is illustrated in Fig. 5.5 for a particular value of the temperature gradient over the growing crystal

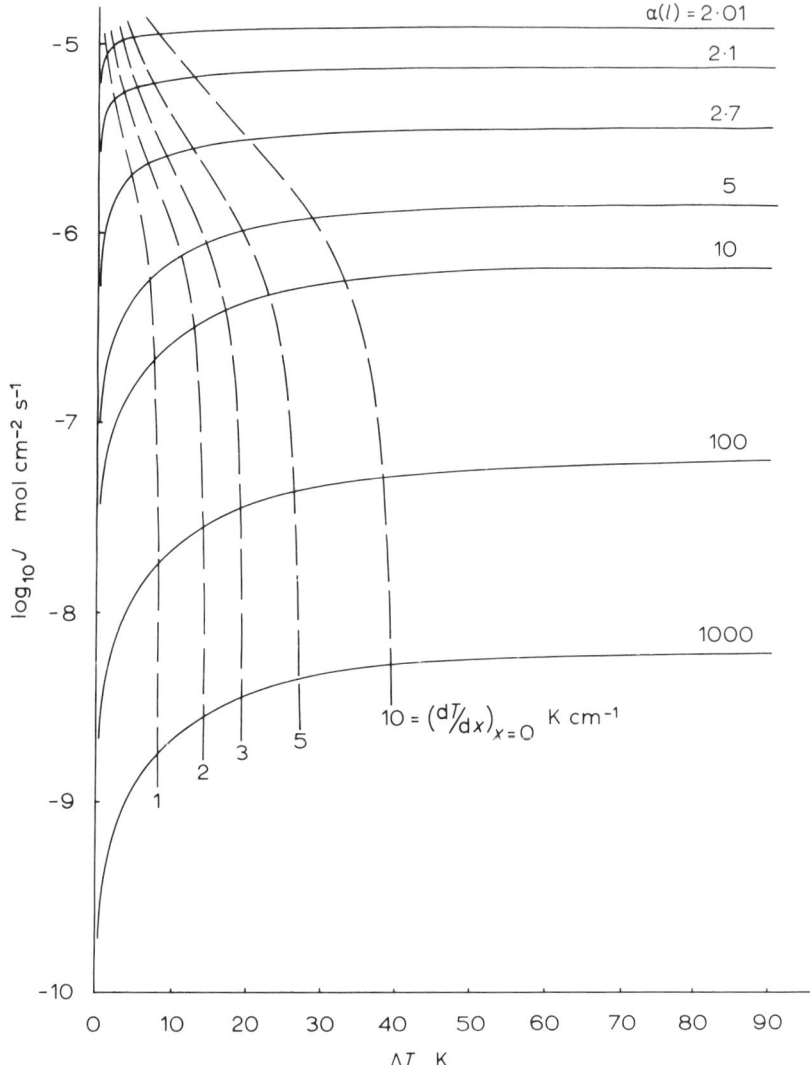

Fig. 5.4 Critical growth rates for cadmium sulphide with various temperature gradients over the crystal: $(dT/dx)_{x=0} = 1, 2, 3, 5$ and 10 K cm^{-1}.

of 1 K mm^{-1}. We see the dramatic effect of even quite small quantities of inert gas on the critical temperature difference for $\alpha(l)$ near m. The critical conditions occur on a region of the J versus ΔT plot which is near to horizontal, so that small fluctuations in ΔT do not produce drastic alterations in transport rate.

In a chemical vapour transport system, there is inevitably a portion of the vapour which plays the part of an inert gas as far as

SEQUENTIAL PROCESSES IN CRYSTAL GROWTH

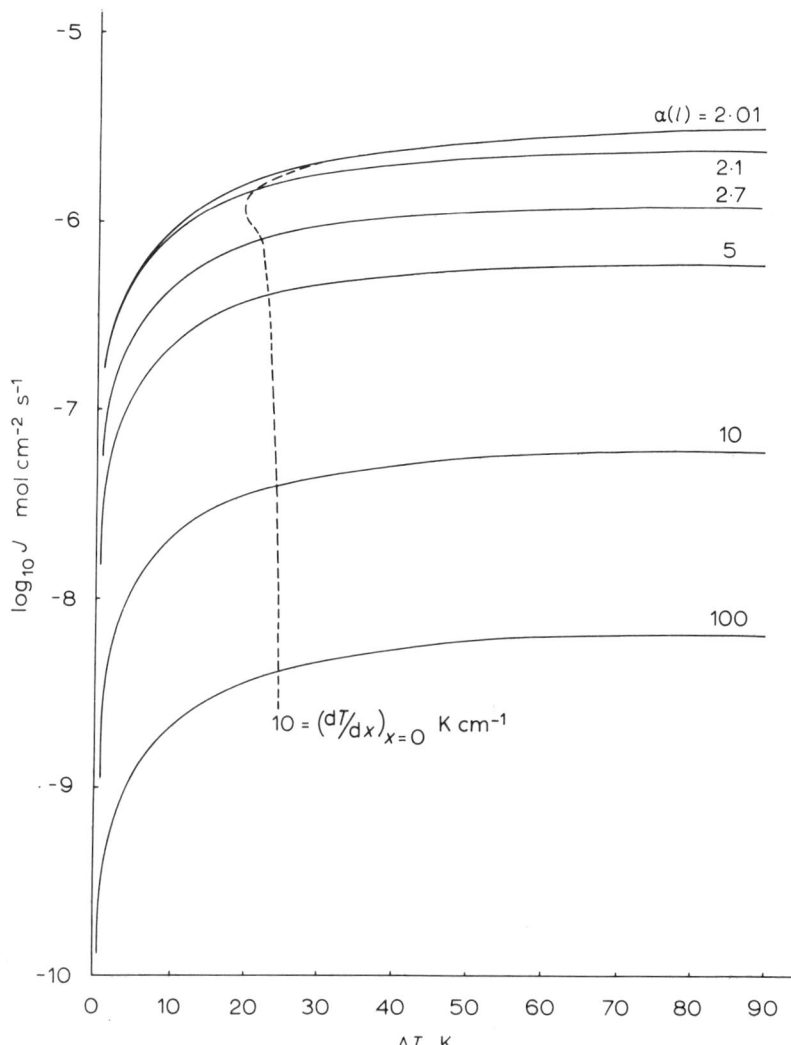

Fig. 5.5 The effect of inert gas on the critical conditions in the growth of cadmium sulphide. The temperature gradient over the crystal $(dT/dx)_{x=0} = 10 \text{ K cm}^{-1}$.

transport rates are concerned. For example, in the simple system:

$$M(s) + HX(g) \rightleftharpoons MX(g) + \tfrac{1}{2}H_2(g)$$

the combination $HX + 2H_2$ has a zero net transport rate, and may be likened to an inert gas in some ways, in particular, in its effect on the critical conditions. The thermodynamic approach to the stability of the surface of a crystal growing by chemical vapour transport has been given in the literature[144,146]; many of the essential features are reproduced in the system of dissociative sublimation with an inert gas present.

If the surface reaction has to overcome a significant activation energy, the vapour over the crystal will be correspondingly supersaturated. For the interface to be stable according to the thermodynamic theory, we require this supersaturation to decrease as one moves away from the surface of the crystal. The stability condition is the same as before, namely:

$$\left(\frac{da}{dx}\right)_{x=0} \leqslant 0$$

but now the activity $a(0)$ at the interface is no longer unity, since the vapour has departed significantly from its equilibrium composition.

The effect of an activation barrier to the surface reaction is to reduce the rate of growth caused by a given temperature difference. This means that the partial pressures of all the vapour constituents vary less rapidly with x, so that critical conditions occur with smaller temperature gradients over the crystal. This can be seen from Equation 5.24, since the effect of the activational barrier is to make $\alpha(0)$ more nearly equal to $\alpha(l)$ (and hence to m). Alternatively, we may say that, for a given temperature gradient over the crystal $(dT/dx)_{x=0}$, critical conditions occur with a larger temperature difference when the surface reaction is hindered.

The dynamic theory of the stability of the surface of a growing crystal has been worked out for a crystal growing from the melt by Tiller[185], and by Mullin and Sekerka.[186] There seems to be no corresponding body of theory describing the interplay of heat and matter flow and its effect on interface stability for a crystal growing from the vapour. The existing theories for crystals growing from the melt do not carry over very well to vapour phase crystal growth because of the large difference in such properties as density, heat conductivity, and diffusion coefficient between vapours and liquids. We do not attempt to supply the theory here, but a description of the problem may be useful.

Consider a single protrusion on the otherwise smooth surface of a crystal. In order for the protrusion to grow more rapidly than the rest of the crystal, it must be supplied with material from the vapour, and be able to get rid of the heat of condensation or reaction. If the protrusion reaches out into a region of the vapour which is more supersaturated than the vapour over the rest of the crystal, it will begin to grow faster. In doing so, it becomes hotter than the rest of the crystal because of the increased rate of production of heat of condensation or reaction. It also reduces the supersaturation in the vapour in its immediate vicinity by withdrawing reactants at a locally increased rate. This in turn sets up local gradients in the partial pressures of all the vapour species, which may be such as to transport extra material to the protrusion, at the expense of neighbouring areas of the crystal. The full solution of the problem must take into account the flows of matter and of heat in three dimensions in a time-dependent way.

That concludes our discussion of morphological stability from the theoretical point of view. It should be realized that an initially smooth crystal surface (for example, a polished substrate on which a thin film is to be deposited) may contain a distribution of growth sites with widely different surface resistances to growth. If parts of the surface are contaminated, further growth on those parts may take place only very slowly, while proceeding more rapidly on clean parts of the surface. The different growth rates on different parts of the surface give rise to surface roughness whether or not the critical conditions, delineated by the thermodynamic theory of surface stability, have been exceeded.

5.5 Separation, purification, and doping by chemical vapour transport

We deal with the questions of separation of different substances, purification of a substance, and incorporating controlled quantities of an impurity ('doping') under one heading, as the three topics are different aspects of one larger topic, namely, simulataneous transport of two or more compounds.

Separation of a mixture

Let us first consider the question of separation of a mechanical mixture of two substances. The simplest case arises if there is a temperature range in which one substance has a significant vapour pressure or dissociation pressure while the other does not, for then the substances may be separated by fractional sublimation. Another simple case arises if a transporting agent can be found which provides a significant rate of transport for one substance but not for the other. The criteria for selecting a good transporting agent were investigated in Section 2.3.1. For separation of two substances by transporting one in a temperature gradient and leaving the other behind, we would look for a transporting agent which reacted with one substance to give an equilibrium constant fairly near to unity (but see Section 2.3.1 for a fuller discussion of this point) while giving a very small equilibrium constant in its reaction with the other substance. The higher the second equilibrium constant, the less separation of the substances will be obtained, in general, although it may be possible to so arrange conditions that one transport reaction is suppressed, even though the equilibrium constant is not very small. For example, transport of a substance using a hydrogen halide as transporting agent may be regulated by adjusting the ratio $p_{HX} : p_{H_2}$. Different transport reactions may depend on different powers of this ratio, so that the relative transport rates may be adjusted.

Note that, although an extremely *high* equilibrium constant for a transport reaction suppresses the rate of transport by that reaction, it also suppresses the rate of other simulataneous transport reactions (a) by lowering the partial pressure (chemical potential) of the transporting agent, and (b) by providing a blanket of more or less stationary vapour, through which the products of the second, simultaneous, reaction have to move.

While it is possible to obtain good separation of a mixture when both components are transported in the same direction (but at very different rates), even better separation can be achieved if the components of the mixture are transported in opposite directions. The requirement is that the transporting agent shall react with one component of the mixture exothermically, and with the other component, endothermically. The mixture is then separated by vapour transport into its components, one of which accumulates at the cold end of the system, and the other at the hot end.[187]

It is interesting to observe what happens to the Stefan flow in the system in which a mixture of components is separated by transport in opposite directions in a temperature gradient. For most transport reactions, Stefan's velocity is in the direction hot → cold (see Section 4.2.3). For the component of the mixture which transports by the endothermic reaction, Stefan's velocity is directed away from the mixed source material, while in the other half of the system, the other component is being transported by the exothermic reaction, and Stefan's velocity is directed towards the mixed source material. Thus the 'wind' blows continuously from the hot end of the system to the cold, with, in general, a different magnitude in each half of the system.

If the 'wind' blows from the mixed source material towards the cold end of the system where the component transported by the endothermic reaction is deposited, it will carry with it all vapour species which exist over the source, i.e. the reactants and products of both transport reactions. On arrival at the cold end of the system, the products of the endothermic transport reaction undergo the reverse reaction to deposit one component of the mixture. The products of the exothermic reaction have no driving force to make them undergo the reverse reaction, and if none of these products is soluble in the depositing solid to a significant extent, they start to accumulate. Accumulation continues until a sufficient concentration gradient is established for the diffusion fluxes of these products away from the cold end to exactly balance the Stefan flow in the opposite direction, or until the concentration of these products builds up to such an extent that they exceed the equilibrium concentration at the temperature of the cold end, and the reverse reaction takes place. In the latter case, separation of the two components is far from complete, since at one end of the system both components are deposited. Re-designing the temperature field is then necessary to suppress deposition of the second component at the cold end.

In either case, the concentrations of the products of the exothermic reaction will be higher at the cold end of the system than over the mixed source material, and although these concentrations may not be in equilibrium with the second component in its pure form, there will be some activity of the second component with which they are in equilibrium Thus the component which is deposited at the cold end will be contaminated to some extent with the second component.

Although the products of the endothermic reaction will penetrate to the hot end of the system by diffusion against the Stefan flow, they will be in lower concentration there than over the mixed source material. Coupled with the temperature change, these factors result in the material transported to the hot end being less contaminated than that transported to the cold end of the system. Furthermore, it is generally observed that larger, more perfect crystals develop at high temperatures than at low temperatures (see Chapter 7). Here, then, are two reasons for selecting an exothermic reaction for separating a wanted component from an unwanted component of a mixture.

It may be possible to choose the endothermic reaction so that a decrease in the number of vapour phase molecules occurs at the mixed source material. In this case, the Stefan velocity is in the direction cold → hot, towards the source. Such a system is illustrated by the well-known example of the separation of copper and cuprous oxide using HCl as the transporting agent, according to the reactions[2]:

$$3Cu(s) + 3HCl(g) \rightleftharpoons Cu_3Cl_3(g) + \frac{3}{2} H_2(g)$$

$$\Delta H^0 = 188 \text{ kJ mol}^{-1}$$

$$\frac{3}{2} Cu_2O(s) + 3HCl(g) \rightleftharpoons Cu_3Cl_3(g) + \frac{3}{2} H_2O(g)$$

$$\Delta H^0 = -92 \text{ kJ mol}^{-1}$$

The existence of Cu_3Cl_3 as the dominant vapour species over cuprous chloride, with Cu_4Cl_4 as the second most important, has been confirmed by spectrophotometric[188], effusion[189], and mass-spectrometric[190] methods. This trimeric molecule fulfills the conditions discussed at the end of Section 4.2.3. for endothermic reactions to involve a decrease in the number of vapour phase molecules, since it is a heavy molecule with large contributions to its entropy from its rotational and vibrational modes.

Purification of a substance

We distinguish separation of the components of a mixture and purification of an impure substance simply on the basis that in the first case the components are present at essentially unit activity, while in the second case the activity of the impurities may be very low and the activity of the host substance may be significantly below unity. The distinction is often not clear cut in practice.

We consider two cases:

(1) The impurity is present at very low activity (and we assume also low concentration) and the activity of the host substance may be taken as unity to a good approximation. The Stefan velocity in the transport system is then determined by the transport of the host substance, with the impurity, if it is transported, causing a negligible perturbation.

(2) The impurity is present in high concentration, so that we must consider two coupled transport systems.

In a closed capsule, the extent to which a substance may be ridded of small quantities of impurity by transport depends on whether the impurities are more easily volatilized than the host substance or less easily. Volatilization may be sublimation or reaction with a transporting agent to form a volatile compound. Impurities which are involatile collect on the source material, so that their activity at the source increases as transport progresses. Since the partial pressure of an involatile impurity over the source is low, it is transported correspondingly slowly. The impurity cannot be incorporated in the growing crystal faster than it is transported, so that good purification results. As the activity of the impurity at the source rises, its partial pressure in the vapour rises, as does its rate of transport, so that the degree of purification falls. The steady state situation occurs when the activity of the impurity at the source reaches a steady value. If this is less than unity, transport of the impurity is keeping pace with transport of the host substance, and no purification is achieved. This steady state may not be reached in the duration of the experiment if the transporting agent is well chosen. If the steady-state activity of the impurity is unity, the impurity accumulates at the source as a separate condensed phase, and purification continues.

A volatile impurity, on the other hand, accumulates in the vapour. It enters the vapour near the source at the rate at which it is uncovered by transport of the host material. This rate may be enhanced significantly by diffusion of the impurity in the source material to the surface, especially if the rate of transport of the host substance is low. The Stefan flow sweeps the impurity towards the growing crystal if the transport reaction is endothermic, or away from the growing crystal if the reaction is exothermic. The partial pressure of the impurity over the growing crystal increases with time, as does the activity of the impurity in the crystal. The steady state situation occurs when the activity of the impurity in the crystal has risen to become equal to that in the source material, and no further purification is achieved. This situation is reached more rapidly if the transport reaction is endothermic, as the impurity is then swept towards the crystal. A large capsule favours purification, as the steady state situation is avoided if the vapour volume can accommodate all the impurity from the original charge without exceeding the vapour pressure which would be in equilibrium with the original concentration of impurity in the source. The necessary capsule volume depends on the properties of the impurity and the transporting agent, on the amount of charge, impurity level, etc. and may well be impracticable.

We see, then, that in a closed capsule, the material transported at the beginning of the run is relatively pure, but the purity declines as the run proceeds, and eventually may be no better than the purity of the starting material.

The accumulation of volatile impurities in the vapour may proceed very rapidly in a closed capsule, if the starting material has a surface film of oxide or contamination. Vacuum degassing is often used to overcome this problem but if the charge material contains pockets of impurity trapped in the solid, these impurities remain in the charge until they are uncovered as transport proceeds. There is no way of removing them from a closed capsule, and so they remain to slow down the rate of transport and to contaminate the grown crystal.

Impurities may be removed continuously if a small hole is drilled in the side of the capsule, transforming it into an 'almost closed capsule'.[146] This arrangement calls for a suitable atmosphere surrounding the capsule, of course, which could be a vacuum in the case of transport by sublimation, or a flowing stream of gas containing a transporting agent in the case of transport by chemical vapour transport. The experimental details of the 'almost closed capsule' are discussed in Chapter 6. For the best purification, the small hole must be placed where the greatest concentration of impurity in the vapour is found, i.e. near the end of the capsule towards which the Stefan flow is directed. The continuous leaking of impurities through the hole alters the nature of the steady state of purification to one in which the grown material is purer than the starting material. The steady state is achieved when impurity is transported at the rate at which it is uncovered, so that no further accumulation in the vapour takes place. Some of the transported impurity leaves the capsule. Correct placing of the small hole ensures that the ratio of impurity to host substance leaving the capsule is far greater than the ratio in the starting material, so that purification is achieved even in the steady state.

When we consider the purification of a substance containing a large quantity of impurity, using a closed capsule system, we make use of the ideas discussed under 'Separation of a mixture', but we must take into account the fact that the substance will not be at unit activity in the starting material, at least initially. We try to arrange for the host substance to be transported rapidly, while the impurity is transported only very slowly or, better still, in the opposite direction. The result is that the impurity accumulates at the source and rapidly reaches unit activity. Thus a starting material consisting of the host substance A and impurity B at activities a_A and a_B becomes two condensed phases which would ideally be a saturated solution of B in A plus almost pure B. As the impurity B accumulates and eventually separates as a second condensed phase, so the activity of A decreases.

Consider as a typical example the purification of a solution of B in A, using a hydrogen halide as the transport agent. If both B and A are transported as the monohalides, the transport reactions are:

$$\text{A (impure solid)} + \text{HX(g)} \longrightarrow \text{AX(g)} + \tfrac{1}{2}\text{H}_2\text{(g)}$$

$$\text{B (deposit)} + \text{HX(g)} \longrightarrow \text{BX(g)} + \tfrac{1}{2}\text{H}_2\text{(g)}$$

The equilibrium constants are:

$$K_A = \frac{p_{AX} p_{H_2}^{1/2}}{a_A p_{HX}} = \frac{p_{AX}}{a_A \lambda}$$

where

$$\lambda = p_{HX}/\sqrt{(p_{H_2})}$$

and

$$K_B = \frac{p_{BX}}{a_B \lambda} \simeq \frac{p_{BX}}{\lambda}$$

when a_B has become effectively unity.

If we arrange for rapid transport of component A by suitable adjustment of the experimental control variables T, ΔT, P and λ (see Chapter 4) we can suppress the transport of B only by ensuring that p_{BX}/p_{AX} is very small. From the equilibrium constant equations, it follows that:

$$\frac{p_{BX}}{p_{AX}} = \frac{K_B}{K_A a_A}$$

which is a function of temperature only, once the transporting agent has been selected. Here we must choose a transporting agent which gives a large ratio of K_A to K_B at a convenient temperature.

If B transports as a trihalide, however, while A transports as a monohalide, we have a little more flexibility. The transport reaction for B is now:

$$B + 3HX(g) \longrightarrow BX_3(g) + \frac{3}{2} H_2(g)$$

for which

$$K_B' = \frac{p_{BX_3}}{\lambda^3}$$

The ratio of B to A in the vapour is now:

$$\frac{p_{BX_3}}{p_{AX}} = \lambda^2 \frac{K_B'}{a_A K_A}$$

which is a function of temperature and λ. We can thus suppress transport of B by choosing a small value for λ. Although this will decrease the transport rate for A, it decreases the rate for B more rapidly. Finally, it may be possible to select a transporting agent which forms a stable condensed phase with the impurity and a volatile compound with the host substance. For example, A may form a volatile oxide or hydroxide in a water vapour/hydrogen mixture, while B forms an involatile oxide. Substance A can then be transported while substance B is fixed at the source in a yet more stable compound.

Purification in an open flow system presents no new concepts so far as the chemistry of the vapour transport is concerned. The reason for using this method is that the continuous flow of gas sweeps away any impurities which are evolved as the source material is volatilized, or by reaction of the transporting agent or any other component of the system with the container material. For example, silica outgasses water on heating, and can be reduced by hydrogen and other reducing agents at high temperatures to form silicon monoxide gas. In a closed capsule, there is no means of getting rid of these gaseous impurities, and so they accumulate until eventually they are incorporated in the crystal at the same rate as they are evolved. In the 'almost closed capsule' system these impurities build up to a lesser extent, but may still be incorporated in the growing crystal at unacceptable levels. The open flow method offers a means of reducing contamination from these sources. Against that must be set the very low yield which results, as most of the starting material passes right through the system and is wasted.

Doping by chemical vapour transport

Doping, or the controlled incorporation of specific impurities to produce certain required physical properties (usually electronic), may be viewed as the reverse of purification. Normally we wish to produce a certain concentration of a particular foreign element in a growing crystal, sometimes within close limits, and accordingly we must arrange to have the corresponding partial pressure of the foreign species, or some volatile compound of it, over the growing crystal. That partial pressure will be a function of the activity of the dopant species in the source material or in a separate reservoir, and of the Stefan flow velocity in the system, and other transport properties.

It should be realized that when crystals are grown in sealed capsules, only a very low degree of control is achieved over their purity, rate of transport, and doping. The worst situation arises if a doped source is used. We have already seen that the concentration of foreign species in the growing crystal increases during the experiment as the foreign species accumulate at the source or in the vapour. Since we expect interaction between one foreign species and another to alter the solubility of each in the host crystal, we would normally expect the concentration of dopant in the growing crystal to depend on the concentration of unintentional impurities, so that the solubility of the dopant species becomes a function of time.

By using a separate source for the dopant species, either mixed with the source material for the crystal or in a separate reservoir at a different temperature, greater control over the vapour composition is achieved. Provided the required doping level is high compared with the concentration of unintentional impurities, so that the effect on solubility is swamped, such variations on the closed capsule method yield fair control.

The open flow method[191]. and to almost the same extent, the 'almost closed capsule' method, provide for continuous control over the vapour composition, as well as producing relatively pure material. In such systems, the dopant species may be included in the furnace atmosphere, and their concentrations varied at will. These systems provide the finest control over doping, purification, and rate of growth, which adequately accounts for their wide-spread use in the electronics industry.

Consider a crystal growing by chemical vapour transport in an atmosphere containing a dopant species. We have already mentioned that the surface energy of each face of the crystal may be altered by adsorption of species from the vapour, for example, the transporting agent or the dopant species. Thus we expect the inclusion of a dopant species in the vapour to alter the state of each surface of the crystal, and thus modify the rate of surface processes. Extreme cases are: poisoning of surface-active sites by trapping dopant atoms, and creation of surface-active sites by adsorption of dopant atoms. The dopant species may affect morphology and growth rate. There is feedback here, since alteration of the growth rate affects the stoichiometry of the crystal, which in turn affects the solubility of the dopant species.

Unfortunately there is little quantitative data available on solubilities of dopants in crystals as a function of stoichiometry, so that a detailed analysis of particular systems may not be possible. We therefore stress the qualitative interactions between transport, surface processes, and adsorption of dopants, so that the crystal grower may appreciate the nature of the problems which are often overome only by lengthy trial and error.

CHAPTER SIX
Experimental Methods of Crystal Growth from the Vapour Phase

We apologize to the reader at the outset for the contents of this chapter. In spite of the title, no detailed account of particular methods of growing crystals will be given. In this chapter we present very brief descriptions of a few typical experimental systems, with no more detail than is needed to discuss them in the light of the theoretical concepts developed in the earlier chapters. This approach may, we hope, lead the reader to exploit fully the desirable features of the various experimental configurations. The reader who wants the experimental details in full for particular practical systems must go to the original papers, or to the well-known books or articles by Schäfer,[2,192] Nitsche,[193] Laudise,[1] Tanenbaum[194] and Tietjen.[191]

6.1 The sealed capsule method

The earliest experiments performed to grow crystals from the vapour phase made use of the sealed capsule or closed tube method. The material to be transported was placed in a capsule of glass, silica, or other suitable material, along with a small quantity of a transporting agent. The capsule was sealed and placed in a temperature gradient in a furnace. It could then be left until a crystal had grown to the required size, which might take hours, days, or months, or in many cases, might not happen within the lifetime of the experimenter. Such experiments are not unknown today.

With the simple theory described in Chapters 4 and 5, we can see very easily why the rate at which crystals grow is quite unpredictable in such apparatus. Even in the absence of surface kinetic barriers, the rate at which material is transported is a sensitive function not only of the rate of change of equilibrium constant with temperature (i.e.

of ΔH, the enthalpy change for the transport reaction) but also of the individual partial pressures. For example, in the transport of a binary compound by dissociative sublimation, we saw (Section 4.3) that the rate of transport is a sensitive function of the ratio of the components of the compound in the vapour. This ratio may vary by orders of magnitude from one experiment to another, unless stringent precautions are taken (see next section). We saw in Chapter 4 that unless this ratio is within, say, a factor of ten of the stoichiometric value, the growth rate becomes extremely small.

Although we are now aware of the reasons why crystals may grow only very slowly in closed capsules, the experimental difficulties remain. In brief, we have to fix the composition of the vapour within fairly close limits, and make sure that the composition does not alter very much during the course of the experiment. It must be realized that the vapour phase often contains only a few milligrams of material, which includes virtually all the transporting agent. There are obvious experimental problems in metering milligram quantities of substances that are frequently volatile and corrosive (e.g. chlorine or hydrogen chloride). The vapour phase must also act as a sink for any excess of one component of the source material and for impurities originally present in the source material. It is not uncommon for a charge of 10–100gm to be loaded into the capsule, and we now see that it is necessary for the total of all impurities, stoichiometric excess of one component, and surface contamination and oxidation to be less than a few milligrams, i.e. 1 part in 10^4.

We must also consider vapour phase constituents arising from the container material. It is well known that silica, which is probably the most widely used capsule material, evolves water and hydrogen on heating, as well as reacting slightly but significantly with many materials to be transported. It thus contributes several extra components to the vapour. At best, these components merely act as inert gas and increase the transport resistance. More often, they dissolve in the growing crystal or adsorb on its surface, and can make a significant difference to its defect-sensitive properties, as well as its rate of growth. Various experimental techniques which go some way to eliminating these difficulties have been devised, and will be described in the next two sections.

In many early closed-capsule experiments, no attempt was made to define a crystallization zone. It was usual for several crystals to form more of less simultaneously, and these crystals would eventually coalesce into a polycrystalline mass, in which there might be a few crystals of useful size.

It is now more usual to define the zone in which the crystal starts to grow by nucleation control or by seeding. Nucleation control involves one of the techniques listed below, or a combination of them:

(a) After the first visible distribution of crystallites is formed, the temperature gradient is reversed and material transported back to the source until only one crystallite remains, when the temperature gradient is changed back again and growth continues. The temperature gradient is often changed by moving the capsule to a different part of the furnace, although of course if a two-zone furnace is used, the temperature can be reversed by altering the relative temperatures of the two zones. The obvious difficulty is that, unless one crystallite is substantially larger than the rest, by the time all but one have been removed, that one will be invisible. It is very hard to know when to start the growth again.

(b) An interesting variation of this method was devised by Scholtz,[195] who imposed a temperature cycle on the growing crystal, so that for some fraction of each cycle the crystal was etching rather than growing. In principle, any sub-microscopic nuclei which form near the growing crystal would have a higher free energy on account of their high surface to volume ratio, would thus be etching for a greater proportion of each cycle, and should be removed almost as soon as they appeared. In practice, the method was not very successful.

Another interesting variation of this method is suggested by a technique used by Bernal and Humphreys-Owen.[196] When growing crystals from solution, they used hot needles to exterminate unwanted crystallites, i.e. to make them dissolve again. Obviously a similar technique should be applicable to vapour growth, not with a hot needle but with a light beam, which could be used at high temperature. For such a technique to be applicable, one would have to be able to see into the furnace. The possibilities and problems here are discussed later on. Then it would be necessary to have a sufficiently intense and well-focused beam of light of a wavelength that would be absorbed by the crystal but not by the capsule material. The possibilities here range from mercury vapour lamps and carbon arcs to lasers, and the experimenter would have to choose the light source appropriate to his particular needs.

(c) A cold spot is established at some point on the capsule, so that the vapour is supersaturated locally and nucleation takes place at this point. Honigman[197] grew large crystals of HMTA at around 80°C in the apparatus illustrated in Fig. 6.1.

Nitsche[198] attached a tube to a capsule, and cooled the end of the

EXPERIMENTAL METHODS OF CRYSTAL GROWTH 229

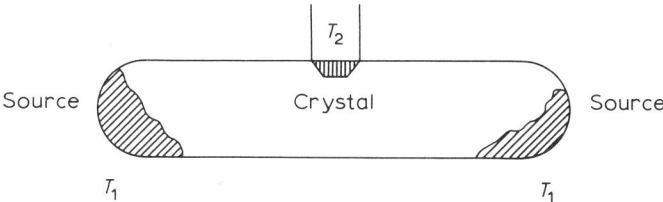

Fig. 6.1 Nucleation control in sublimation growth.

capsule by circulating cold air through the tube. In this way he grew large crystals of zinc sulphide at 950°C.

(d) If the capsule is moved through the furnace at the same rate as the crystal grows, the temperature distribution between the source material and the surface of the growing crystal remains unaltered. This method was used by Pizzarello[199] in 1954 to grow crystals of lead sulphide, by Rabenau et al.[200] for gallium arsenide, and by Kaldis[201] for zinc selenide. It is now a very commonly used technique.

The apparatus depicted in Figure 6.2. was used by Hanak and Berman[202] to grow Nb_3Sn.

It should be emphasized, in connection with nucleation control, that even if a single crystal is growing steadily, appreciable supersaturation can occur in the vapour if the critical conditions discussed in Chapter 5 are exceeded. The supersaturation reaches a maximum at some distance ahead of the surface of the growing crystal. The distance is a function of the distribution of temperature in the system and of the gas transport. If the supersaturation is sufficient to cause nucleation, extraneous crystallites will form ahead of the growing crystal.

If the experimenter wishes to remove extraneous crystallites by reversing the temperature gradient, he will need to be able to see into the furnace. A cylindrical furnace open at the ends permits one to observe the crystallites which are first formed on the end of the capsule by peering into the furnace from the end. Once the crystal has grown to fill the width of the capsule, however, this line of vision is blocked off. Further crystallites forming ahead of the growing crystal cannot be seen and so cannot be removed except by blind guesswork.

The furnaces devised by Reed et al.[203] and by Moss et al.,[204] permit continuous observation of the growing crystal throughout the experiment by incorporating a window running along the length of the furnace. It is a great advantage and very interesting to be able to see the crystal while it is growing. It is then possible to

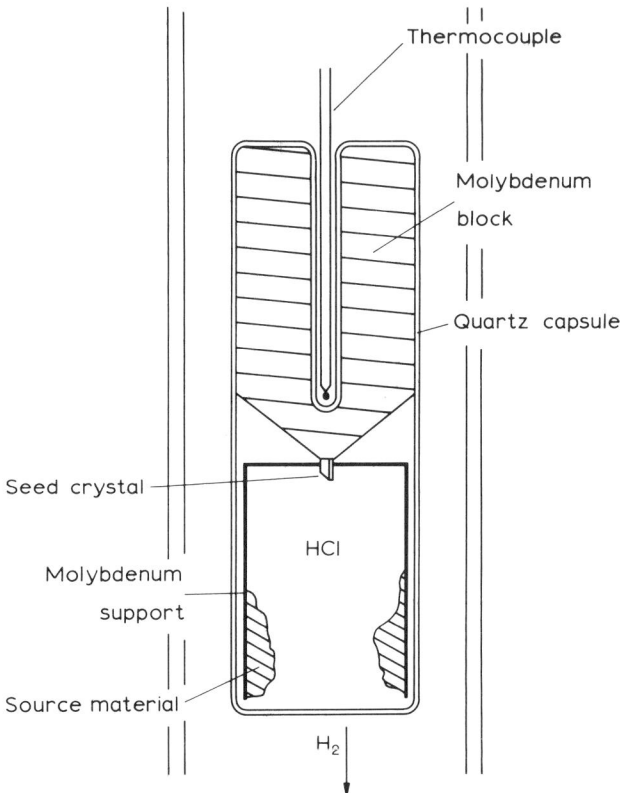

Fig. 6.2 Hanak and Berman's apparatus [202] for growing Nb_3Sn.

detect any new crystallites which form and also to monitor the rate of growth continuously by viewing the advancing front of the crystal through a cathetometer. One can then see if the rate of growth is uniform, oscillatory or irregular. The temperatures of the source material and of the crystal surface can be measured with an optical pyrometer, and the way in which the rate of growth varies with temperature difference may be explored.

Perhaps most important of all, one can verify that the crystal is still growing. It was not uncommon for experiments to be continued for weeks or months in furnaces without some means of seeing the crystal, only to find that very little material had been transported. The expenditure of time, effort and money on such abortive experiments is considerable.

For various reasons, the temperature field in a typical cylindrical

EXPERIMENTAL METHODS OF CRYSTAL GROWTH 231

furnace is not radially symmetrical. If the axis of the furnace is horizontal, then the top part is usually a few degrees hotter than the bottom. If the furnace is set with its axis vertical, this effect is removed as far as the radial distribution is concerned, but the heat field still depends on the configuration of the heating elements, lagging, etc.. If there is a window in the furnace for the growing crystal, some heat is inevitably lost through it, and this may cause a serious disturbance in the temperature distribution. The amount of heat lost through the window can be appreciably reduced by (a) using a double wall to the furnace, e.g. two concentric silica tubes, and evacuating the space between, to eliminate conduction and convection of heat, and (b) coating one surface of the window with a layer of gold, thin enough to be transparent (greenish) yet thick enough to reflect most infra-red radiation. Indium oxide[205] may also be used as an infra-red reflector, and is transparent to visible light. With these precautions, the disturbance to the temperature field is probably only a few degrees at $1000°C$, yet such an inhomogeneity in the temperature of the crystal can result in an inhomogeneity in growth rate of a factor of ten or a hundred in unfavourable cases.

It is common practice to smooth out the radial inhomogeneity of temperature by rotating the capsule at a rate of one revolution every few seconds. Steady and continous rotation is accomplished by using a small electric motor and a gear train to produce the required rate of rotation. The motor drives a rod from which the capsule hangs. Alternatively, the driven rod may enter the furnace from below, and the capsule be supported on the end of the rod.

The effect of changing the shape of the end of the capsule has been studied by Bulakh[206] and in the authors' laboratory, as in many others, no doubt. The evolution of capsule shape used in the authors' laboratory is shown in Fig. 6.3. A rounded end (Fig. 6.3a.) usually results in a distribution of crystallites over several centimetres. A tapered, pointed end defines the cold region much better. Although crystallites still form over much of the end section, it is usually possible to remove unwanted ones by reversing the temperature gradient until a single crystallite remains. However, it is found that when the temperature gradient is changed back, crystallites soon form again in exactly the same places, indicating that there are active sites on the capsule wall with low over-potentials for nucleation.

The next stage was to extend the end of the capsule by a narrow tube. Nucleation takes place inside the tube in several places, and after a time the crystallites grow together. By the time the growing mass emerged into the capsule, it often happened that one crystal occupied the full width of the narrow tube, and this crystal continued to grow into the capsule.

The final stage of the evolution was to provide a crystallite or seed, rather than wait for nucleation. A convenient way of doing so is to trap a crystallite above a constriction in the end of the capsule, and insert a stopper above it (Fig. 6.3d). The stoppered joint is conveniently made of a standard B7 or B10 cone joint, which is sufficiently gas-tight. A slight leak is, in any case, more to be desired than avoided (see Section 6.2.)

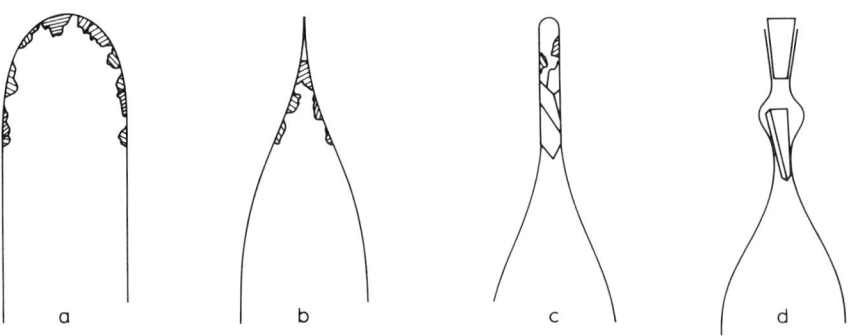

Fig. 6.3 Nucleation control in vapour growth.

(a) A rounded end to the capsule usually results in a wide distribution of nuclei forming simultaneously.

(b) A long, tapered end decreases the distance over which nuclei are formed.

(c) A long, narrow tube, joined to the top of the capsule, often results in a single crystal growing into the capsule.

(d) Seeding arrangement.

Bulakh[206] has concluded that a rounded end to the capsule gives the lowest possibility of developing unwanted nuclei, assuming that there is initially just one nucleus placed centrally in the end of the capsule.

Instead of defining the zone in which the initial nucleation is to take place, one can go one step further as we mentioned, and introduce a nucleus or *seed* at some place in the capsule. This seed removes the supersaturation in the vapour by growing, so that, with sufficient control over the sytem no further nuclei form. Kaldis[201] used this method for growing zinc selendide by transport with iodine. His capsule design is sketched below (Fig. 6.4).

Cadmium sulphide has been grown in the authors' laboratory, using a seed trapped between a constriction in the capsule and a removable stopper (Fig. 6.5.)

EXPERIMENTAL METHODS OF CRYSTAL GROWTH

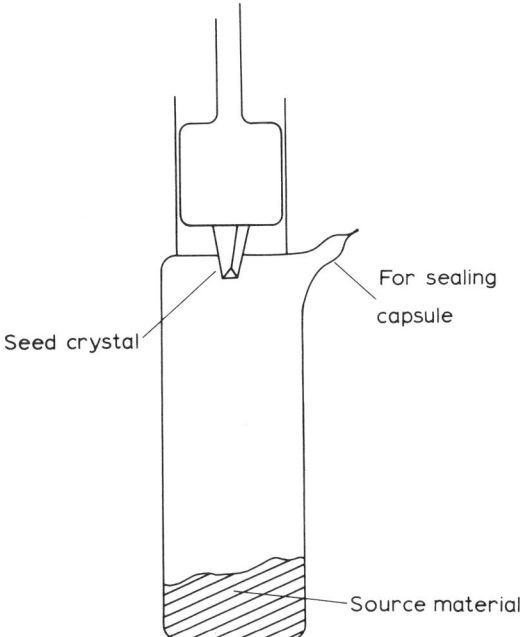

Fig. 6.4 Seeding in crystal growth (after Kaldis[201]).

As mentioned earlier, the temperature in a cylindrical furnace usually varies radially, being lowest at the axis and higher at the walls. Rather than struggle to overcome this effect, Scholtz[207] devised a crystal growth apparatus to exaggerate and make use of the effect. His apparatus is shown schematically in Fig. 6.6. The capsule is heated from the sides and from above, and cooled in the centre at the bottom to exaggerate the radial temperature gradient which drives the transport reaction. The capsule is rotated in the furnace to make the radial temperature gradient cylindrically symmetrical. Scholtz used this apparatus for growing ferric oxide.

A most sophisticated version of the sealed capsule method was used by Kaldis[208] to grow oxides and sulphides of the rare earths. These are notoriously recalcitrant materials for the crystal grower. In a herioc and successful attack on these substances, Kaldis used welded tungsten capsules and an operating temperature near 2000°C to produce crystals. His apparatus is shown in Fig. 6.7.

The seed is usually a small crystal, but there are advantages in using an extended area of seed, such as a plate. Because a small seed has only a small surface area compared with the cross-sectional area

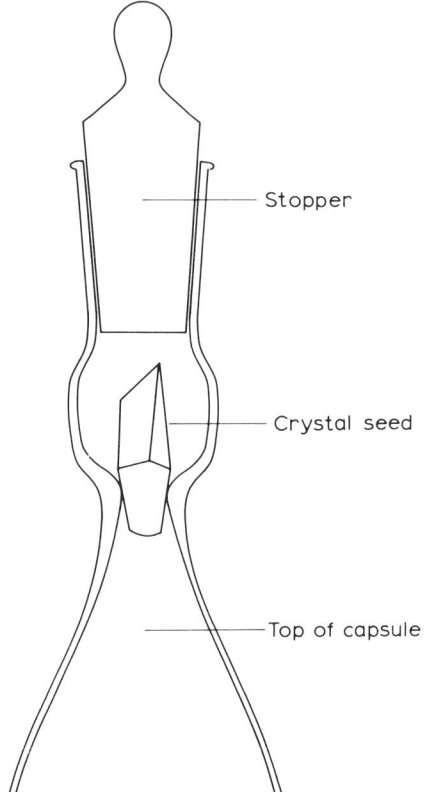

Fig. 6.5 Seeding in crystal growth. The method used in the authors' laboratory.

of a typical capsule, it must grow at a greater rate, per unit surface area to remove supersaturation in the vapour ahead of it than a plate occupying most of the width of the capsule would have to. If there is a surface kinetic barrier to growth, a high growth rate per unit area can only be achieved by imposing a large chemical overpotential at the growing surface, i.e. by causing the vapour to be supersaturated. It is clear why seeding does not always prevent extraneous crystallites developing.

A surface kinetic barrier to growth can develop during a growth experiment if contamination builds up on the surface. The overpotential required to drive the surface processes increases as the contamination builds up, and less potential is available for the gas transport process. The growth rate decreases, and the supersaturation of the vapour ahead of the crystal increases. There comes a point

EXPERIMENTAL METHODS OF CRYSTAL GROWTH 235

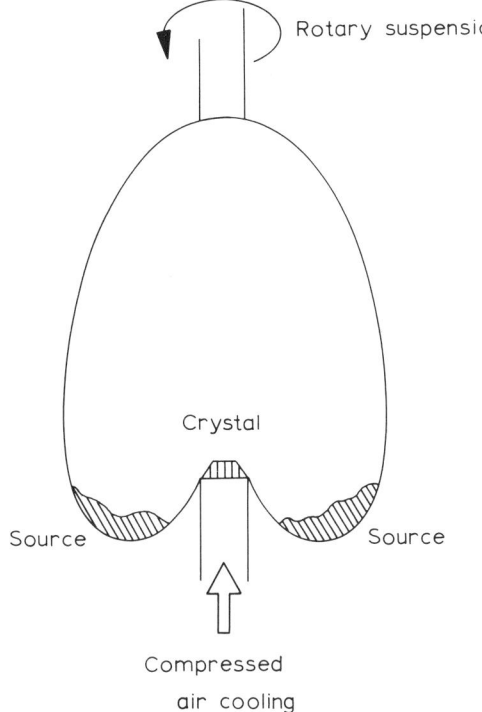

Fig. 6.6   Growth of ferric oxide [207].

when the supersaturation is sufficient to cause incoherent nucleation, either on the walls of the capsule or on the contaminated surface of the crystal, and more rapid growth starts again.

If the capsule remains in the same place in the furnace, the temperatures of the surface of the crystal and of the source are determined by their position in the furnace, and will normally change as the crystal grows and the source material is transported away. The amount of temperature change depends, of course, on the temperature gradient into which the crystal is growing. Consider the temperature profile depicted in Fig. 6.8., and imagine a crystal growth capsule placed in this temperature field in the position shown. At some early point of the experiment, the situation denoted by the upper picture of the capsule obtains, with a small crystal growing at temperature $T_1(0)$, and a lot of source material at temperature $T_1(l)$.

As the experiment progresses the crystal grows into a hotter region, while the source retreats into a colder region. The rate of growth decreases accordingly. After a very long time, when the

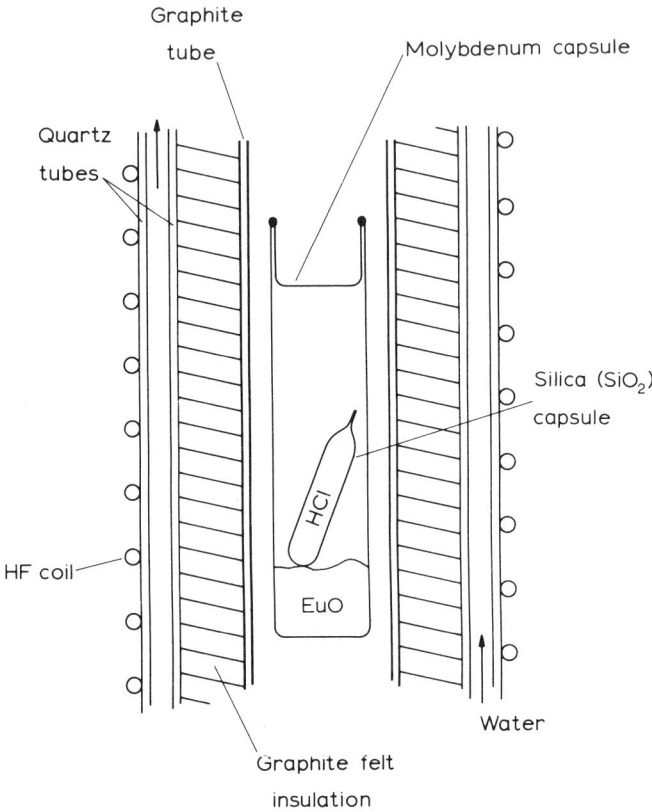

Fig. 6.7 High-temperature crystal growth system [208].

crystal and source surface temperatures have become very nearly equal, growth practically stops, as depicted by the lower drawing of the capsule, with $T_2(0) = T_2(l)$.

The temperature profile shown in Fig. 6.8 with a maximum between the crystal and the source, is typical of those used in practice to achieve a steep temperature gradient at the surface of the growing crystal, to avoid constitutional supercooling of the vapour (see Section 5.4). If the temperature profile increased from the crystal to the source and beyond, then both the crystal and source temperatures would increase as the growth experiment progressed. The temperature increase might affect the growth rate, or it might not. Of course, if both the crystal and the source were placed in regions with no temperature gradient, then their temperatures would remain unchanged. Such an arrangement is unusual and perhaps hard

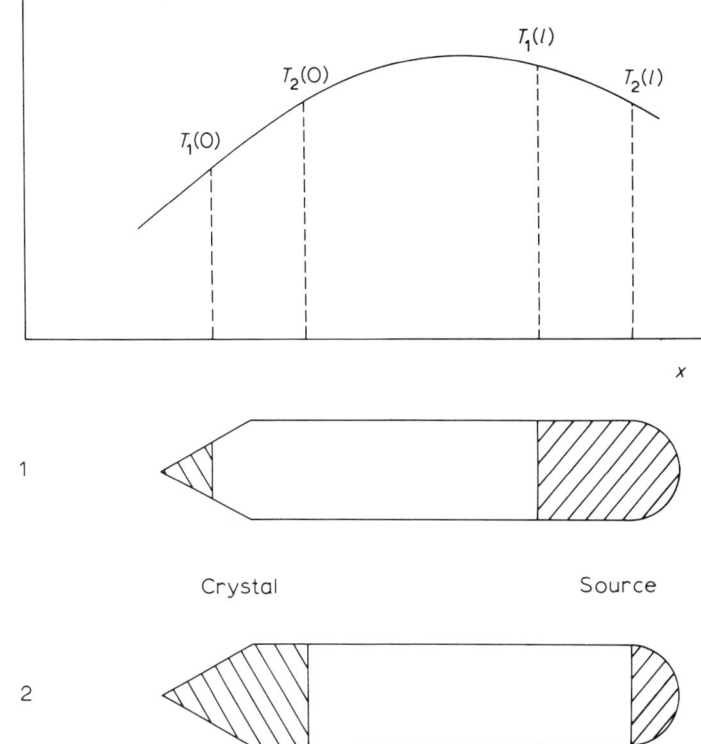

Fig. 6.8 Schematic temperature profile in a crystal growing furnace.

to achieve. It would certainly cause constitutional supercooling in the vapour.

To maintain control of the rate of growth and composition of the crystal it is necessary to keep both the crystal and the source at constant temperatures. This is achieved by 'pulling', i.e. moving the capsule through the furnace continuously, so that the crystal surface and the source are always in the same place in the furnace. It may seem that very fine control over the pulling rate is required, but that is only true if one is working under conditions where the growth rate is not sensitive to the temperature difference applied. If the growth rate can alter as the temperature difference alters, there is a tendency for the system to be self-correcting. If the pulling rate is initially faster than the growth rate, then a temperature profile such as that sketched in Fig. 6.8 ensures that the temperature difference increases until the growth rate matches the pulling rate.

We have included in this section a description of some important aspects of experimental technique for growing crystals in closed capsules. Many of these techniques, such as rotating the capsule, and pulling, are equally important when growing crystals by the 'almost closed capsule' methods discussed in the next section.

6.2 Almost sealed capsule methods

The initial production of gaseous components from surface contamination or oxidation of the source material, the continuous outgassing of impurities, and the lack of control over the vapour composition (and hence the composition of the growing solid) are serious shortcomings of the closed capsule method which have been overcome to some extent by various types of 'almost closed capsule'.

Piper and Polich[209] used a capsule that was initially open at the end where the crystal was to grow, so that gaseous contamination

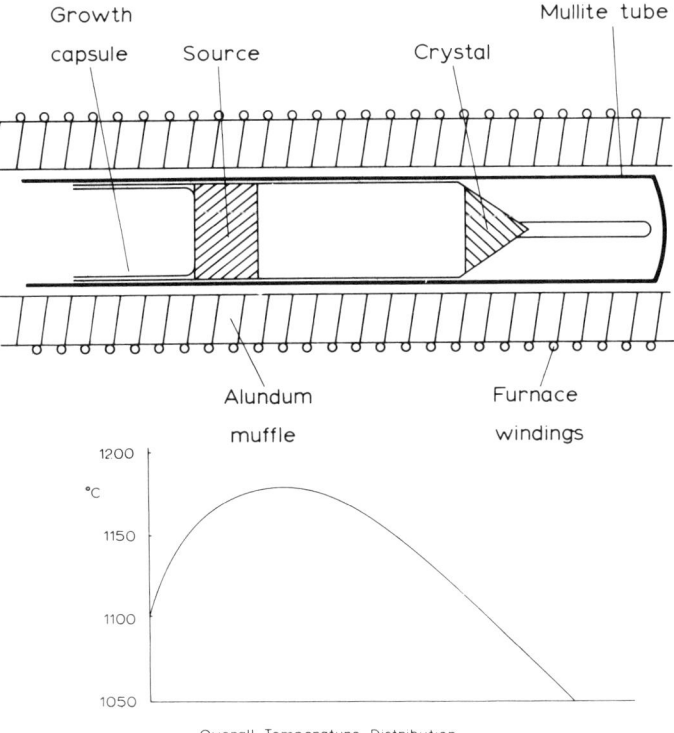

Fig. 6.9 Furnace cross section and temperature profile (drawn to the same scale) used by Piper and Polich[209] for growing cadmium sulphide.

produced during the early part of the experiment could escape. (Fig. 6.9) A cooled stopper was then inserted, or in some variants of the method, the capsule was sealed by material growing over the opening. This approach removes volatile impurities if they diffuse rapidly out of the source material and has been used to grow some of the largest crystals of the II–VI compounds in particular. It cannot be used for chemical vapour transport in its simple form although it would be possible to arrange to meter in some volatile transporting agent immediately before stoppering the capsule. Once the capsule is stoppered, impurities or excess of either component can build up over the growing crystal, reducing the rate of growth and causing a steady change in the composition of the growing solid. However, the Piper–Polich method is far more reliable and reproducible in general than the closed-capsule method.

To improve control over the composition of the vapour, and hence of the solid, Prior[210] fixed the activity of one component by connecting the growth capsule via a tube to a reservoir containing that component as a condensed phase. (Fig. 6.10) The temperature of the reservoir is controlled independently. The connecting tube is made fairly long and narrow, to reduce diffusion of other components into the reservoir. Woods[83] used this method to grow crystals of zinc selenide, and he found it necessary to heat his source material carefully in vacuum beforehand to remove volatile impurities. With these techniques he was able to grow large crystals, of well-controlled conductivity, quite rapidly.

Faktor and co-workers[141, 143, 146] have used a third variety of the almost closed capsule, the 'perforated capsule'. Their capsules were drilled with a tiny hole, 0.2 mm diameter, through the capsule wall near the source material. Such capsules were used to grow cadmium sulphide by dissociative sublimation, and in these experiments, a vacuum (10^{-5} Torr), was maintained outside the capsule. The leak hole serves two main purposes. It allows impurities, which would otherwise accumulate in the vapour inside the capsule, to leak away continuously. This aspect has been discussed at length in Section 5.5. Also, it allows a small amount of the source material to escape, and in doing so, fixes the ratio of the components in the vapour near the source. In the capsule, the pressure is around 10 Torr, so that flow through the leak hole is molecular flow. Since the components leak out in stoichiometric proportions if the material sublimes congruently, their partial pressures inside the capsule near the leak hole are in a ratio determined by their molecular weights. This ratio is about 2.7 for cadmium sulphide (i.e. $p_{Cd}/p_{S_2} \simeq 2.7$). By thus fixing the ratio of components, it is possible

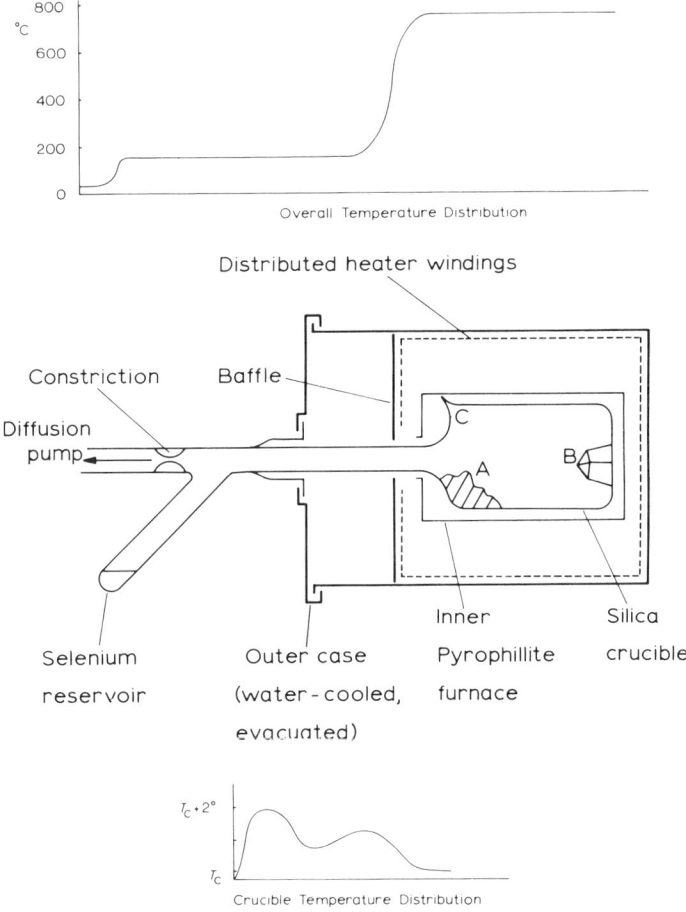

Fig. 6.10 Control of vapour composition in growing group IIb selenides[210]. A: source, B: crystal, C: sealed charging port.

to grow crystals of fairly uniform composition. The rate of loss of material is negligible, e.g. 10 mg h^{-1} when the capsule temperature is near 1000°C, which is less than 1% of the attainable transport rate.

The composition of the vapour inside the capsule can be regulated by substituting a controlled atmosphere for the vacuum surrounding the capsule. Thus extra of either component may be introduced either as the vapour of the pure component or as a volatile compound. The presence of a gas at some finite pressure outside the capsule reduces the rate of loss via the leak hole.

The same type of capsule may be used for chemical vapour transport. Instead of a vacuum surrounding the capsule, a flow of gas containing the transporting agent is used. Such experiments are conveniently carried out at a pressure of 1 atm, and flow through the leak hole is no longer molecular flow but a combination of diffusion and Stefan flow (see Chapter 4). The ratio of components inside the capsule is governed by the composition of the gas outside the capsule, and by the condition that the net fluxes of the different components leaving the capsule must be in the correct stoichiometric ratio. Such a capsule system has been used in the authors' laboratory for growing large crystals of gallium arsenide, and for depositing epitaxial layers of gallium arsenide on substrates. The various aspects of purification and doping in such a process have been discussed in Section 5.5.

The apparatus used for growing large crystals of gallium arsenide is illustrated schematically in Fig. 6.11. The capsule is in two parts with a ground glass joint between, so that it can be loaded and emptied

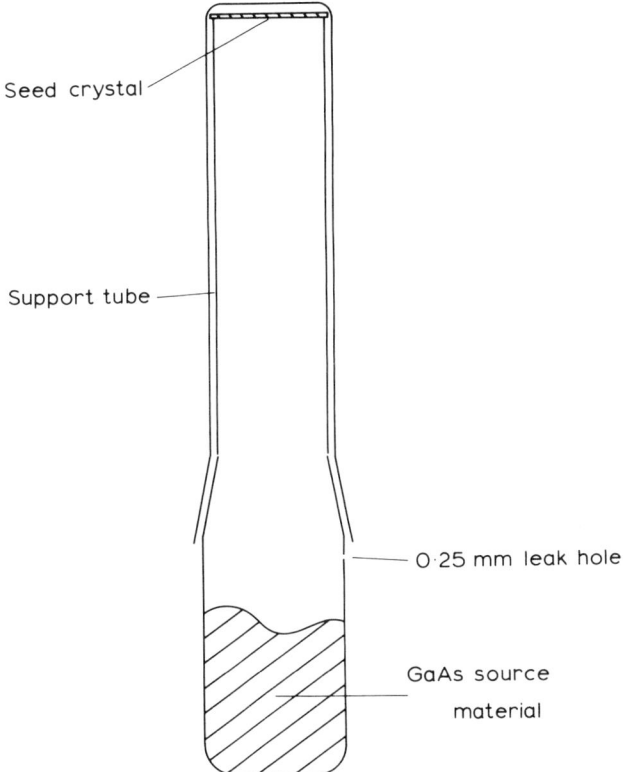

Fig. 6.11 Capsule for growing crystals of gallium arsenide, used in the authors' laboratory.

easily. The ground glass joint is gas tight, compared with the conductance of the leak hole. About 30 g of charge are loaded into the bottom, and a thin single crystal plate is supported at the top on a silica tube which rests on the ground glass joint. The capsule is supported in the furnace on the bottom end of a silica rod, which is rotated by a motor mounted above the furnace. A second motor drives a screw which moves the silica rod (and hence the capsule) at a steady rate through the furnace. The gas flowing through the furnace tube is hydrogen, saturated with arsenic trichloride by passing it through a bubbler. On entering the furnace, the arsenic trichloride is reduced by the hydrogen to arsenic vapour and hydrogen chloride.

After the capsule has been loaded with the charge material and seed plate and placed in the furnace, the furnace and capsule are evacuated and flushed out with pure hydrogen several times, first at room temperature and then while the furnace is heating to the operating temperature of 750–800°C. In this way, residual gases and surface contamination are largely removed. The final filling is with the hydrogen/transporting agent mixture. Repeated flushing and evacuating is a far more efficient way of getting rid of residual gas than evacuating to low pressure.

The vapour pressure of HCl in the capsule is determined by the temperature of the arsenic trichloride bubbler. In Section 4.4.5 we saw how the transport rate of gallium arsenide varies as the temperature of the bubbler is changed. For convenience, the temperature is controlled at around 20°C, which results in a growth rate of a little under one millimeter a day (10–15 mg h^{-1}). Faster transport could be obtained by heating the bubbler, but it would be necessary to heat all the pipework leading from the bubbler to the furnace as well, to stop arsenic trichloride condensing out.

This process is used to purify material for later use as a source in epitaxial deposition experiments. We have not attempted to optimize any parameters except the purity of the transported material, yet it is not uncommon for the experiment to yield a large single crystal.

The 'almost closed capsule' system[146] used in the authors' laboratory for growing epitaxial layers of gallium arsenide is shown in Fig. 6.12. In this apparatus, the source material is held in the top part of the capsule. The substrate, which is a single crystal wafer of gallium arsenide, rests on the optically flat end of a silica rod which enters the furnace from below. The capsule also sits on this rod, as illustrated in Fig. 6.12. The rod is rotated by a motor below the furnace and can also be moved vertically, though no arrangements are provided for continuous pulling in this apparatus.

The aim here is to grow epitaxial layers of thickness 1 μm or less,

EXPERIMENTAL METHODS OF CRYSTAL GROWTH

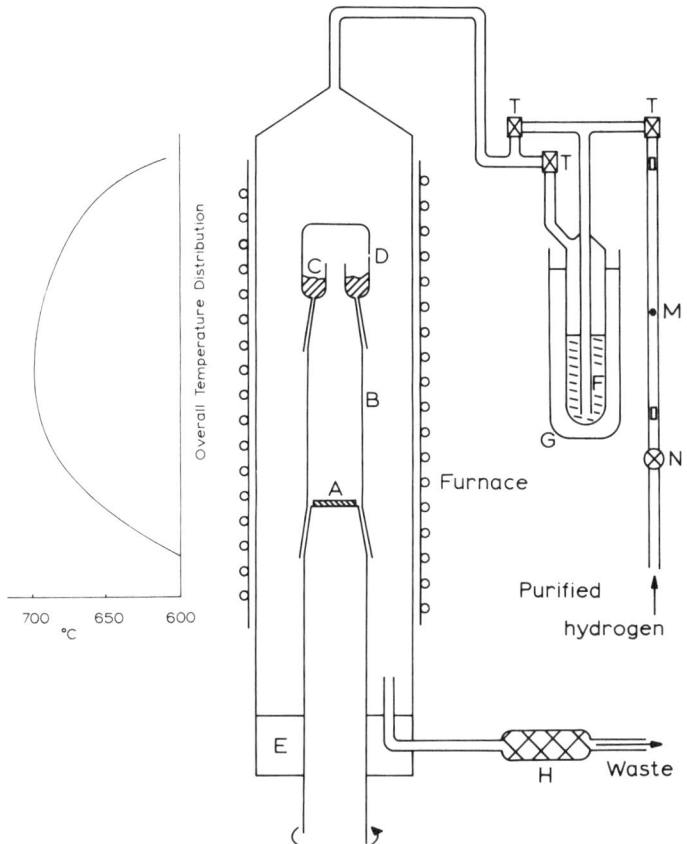

Fig. 6.12 Capsule apparatus for growing epitaxial layers of gallium arsenide, used by the authors.

- A: substrate
- B: capsule
- C: source material
- D: leak hole
- E: P.T.F.E. block
- F: Arsenic trichloride
- G: thermostatic bath
- H: soda-lime filter
- M: flow meter
- N: needle valve
- T: taps

and to this end we have designed our experimental conditions so that the rate of growth is controllable down to a fraction of a micron per hour. In this apparatus, the gas inlet system is basically the same as in the bulk transport apparatus, except that here we control the arsenic trichloride temperature at around 0°C. In addition, there is a separate gas inlet line for introducing dopants.

Epitaxial layers of gallium arsenide have been grown in this apparatus at substrate temperatures as low as 650°C, compared to the 750°C usual in the open flow method when growing gallium arsenide. The technique is not without its problem, however. Because of the radial symmetry of the apparatus, the grown layers frequently show radial variations in thickness, surface morphology, and electrical characteristics. What may be happening is that convection currents are set up in the capsule, and the convective velocity of the vapour predominates over the Stefan velocity and diffusion. Assuming that there is equilibrium between the vapour and the solid over the source and over the substrate, we can calculate the composition of the vapour and hence its density. Although the hot end of the capsule is uppermost, the vapour density is greatest at the top, because the density difference due to composition far outweighs that due to temperature. Thus gas transport results in a driving force for convection. In Section 4.7 we discussed one convective flow pattern that could possibly arise in our apparatus.

An interesting experimental technique that has some features in common with the 'almost closed capsule' methods is the 'close-spaced' transport method, described by Nicholl.[211] This is essentially a method for growing thin films on a substrate. The source material is in the form of a plate, and is positioned very close to the substrate, typically 0.1–0.4 mm. The vapour transport resistance is thus drastically reduced. The necessary temperature gradient is produced by heating the source and substrate independently with infra-red lamps. With the very close spacings used, the temperature difference between the two adjacent surfaces is difficult to ascertain, but is almost certainly less than that measured between the outer surfaces of the substrate and the source, because of heat transfer between the inner surfaces.

The substrate-source assembly is enclosed in a suitable chamber through which a gas containing the transporting agent is passed. Robinson[212] has grown epitaxial layers of gallium arsenide on germanium in this way, using residual water vapour as transporting agent in an atmosphere of hydrogen. He also reported transport using residual water vapour in a vacuum of 10^{-6} Torr.

This method has considerable advantages over the capsule method for growing epitaxial layers. The radial variation of transport rate and surface stability are to a large extent overcome. It is possible to alter the composition of the vapour between the source and the substrate more quickly than in the perforated capsule method, so that better control over doping profiles should be obtainable. In common with the 'almost closed capsule' methods, contaminants and impurities are allowed to escape, while the loss of material is kept low – typically a few percent of the material transported. A fairly uniform distribution of temperature over the substrate and over the source was obtained by Gottlieb and Corboy[213] by mounting the source and substrate on molybdenum plates. The apparatus has the advantage of simplicity and would seem to be an interesting possibility to consider for the crystal grower who is not pursuing the ultimate in flexibility and purity but who wants to produce epitaxial layers with a relatively small expenditure of capital.

6.3 Open flow systems

For much of the theoretical discussion in Chapters 4 and 5, whenever it was necessary to use a physical model of a crystal growth system, we chose a model more akin to a closed capsule system than to an open flow system. The reader might be excused for thinking, at first sight, that the discussion is not relevant to the open flow system. We indicate later in this section how the concepts of Chapters 4 and 5 have been carried over almost unchanged to describe crystal growth in an open system.

The important characteristic of the open flow method is that the reactant gases are passed through a reactor tube by the action of an external force (usually the pressure in a gas cylinder), and the crystal seed or substrate is supported in the flow of gas. The reactants which go to form the crystal may be produced in various ways to be described later. They are usually swept along in a carrier gas, which may be an inert gas, but which is very often one of the reactants (e.g. hydrogen), usually in considerable excess.

The reactor tube may be made of any material which does not melt, soften, or react appreciably with the species in the system at the operating temperature. At temperatures up to about 500°C, pyrex glass is suitable in many systems. At higher temperatures, silica is very commonly used, but alternatives are alumino-silicates, sintered oxides of aluminium or magnesium or both, graphite vitreous carbon, refractory carbides, or refractory metals.

The type of reactor which is probably most widely used for growing crystals by the open flow method consists of a silica reactor tube in a cylindrical open-ended furnace. However, at about 700°C and above silica is a source of contamination (largely water vapour) which may be unacceptable when material for sophisticated electronic devices is being prepared. Furthermore, silica is reduced by hydrogen:

$$SiO_2(s) + H_2(g) \rightleftharpoons SiO(g) + H_2O(g)$$

$$K_p = \frac{p_{SiO} \, p_{H_2O}}{p_{H_2}}$$

At a given temperature of operation the partial pressure of silicon monoxide is inversely proportional to the water/hydrogen ratio. The silica monoxide may be adsorbed on the surface of the growing crystal and may dissolve in it. The silicon and oxygen impurities may result in traps for current carriers, as unwanted donors or acceptors, or may act as scattering centres, thus lowering the mobility of the current carriers.

Silica, or silica in a reduced form, may also be transported onto the growing crystal by reaction with transporting agents such as chlorine or hydrogen chloride.

The popularity of silica is no doubt largely due to its excellent fabrication properties, coupled with the wide range of standard shapes and sections available. Here it compares favourably with the refractory metals and with high-temperature materials such as graphite and vitreous carbon.

The problems which arise with silica at high temperature have been overcome to some extent by using a 'cold-walled reactor'. Here the silica acts only as an envelope to contain the reactive gases and is usually water-cooled. The high temperature zone is produced by radio-frequency heating, using a graphite or vitreous carbon susceptor. In a chemical vapour deposition system using an exothermic reaction with an extreme equilibrium (e.g. decomposition of silane at 1200°C to deposit silicon)[174] the susceptor is merely a carbon support on which the substrate rests. The cold-walled reactor has been used for chemical vapour transport, using an endothermic reaction with an equilibrium constant near unity at the temperature of operation. In this system an inner reactor tube of vitreous carbon was used, to contain the reactive gases as much as possible. [214]

The open flow method is, of course, wasteful of material. Typically, only a few percent of the starting material is deposited in useful form as a single crystal or monocrystalline layer. The rest passes to waste, and is not usually recovered. However, the cost of the starting materials is often negligible in comparison with the value of the grown crystal or layer. An example discussed later in this section emphasizes this cost comparison. Ruby for lasers has been grown, using aluminium, chlorine, carbon dioxide and hydrogen as starting materials.[215]

Against this minor drawback, the open flow method has several advantages to offer. It is a dynamic system, so that impurities do not accumulate in the vapour. They may be introduced with the gas stream, of course, but this source of contamination can usually be brought under control. In the same way, it is not possible for a large excess of one component to build up. One exception to this

EXPERIMENTAL METHODS OF CRYSTAL GROWTH 247

statement can occur in systems such as the Effer/Knight[216, 217] process for depositing gallium arsenide. The source material is often liquid gallium which has been saturated with arsenic by passing arsenic vapour over it, until a crust of gallium arsenide forms over the gallium surface. The thickness of the crust varies over the gallium surface in a way which depends on the temperature distribution. If inadequate control is exercised, the crust may be very thin in places, so that the least shake or vibration cracks it, exposing liquid gallium. The activity of gallium rises abruptly to near unity, with a corresponding decrease in the arsenic activity. The ratio of gallium to arsenic species in the vapour rises greatly, and remains high until the crack is closed up with the formation of a new crust. The composition of the vapour may vary rapidly over a wide range. This problem has been investigated in detail by Shaw.[218] In spite of this shortcoming, liquid gallium sources are used when the greatest purity is required, because the gallium scrubs the gas stream of impurities, such as copper, which are very soluble in liquid gallium.

Being a dynamic system gives the open flow method great flexibility. The composition of the input gas stream may be varied at will during an experiment, to obtain crystals or layers of graded composition, conduction type or doping level. It is well adapted to the growth of alloys. Mixed III–V compounds such as Ga – As – P or In – Ga – P, or mixed II–VI compounds such as C – dS – Se have been grown in this way.[191] The composition of the crystal can be changed as rapidly as the necessary change in vapour composition can be brought about. If the dead volume of the apparatus is small, and if the crystal or epitaxial layer is grown relatively slowly, the abruptness of change from one composition of solid to another may be limited only by solid state diffusion at the operating temperature.

The open flow system has been used for producing crystals of anomalous shape, for example, the large, dislocation-free platelets of α-alumina grown by White.[219] While such anomalous structures are an interesting area of study as well as being attractive from the practical point of view, we will not concern ourselves with them, as the growth mechanisms which lead to the extreme shapes are rarely well established.

Let us consider a typical open-flow horizontal reactor, as illustrated schematically in Fig. 6.13. The gas stream containing the transporting agent enters the reactor from the left, and encounters the source which is at temperature T_1. The transporting agent reacts with the source material to form volatile products, which are carried in the gas stream towards the substrate.

Because of the dynamic nature of the system, equilibrium cannot

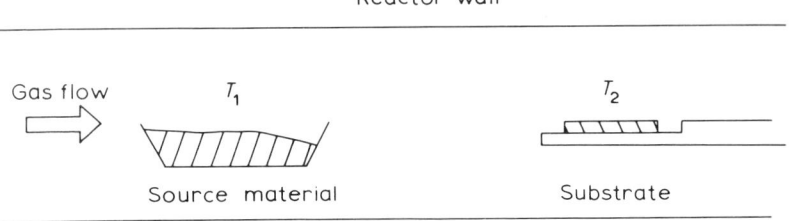

Fig. 6.13 Schematic open flow system for growing epitaxial layers.

be established between the source material and the gas stream, and in some experimental systems it is not approached very closely. Thus the composition of the gas stream yields a partial pressure product K_1^* (see Section 5.2) which is less than the equilibrium constant K_1 appropriate to the temperature T_1 of the source. The departure from equilibrium is inevitable (though it may be made small). The products of the reaction have to diffuse across the stream of flowing gas, since the reaction can only take place at the surface of the source.

The equilibrium of the vapour with the source can be assisted by increasing the total reaction rate (i.e. by having a large area of source material and by operating a high temperature) and by speeding up mixing processes in the gas stream by producing local turbulence. There are two advantages to be gained by approaching closely to equilibrium; the composition of the vapour may be calculated from thermochemical data, if they are available, and will be less sensitive to slight fluctuations in the gas flow pattern or velocity.

The gas stream passes down the furnace tube to the deposition zone, where there may or may not be a seed or substrate. Let us imagine that an epitaxial layer is to be grown, so that there is a substrate in the deposition zone, at temperature T_2. If the transport reaction is endothermic, we make $T_2 < T_1$. The gas stream reacts at the substrate to deposit material, and again equilibrium may be approached closely, or it may not. The partial pressure product K_2^* will be greater than the equilibrium constant K_2 at temperature T_2, though less than K_1^*.

Various workers have described theoretical models of the hydrodynamics and thermochemistry in the region over the substrate. Eversteyn and Severin[174] who deposited silicon by decomposing SiH_4 and $SiCl_4$ in a hydrogen atmosphere at about 1200°C, have described a successful model which accounts well for the performance of their apparatus. They performed flow-visualization experiments using smoke, and found that the gas flow was turbulent over

EXPERIMENTAL METHODS OF CRYSTAL GROWTH

their substrate, except in a boundary layer some 4 mm thick next to the substrate. The turbulence undoubtedly arises when the cold gas encounters the substrate holder at around 1200°C, and is caused by thermal convection currents.

The boundary layer model has also been applied to systems which do not have the sudden temperature changes associated with the silicon deposition process mentioned above. It has been assumed by some workers[220] that the mass transport boundary layer over the substrate is the same as the viscous boundary layer over a flat plate immersed in an infinite flowing medium. The theory for the viscous boundary layer in an infinite medium is well developed,[221] and shows that the thickness $\delta(x)$ of the layer at a point a distance x from the leading edge of the plate is given by:

$$\delta = 4.64 \sqrt{\frac{\nu x}{U_s}} \qquad (6.1)$$

where U_s is the stream velocity and ν is the kinematic viscosity (see Section 4.6 and Equation 4.96). This theory is perhaps more applicable to fast flows than to the relatively slow flows used in crystal growth systems. To begin with, the theory predicts a boundary layer of zero thickness at the leading edge of the plate, suggesting that the phenomena of viscous forces and diffusion are unable to make themselves felt in the upstream direction. This is obviously an approximation that becomes increasingly justifiable as the free stream velocity is increased. Equation 6.1 shows that the boundary layer builds up rapidly if the free stream velocity is small. If we insert the value of ν appropriate to hydrogen at 1000°C (about 8 cm^2 s^{-1} from Table 45), we find that for U_s = 1 cm s^{-1}, typical of flow rates used in practice, $\delta \cong 13\sqrt{x}$ cm The boundary layer becomes 1 cm thick a distance 6 x 10^{-3} cm along the plate. Since the separation of the substrate from the wall of the reactor tube is usually a centimetre or two, we see that the boundary layer over the substrate feels the effect of the finite bounds of the system almost immediately. We conclude that the gas flow is fully-developed laminar flow very nearly everywhere, (except when it becomes locally turbulent over small protrusions, etc.).

In a useful model for this situation,[222] the flow over the substrate has been pictured as illustrated in Fig. 6.14. After a very short distance along the substrate or substrate holder, the flow has become fully-developed laminar flow, with a lateral velocity variation which is roughly parabolic, though its precise form depends on the geometry of the apparatus. The line of maximum stream velocity is at some distance above the substrate, which will be roughly half the distance between the substrate and the wall.

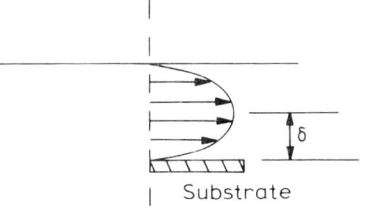

Fig. 6.14 Velocity distribution over the substrate in the open flow system[222].

Because the gas flows fastest along this streamline, its composition will most closely approximate that of the free stream in the region between the source and the substrate. The composition changes smoothly in a direction normal to the substrate, being such as to give a partial pressure product of K_2^* at the substrate surface, and approximating to K_1^* along the fastest streamline. The region around this fastest streamline (the 'core') acts as a virtual source. The problem is now reduced to describing mass transport from the core to the substrate.

As a simple approximation, it was assumed that the one-dimensional flow equations, developed in Chapter 4 could be applied to this system. The length of the flow path l was taken as being the shortest distance between the core and each point on the surface of the substrate. The length of the flow path is therefore least for points along the axis of the substrate, and greatest for points along its sides. Fig. 6.15 makes this clearer.

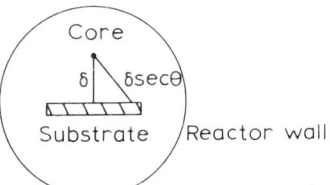

Fig. 6.15 Cross-section of open flow system: length of flow path from core to substrate[222].

In Chapter 4, we showed that the growth rate J is inversely proportional to the length of the flow path l. We therefore expect maximum rate of growth at the centre, falling off towards the edges.

EXPERIMENTAL METHODS OF CRYSTAL GROWTH

If z is the horizontal co-ordinate perpendicular to the axis of the reactor, we have:

$$\tan \theta = z/\delta$$

so that

$$l = \sqrt{(\delta^2 + z^2)}$$

or

$$J(z) = J(0)/(\delta^2 + z^2)^{\frac{1}{2}} \qquad (6.2)$$

The variation of growth rate with z is shown in Fig. 6.16. This distribution agrees quite well with experimental determinations of layer thickness, and with an exact numerical analysis of a silicon deposition system by Takahashi et al. [223] If the width of the substrate is 4δ, the average integrated rate of growth is about $0.88 \, J(0)$.

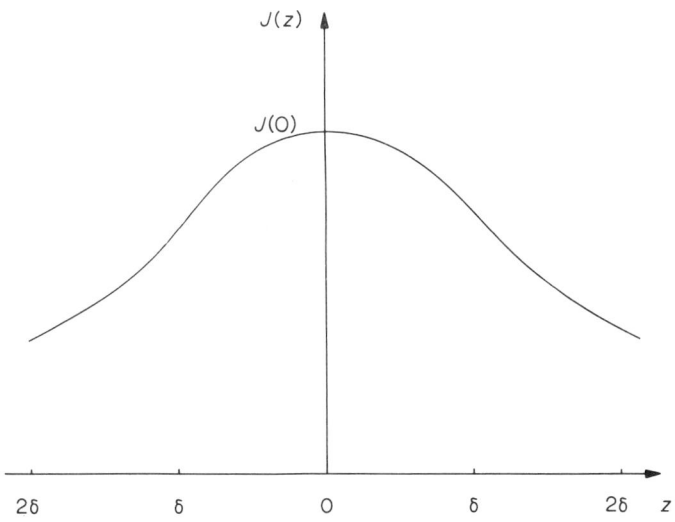

Fig. 6.16 Variation of growth rate J across substrate [222].

This model cannot account for the variation of growth rate with free stream velocity without some modification. In fact, it would appear that until the stream velocity has risen to a value where it is reasonable to consider a viscous boundary layer over the substrate of the type described by Equation 6.1, there should be no change in the rate of growth. This is in agreement with Shaw's observations.[156] Of course, if a solid or liquid source for any of the components is used, increasing the flow rate may cause the vapour composition to depart further from the equilibrium composition at the source temperature (i.e. make K_1^* smaller), so that the observed growth rate decreases.

The use of gaseous components as the source obviates this problem, and enables us to use the growth system for studying both the thermodynamics and the chemical kinetics of vapour growth. The equilibrium constant for the growth reaction may be found by adjusting the gas composition so as to give zero growth rate at a particular temperature, or by adjusting the temperature so as to give zero growth rate for a particular composition of vapour. It is not easy to ascertain that the growth rate is zero, so in practice one would vary the temperature, say, to obtain a range of positive and negative growth rates, plotting growth rate versus temperature, to find the temperature at which the plot crosses the axis corresponding to zero growth rate. The value of using both positive and negative points for this interpolation cannot be over-emphasized. Although the growth rate varies linearly with temperature near zero rate, the gas composition at the surface of the substrate is equal to the input composition only if the growth rate is zero. Attempting to extrapolate to zero rate from a series of positive rates may be unreliable.

Once the thermodynamics of the transport or growth reaction are well established, by this method or by any other (see, for example, our modified entrainment method[180,181] described briefly in Chapter 7), the experimenter is in a position to study the gas transport and chemical kinetics. Knowing the equilibrium constant K_2 at the temperature of the substrate, and having a model of vapour transport, the experimenter can calculate what vapour composition K_2^* is produced at the substrate surface by the observed rate of transport J, and hence determine the affinity $A, = RT \ln (K/K^*)$ for the chemical reaction at the surface. We take this line of thought a few stages further in Chapter 7.

We conclude this chapter with a short description of some of the different types of open flow crystal growth systems.

(1) Solid source plus carrier gas, with no transporting agent. Such a system has been used for growing crystals of many single component substances and for binary compounds such as the II–VI compounds where both components are fairly volatile. Here one may use the substance itself as the source material, or its components as condensed phases. In the latter case, if the temperatures of the components are independently controlled, the composition of the vapour may be varied at will to produce crystals of different

properties (Section 5.3). We illustrate this type of system with the apparatus used by Bulakh[224] for growing crystals of cadmium sulphide. He obtained a good yield of large platelets in the apparatus shown in Fig. 6.17, in which the flow of inert gas over each component element may be controlled independently.

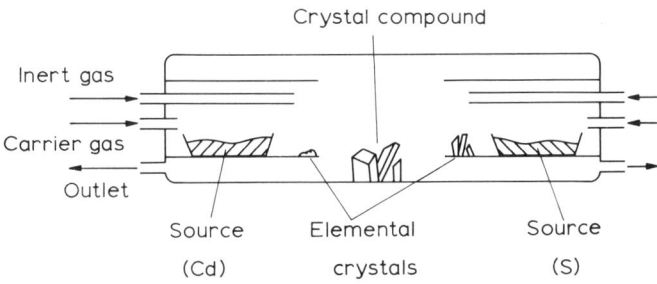

Fig. 6.17 Bulakh's apparatus for growing platelets of cadmium sulphide.

(2) Solid source, transporting agent, and carrier gas of some sort, which is often one gaseous component of the transport reaction, such as hydrogen, in considerable excess. This is perhaps the most widely used open flow system for chemical vapour transport and has been used for a very great number of compounds. We take the Knight – Effer – Minden process[191] for growing crystals or thin films of the III–V compounds as being a common example and quite typical. Fig. 6.18 illustrates one form of the apparatus. The source material may be the solid which is to be transported, or the liquid group III metal which has been saturated with the group V metalloid until a crust of the compound has formed on top.

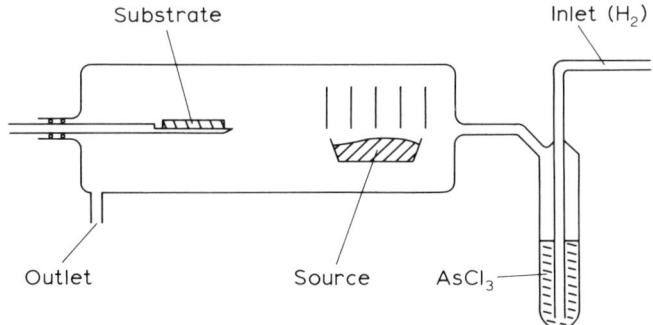

Fig. 6.18 Schematic chemical vapour transport system.

The transporting agent is most commonly hydrogen chloride, although other hydrogen halides have been used. One way of obtaining relatively pure HCl is shown in Fig. 6.18. The purified hydrogen stream is bubbled through the liquid trichloride of the group V element. Alternatively, gaseous HCl, mixed with hydrogen, may be fed in over the source.

(3) Condensed phase source for one component, gaseous sources and transporting agents otherwise. This is also a very widely used type of system. For example, Reisman[225] *et al.* have deposited thin films of ZnO on α-alumina and spinel substrates at 750–900°C, using the reaction:

$$Zn\ (gas) + H_2O\ (gas) \rightleftharpoons ZnO\ (solid) + H_2\ (gas)$$

The zinc vapour is obtained by passing helium over zinc at a controlled temperature, and the water vapour by bubbling helium through water at a controlled temperature. Hydrogen may be used in place of part or all of the helium to modulate the equilibrium. The zinc and water vapours are mixed at 1100°C, where the activity of ZnO is below unity. The mixture is passed over the substrate, at a temperature of 750–900°C, where the deposition reaction takes place. This experimental system gives the operator control over a wide range of input partial pressures, since the partial pressure of zinc, water, and hydrogen at the input are independently variable.

Schaeffer[215] has used a transport system in this category to grow ruby crystals for lasers, using the reaction:

$$2\ AlCl_3(g) + 3\ CO_2(g) + 3H_2(g) \rightleftharpoons Al_2O_3(solid)$$
$$+ 3\ CO(g) + 6\ HCl(g)$$

The aluminium trichloride was obtained by passing chlorine in argon over high purity aluminium. The gaseous reactants are passed over a substrate of Al_2O_3 at 1550–1800°C. Growth rates of up to 76.5 mg cm^{-2} h^{-1} were obtained, and single crystals of 80 g could be produced (Fig. 6.19).

Tietjen[191] has grown crystals of III–V compounds and their alloys using a very flexible system in which the group III metal is present as a liquid source and is transported as the sub-halide with hydrogen chloride, as transporting agent, while the group V metalloid is fed in as the gaseous hydride. His method has some advantages over the Effer – Minden process described under (2). His apparatus is shown schematically in Fig. 6.20.

Two source zones are shown, which would contain different group III metals at different temperatures if a mixed III–V compound were to be grown. The vapour pressures of the group III

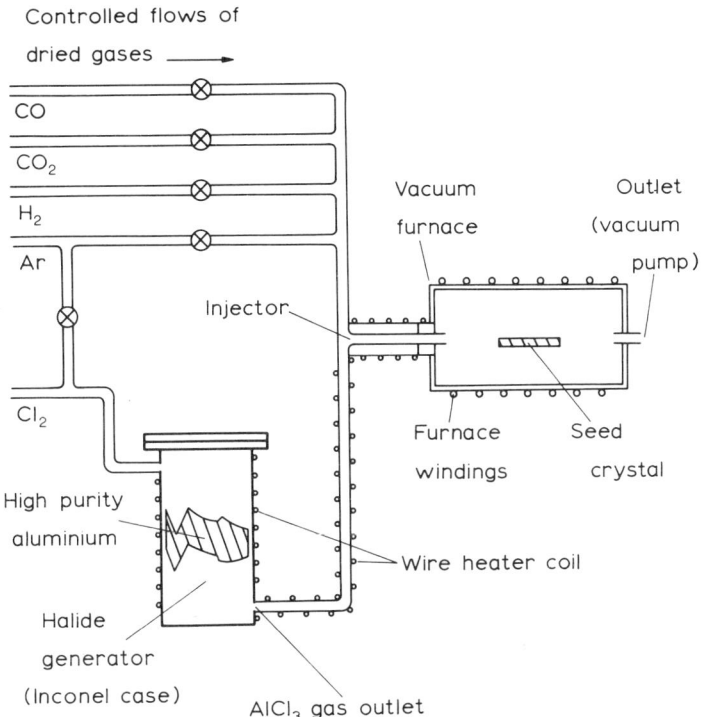

Fig. 6.19 Apparatus for growing ruby crystals (after Schaeffer[215]).

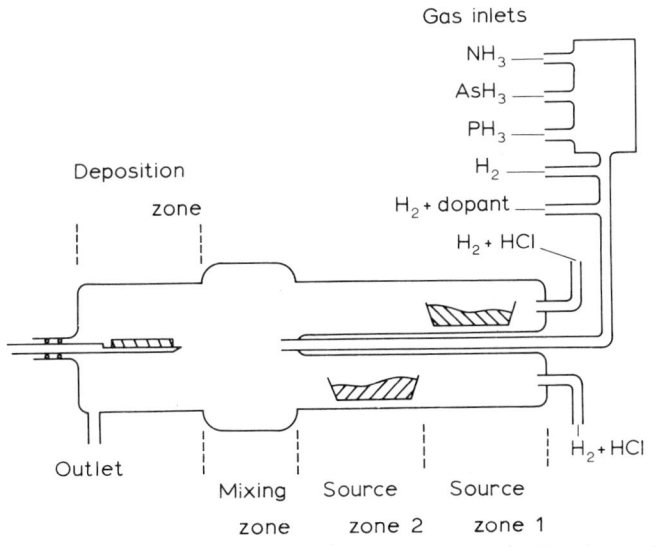

Fig. 6.20 Growth of mixed III-V compounds[191].

species and the group V species are under independent control, so that their ratio may be varied at will over a wide range, instead of being fixed as in the Effer – Minden process. The crystal grower thus has control over the stoichiometry of the growing crystal. The group V hydrides are unstable with respect to the component element gases at the experimental temperatures, but according to Ban,[155] AsH_3 and PH_3 are not completely decomposed, while NH_3 is hardly decomposed at all, during their passage through a hot silica tube under conditions similar to those used in the epitaxial deposition apparatus. The vapour thus arrives at the substrate with a certain amount of built-in overpotential. Furthermore, the group V element is present in the vapour species as one atom per molecule instead of two or four (as in N_2, P_2, As_2, As_4, etc.). Let us suppose that if one attempts to grow crystals of III–V compounds from a vapour in which the group V element is present as the dimer or tetramer, the process with the lowest conductance L (see Chapter 5) is the dissociation of the group V molecules on the crystal surface to form adatoms. The concentration of adatoms will be low, so that the exchange current J_0 for all subsequent process is restricted. If an unstable molecule such as NH_3 or AsH_3 is used as the source of the group V element, it appears that the crystal surface may act as a catalyst for decomposing these molecules, so that the concentration of adatoms of the group V element is increased considerably. The exchange current for all subsequent processes can now be much larger, although the adatoms may remain for only a short space of time on the surface, before forming dimers or tetramers and desorbing. This hypothesis seems to be a possible explanation for the success of Tietjen's method,[191] particularly for GaN, AlP, and AlN which are hardly amenable to vapour growth by other methods.

An interesting modification has been used by Shiloh and Gutman[239] to deposit gallium nitride. The gallium was transported as the monochloride, obtained by passing the trichloride over liquid gallium at 800°C. The nitrogen source was dried, purified nitrogen gas. The process was carried out at reduced pressure (53 Torr), and a discharge produced by passing the nitrogen through a microwave cavity. The reaction between excited nitrogen and gallium monochloride to form gallium nitride took place at around 600°C. Attempts to deposit gallium nitride using unexcited, molecular nitrogen have failed, and it is thought that the large bond energy of the nitrogen molecule presents an activation barrier to the chemical reaction which effectively prevents growth at moderate temperatures.

(4) Gaseous sources only. Elements such as silicon and germanium

have been deposited as thin films by pyrolysis of their tetrahydrides on a heated substrate. Typically, silane or germane mixed with hydrogen is passed through the silica reactor tube which is water-cooled. The substrates rest on a graphite susceptor which is heated by radio-frequency heating. On contact with the hot substrate, the silane or germane decomposes to deposit silicon or germanium, liberating hydrogen. The resulting films are of very high purity, since the source gases can be obtained free of any impurities that can be pyrolysed.

Compounds have been deposited as thin single-crystal films by using gaseous compounds as source materials. Manasevit et al.[226] have used organometallic compounds of the group II and III elements (e.g. trimethyl gallium) to grow II–VI and III–V compounds. The group VI elements may be used as elemental vapours or as hydrogen chalcogenides. The group V elements were obtained from the trihydrides. In the case of the AlN and GaN, the gas mixture undergoes pyrolysis in several stages at different temperatures, so the gases are led separately into the vicinity of the substrate before mixing. Substrate temperatures of 700–1200°C are used, depending on the compound to be deposited. The process showed some promise, and with further development might produce material suitable for electronic devices.

(5) Ternary and higher compounds. Chemical vapour transport in open flow systems has been used for growing crystalline films of magnetic garnets by Pulliam[150] et al. and by Taylor and

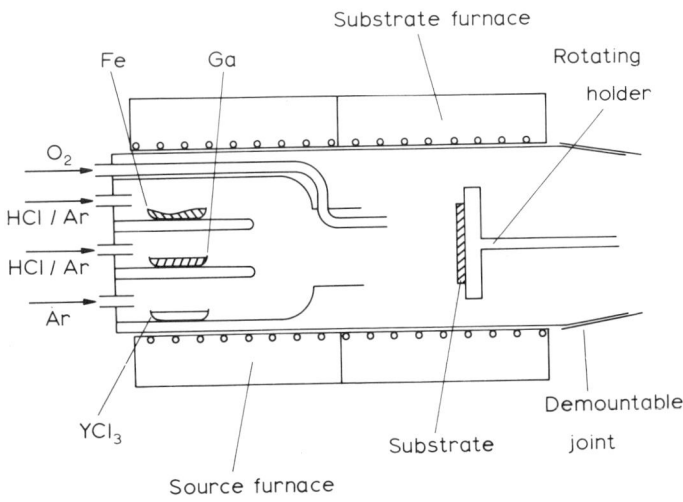

Fig. 6.21   Growth of rare-earth garnets[227].

Sadagopan.[227] Spinel-type ferrites have been grown by Takasu,[228] and by Pulliam *et al.*[150] In Fig. 6.21 we depict the apparatus used by Taylor and Sadagopan, for growing films of gallium-substituted yttrium iron garnet. Iron and gallium are transported as lower chlorides, using hydrogen chloride in argon as transporting agent. Yttrium is transported as the trichloride by evaporating the trichloride in an argon stream. Oxygen is fed in separately. A substrate of dysprosium-substituted gadolinium gallium garnet is used, held at 1200°C. It is necessary to have good control over the temperatures of the various sources and over the gas flows, as these determine the composition of the vapour over the substrate, and hence the phase which grows. Thus, besides the desired composition of Ga:YIG, deposits of polycrystalline $Gd_3(Ga, Fe)_5 O_{12}$, garnet/$Fe_2 O_3$ mixtures, and $Y_2 O_3$ have been produced by adjustment of the HCl and O_2 flow rates.

In principle, an analysis of such a system along the lines of the treatment given in Chapter 4 is only a little more difficult than for chemical vapour transport of a binary compound. It is unlikely that all the necessary thermochemical data will be found, but reasonable estimates can be made, and calculations can then be carried out, to find the partial pressures of the various components at the substrate in terms of the rate of growth of the compound. While such calculations are unlikely to give accurate values for these partial pressures, they provide a basis for investigating the interactions between the experimental variables. To carry out such an investigation empirically, with around a dozen variables, is a Herculean task, and a poor way to learn about the crystal growth system.

CHAPTER SEVEN
Areas for Further Study

7.1 Exploration of concepts

In this chapter, we propose to explore certain twilight areas of crystal growth from the vapour, and to present some ideas which, while plausible (or so we would hope), have not been directly tested. We are entering the realm of hypothesis, i.e. of reasonable speculation. We would ask the reader to bear this is mind as he reads on.

We view crystal growth from the vapour as two distinct processes:

(i) Vapour transport
(ii) Surface reactions on the growing crystal.

If a solid or liquid source of material is used, we also have:

(iii) Vaporization at the source.

Since much of what follows applies to the vaporization process, as well as to crystallization, we leave the third process implicit in our discussion. Of course, the vaporization process is less critical than the growth process in some ways, in that the restrictions on temperature and surface area are usually less rigid, and we do not have to worry about crystallinity. We leave the reader to apply the remarks in this chapter to vaporization as the need arises.

The second process, surface reactions on the growing crystal, includes many possible sub-processes, as mentioned in Section 5.1, forming a series of processes, with other possible processes in parallel (Fig. 5.1). It is to the interactions between the first and second of these processes that we direct our attention in this chapter.

Our interest is in growing crystals, which normally means either large single crystals or flat, thin films of single crystal on a substrate. Occasionally we may want to grow extreme forms such as platelets or whiskers. In general, however, we know nothing of the atomic or molecular processes of crystal growth. We may mention here two theories which have been developed. The first is the 'complete condensation' mechanism. Atoms from the vapour strike a very cold surface, and stick where they hit. There is no desorption and no migration on the surface. This is the simplest picture of growth from the vapour, under extreme conditions which are readily achieved in practice. There is little in the way of conceptual difficulty in this mechanism. Substances grown under 'complete condensation' conditions are usually not single crystals, and may be glasses, so that a study of this mechanism yields little insight into the atomic and molecular processes of crystal growth.

Secondly, we mention the 'atomic steps' mechanism described by Burton, Cabrera and Frank.[124] The crystal surface is imagined to contain a number of atomic steps, and the step edges contain kinks. Atoms or molecules are strongly adsorbed at a kink, less strongly adsorbed on a flat region of the surface (a 'terrace'). Imposing a supersaturation in the vapour causes an increase in the number of adatoms on the surface, and hence an increase in the number arriving at edges and kinks. Since there is no corresponding increase in the rate at which atoms leave the edges and kinks, the steps spread across the crystal surface and the crystal grows. We may ask: 'where do the steps originate?'. They are not thermodynamically stable surface defects except when the supersaturation is great — great enough for two-dimensional nucleation to be facile. Of course, a polished substrate or a seed crystal which has been cleaved from a larger crystal probably has a stepped surface, but these original steps grow out when a few atomic layers have been added to the crystal.

The 'screw dislocation mechanism', proposed by Frank,[26] has been described in Chapter 3. A single dislocation with a screw component, emerging at the surface, results in the growth of a spiral step, while a pair of such dislocations with screw components of opposite signs generates a series of steps in the form of closed loops. These phenomena have been observed by interference contrast microscopy and by electron microscopy using the gold decoration technique. Crystals of sodium chloride grown from aqueous solution,[28] and of garnet[229] and silicon carbide,[230] among others, show these effects. On the other hand, a vast number of substances grown from the vapour by chemical vapour transport do not show evidence of growing by the surface step mechanism, whether the steps are produced by screw dislocations or otherwise. Yet these substances grow at useful rates under quite negligible overpotentials. Clearly there are atomic mechanisms, other than two-dimensional nucleation and the surface step mechanism, by which crystals can grow with very small supersaturations. We do not have any picture of such a mechanism with detailed evidence in support of it. We would suggest tentatively that the general idea of surface roughness, as a controlling factor in crystal growth, developed by Jackson[92] for growth from the melt, may be extended to include growth by chemical vapour transport. Surface adatoms should be stabilized by the presence of an adsorbed transporting agent.

A substance M being transported by X_2 as the volatile species MX will have a certain coverage of adatoms M and molecules X_2 and MX. If the transporting agent is well chosen, the partial pressure of MX may be near its saturated vapour pressure, so that the molar free energy difference between MX vapour and MX as an adsorbed layer is small. The total coverage of M as M adatoms and MX molecules will be very much greater than for substance M in equilibrium with its vapour alone. Thus one possible role for the transporting agent is to convert an atomically smooth surface (i.e. very few adatoms) to an atomically rough surface. On the rough surface, growth can take place by the continuous process of adding more adsorbed M, without the need for two-dimensional nucleation. On the smooth surface, two-dimensional nucleation is required unless there are atomic steps already existing (e.g. because of dislocations with a screw component intersecting the surface).

This is an aspect of crystal growing that has received little attention in recent years. Yet if much progress is going to be made towards bringing crystal growth under control, without expending great effort and much money on each new compound, it will be necessary to gain insight into the atomic mechanisms on surfaces which produce growth of crystals. To this end, we need to have some knowledge of the following:

(1) What vapour species are present under the conditions of growth?
(2) What is the equilibrium composition of the vapour over the crystal?
(3) How far does the composition depart from equilibrium during growth, and what is the relation between growth rate and supersaturation?
(4) How do we alter the surface kinetics by altering the individual partial pressures, or by changing the vapour species (e.g. changing the transporting agent)?

(5) What species are present on the crystal surface? What are their binding energies, their mobilities on the surface, their concentrations?

(6) What is the state of the surface, i.e. its defect state, stoichiometry, dislocation density, orientation?

This list is not exhaustive. There is probably not a single crystal growth system for which all this information has been obtained.

Items (1) and (2) are essential information which we must obtain before we can begin to talk sensibly about the surface kinetics. As regards item (3), it must be remembered that if the rate of growth is observable, transport in the vapour will change the vapour composition between the source or vapour inlet and the growing crystal, and this change will depend on the growth rate. Item (4) is the natural extension of items (1)–(3). The elucidation of items (5) and (6) is a difficult exercise in surface chemistry, and some aspects are probably a little beyond the reach of the present generation of techniques for surface analysis. We will not have anything more to say about them here.

In Chapter 4 we saw how the vapour transport may be described by a simple model, and how we may obtain either of two kinds of information:

(1) If we know the partial pressures at two points in the system, we are able to calculate the rate of transport. Using thermodynamic data for the transport reaction, we can calculate the equilibrium composition of the vapour over the source and over the crystal, and hence calculate a maximum rate of transport, i.e. the rate that would be achieved in the absence of surface resistances to growth. Such calculations are useful in designing a crystal growth system. They may profitably be carried out using estimated or inferred thermochemical data, as they provide information as to which variables must be brought under close control and which can be allowed more freedom. We discuss the question of maximizing the growth rate later. If accurate thermodynamic data are available, these calculations may be sufficiently reliable to provide evidence for appreciable surface resistance to growth when the observed growth rate falls significantly below the calculated one.

(2) If the partial pressures are known or measured at one point in the system, and the rate of transport is measured, we can calculate the actual composition of the vapour adjacent to the growing crystal. Methods of measuring partial pressures will be discussed later. Knowing the partial pressures immediately over the growing crystal, we can study the influences of gas composition and growth rate on

AREAS FOR FURTHER STUDY

such crystal properties as stoichiometry and electrical conductivity. Furthermore, if we have to hand some accurate thermodynamic data, we are in a position to say how far the vapour composition departs from equilibrium with the solid.

In addition to accurate thermodynamic data, we would also require knowledge of diffusion coefficients for all the species in the vapour. This is a tall order, since it often happens that the vapour species are not stable gases, but only exist because of the presence of other gases. For example, we cannot obtain a binary mixture of, say, hydrogen and a metal halide, since some reaction to form hydrogen halide is inevitable if the metal halide is to be the transporting species in a system containing hydrogen. There are various ways around this problem. One is to use an inert gas as the majority species and to measure binary diffusion coefficients for the reactive gases in the inert gas. Relatively unstable gases, such as lower oxides and halides of the metallic elements may be generated by reaction between the pure metals and their higher oxides or halides in solid or liquid form. Alternatively, we may avoid the need for knowledge of the diffusion coefficients by measuring the partial pressures at several points along the transport path (in principle, two points are sufficient) and fitting the observed partial pressures to exponential functions of the type given by Equation 4.47, for example. This curve can then be extrapolated to find the partial pressures at the crystal surface. Coupled with accurate thermodynamic data, this information enables us to study the kinetics of the surface processes. A third alternative is to work in such a region of temperature and vapour composition that the transport rate is small, the growth process being limited almost entirely by the rate of the surface processes. The variation in vapour composition is then quite small, so that uncertainties in diffusion coefficients are less important. Note that it is not possible to eliminate the vapour transport process completely, as nothing can be learned about a process which procedes at a zero net rate (zero entropy production).

Methods of measuring partial pressures fall into two groups: those which measure some property of the individual components (e.g. mass spectrometry and absorbtion spectroscopy) and those which measure a property of the vapour mixture, such as density, from which the individual component contributions must be inferred.

The mass spectrometric method seems at first sight the most useful. It is sensitive to a few parts per million or better, and is a well-established analytical technique. There are several problems which may arise when we try to use a mass spectrometer for measuring partial pressures in a crystal growth system. The central problem is that often we are dealing with reactive gases at several hundred degrees centigrade and at a pressure of around one atmosphere. We are faced with the problem of extracting a sample so as to cause no significant perturbation to the system which is being sampled. We would ideally like not to perturb the sample, but if the pressure of the vapour is above about 10^{-6} Torr in the crystal growth system, it must be reduced to around this pressure before entering the ion source of the mass spectrometer. In some cases, this pressure reduction may be possible without any other perturbation. For example, a monatomic vapour issuing from a Knudsen cell with a sharp-edged orifice may be taken as a directly representative sample of the vapour in the cell. If the vapour is a mixture, the composition leaving the cell is perturbed, in that the conductance of the orifice is inversely proportional to the square root of the atomic or molecular weight. Such mass discrimination is well understood and can be allowed for.

Any more complicated vapour (i.e. a vapour of molecular species) may introduce unknown perturbations. As the pressure is reduced, there is a possibility that the molecular species will dissociate, at an unknown rate.

The greatest problem in sampling for mass spectroscopy is probably that posed by a vapour at a pressure which is too high for practical work with a Knudsen orifice, i.e. at about an atmosphere pressure or above. To drop the pressure to the region of 10^{-6} Torr, it is necessary to resort to fine capillaries or porous plugs. Here we are introducing an extended surface area with a pressure gradient over it. The possibilities for adsorption, surface reaction and migration along the surface are innumerable and all poorly characterized.

Having directed our sample into the mass spectrometer, we need to know the relationship between a given partial pressure in the crystal growth system and the corresponding signal on the mass spectrometer. Calibration for gases which are relatively unstable products of reactions at high temperature is not straightforward. Then there are problems with contamination of the mass spectrometer resulting in spurious readings. In summary, the mass spectrometric method is a powerful technique that requires a deal of taming. It has been successfully used to study high temperature equilibria at low pressures using Knudsen cells. Even here, considerable thought must be given to ways of investigating the dissociation of molecular species on leaving the cell.

One area in which the mass spectrometer has great power is in establishing what species are present in a crystal growth system. For those molecular species which do not dissociate on ionization, identification on the basis of mass number is usually straightforward, particularly if one or more of the elements in the molecule has more than one stable isotope so that a characteristic isotopic pattern of peaks is produced. The presence of species which dissociate on ionization may often be inferred from the shape of the ionization efficiency curve of the fragment peaks. This subject is discussed critically by Grimley[231]. One is always faced with the problem of polymeric species which may dissociate completely in the sampling system, and hence will not be detected.

Absorption spectroscopy[232] and Raman spectroscopy[233] are two further techniques that may be used for measuring the composition of a gas at high temperature. Here again it is by no means a straightforward analysis. There are problems in calibrating the spectrometer for species which are stable only under the conditions which prevail in the crystal growth system. The advantage of these techniques is that it is not necessary to bleed off a small sample of vapour and measurements can be made on a system which is virtually unperturbed. Raman spectroscopy permits investigation of a very small region of the vapour, since the Raman emission is usually observed at right angles to the primary beam. Under favourable conditions, Raman spectroscopy permits measurement of the local temperature as well, since the interval between the Stokes and the anti-Stokes frequencies is temperature dependent. It should also be possible to obtain some information about unknown vapour species with these techniques[233].

For investigating the composition of the vapour along a crystal growth system, there would seem to be much in favour of Raman or absorption spectroscopy, although the problems here are by no means slight. If we wish to investigate the vapour composition in an open flow type of system the problems become more acute, in that most of the change in vapour composition occurs in a relatively small distance over the crystal (and over the source if solid or liquid sources are used). The difficulties in sampling for mass spectrometry from such a system cannot be lightly discarded.

Moving from methods of analysis which give direct indications of the partial pressures of the individual species at a more or less well defined point in a crystal growth system, we now turn to indirect methods of analysis which depend on some average property of the vapour as a whole. The usual property to work on is vapour density, which may be measured by various methods, such as those described below.

(1) Buoyancy. A thin-walled evacuated bulb is suspended from a sensitive balance in the vapour under investigation, and the apparent weight of the bulb in the vapour is compared with its weight in a standard atmosphere (e.g. dry air). For a given size of bulb, the sensitivity of this method depends on the minimum weight difference that can be measured with the balance. At room temperature, a bulb of 10 cm^3 volume changes its apparent weight by 10 μg if the mean molecular weight of the vapour is changed by 4×10^{-3}. The method can be made very sensitive, by using a large bulb, but the result is, of course, an average of the vapour density over the volume of the bulb. At high temperatures, and with reactive vapours, there are problems in choosing a suitable material for the bulb.

(2) Velocity of sound. The velocity of sound in a gas is inversely proportional to the square root of the vapour density, and can be measured accurately by measuring the phase change introduced by passing the sound through the vapour under study, compared with a reference signal. By such a method it should be possible to measure sound velocity to one part in 10^4 or one part in 10^5, and hence only twice that error in the vapour density.

(3) Torsion effusion[234]. A capsule or cell is suspended with its axis horizontal on a fine fibre. The capsule has a hole near one end, so that the vapour effuses from the capsule in a horizontal jet at right angles to the axis of the capsule. The momentum of the jet applies a torque to the capsule, depending on the mean vapour density inside the capsule. The vapour density can be found by measuring the amount of twist of the fibre suspension. Usually two holes are used, at opposite ends of the cell. This method has been fully described in the literature. Flow in the jet must be molecular flow, so that the method is only applicable at pressures below a fraction of an atmosphere.

A method for studying the thermodynamics and reaction kinetics of transport reactions has recently been developed in the authors' laboratory[180, 181]. It is a modification of the well-known entrainment or transpiration method. The important modification is that the region of the vapour in which the composition changes is confined to a narrow channel of accurately known geometry. The effects of vapour transport can then be allowed for. The method is illustrated schematically in Fig. 7.1.

The cell containing the solid or liquid phase is suspended in a furnace from a microbalance, and a stream of transporting agent and carrier gas passed through the furnace. The method may also be used to measure vapour pressures and sublimation pressures, in which case an inert carrier gas would be passed through the furnace. In the

AREAS FOR FURTHER STUDY

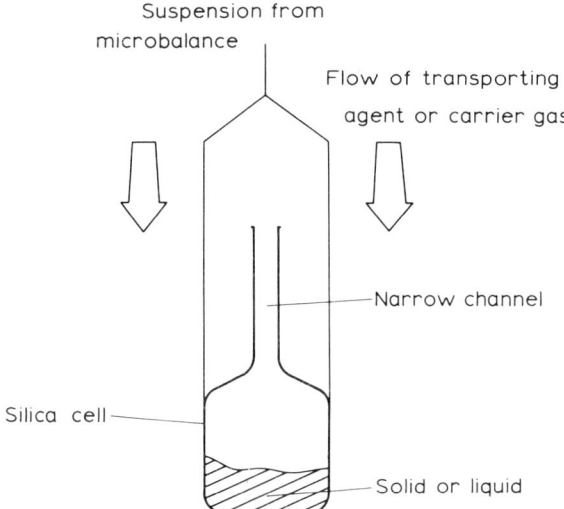

Fig. 7.1 Capsule used in the modified entrainment method [180, 181].

general case of a transporting reaction, the transporting agent enters the cell via the channel, and the vapour products leave by the same route. The vapour transport is well described by the formalism of Chapter 4, so that the measured rate of loss of weight can be related to the partial pressures at the ends of the channel. By careful design, one can ensure that at the open end of the channel the reaction products are quickly diluted and swept away by the gas stream, so that their partial pressures are effectively zero. It is then possible to calculate the partial pressures inside the cell. If these pressures are insensitive to the dimensions of the channel, which will always be so if the channel is small enough, it is reasonable to assume that the vapour in the cell is in equilibrium with the solid or liquid phase. Thus it is possible to study the equilibrium. By making the conductance of the channel to gas transport progressively larger until the calculated partial pressures within the cell show some dependence on the channel dimensions, it is possible to study the reaction kinetics. The two situations may be compared with Knudsen and Langmuir experiments. The composition and pressure of the reactant gas stream may be varied at will, so that a wide range of conditions is open to investigation. While it is obviously not possible to identify the vapour species by this method, a careful experiment may suggest the participation of hitherto unsuspected species. Their existence can then be checked using some other technique.

Establishing the thermodynamics and investigating reaction mechanisms is likely to be particularly rewarding in the case of the 'ungrowables', such as boron, graphite, diamond, zinc oxide, cadmium sulphide, boron nitride, aluminium nitride and others. While these substances may be grown from the vapour by one means or another, usually extreme conditions of temperature, pressure, or supersaturation are used.

In Chapter 5 we derived the equation 5.10:

$$J = J_r \left\{ \frac{K}{K^*} - 1 \right\}$$

where J_r is the reverse or exchange current, also written J_0, K is the equilibrium constant, and K^* is the partial pressure product. For the growth rate to be sizable either J_0 or $(K/K^*) - 1$ must be large. It is often possible to produce a large overpotential $(K/K^*) - 1$ and hence obtain a finite growth rate when J_0, the reverse current, is small. This procedure brings its attendant problems, since the large overpotential may support rapid nucleation of new crystallites, or stabilize new phases or a glassy structure. It is often possible to increase J_0 by increasing the temperature at which the crystal is grown, since J_0 contains an exponential temperature factor. Again, such a procedure may introduce problems. It may be necessary to use different container materials at the higher temperature, or it may be that the desired solid phase is not stable above a certain temperature. In depositing crystalline layers on a substrate, it is often necessary to achieve a certain dopant profile in going from the substrate to the deposited layer, and raising the temperature may make solid state diffusion of the dopant species so rapid that the required profile cannot be maintained. Yet considerable success has been achieved in crystal growth at high temperatures, e.g. the silicon epitaxy technology (1200°C)[174], vapour transport of ruby crystals (1800°C)[215] and rare-earth oxides (2000°C)[208].

Besides temperature, there are other variables which enter into J_0, and in some cases it may be possible to obtain a sizable growth rate when the dependence of the reaction kinetics on the individual species is understood. On the other hand, once the reaction kinetics have been elucidated, it may be possible to find a way of changing the reactants so as to increase J_0 very greatly, by reducing the activation energy for the surface reaction. In some cases this has been achieved on intuitive grounds. Thus the reaction to deposit gallium nitride:

$$GaCl\ (g) + \tfrac{1}{2}N_2\ (g) + \tfrac{1}{2}H_2\ (g) \rightleftharpoons GaN(s) + HCl\ (g)$$

AREAS FOR FURTHER STUDY

is hindered by the exceptionally high dissociation energy of the nitrogen molecule. Replacing the nitrogen by ammonia has lead to a facile reaction, by introducing a reaction path which is less hindered.

It is appropriate at this stage to investigate the 'exchange current' J_0 in more detail. For each stage of the chemical vapour transport process, there are at any instant some molecules going one way through the process and some going the other way. The difference between these two 'fluxes' is the net flux which we have called J and which is the same at each stage. The 'exchange flux' or exchange current J_0 is the smaller flux, i.e. the reverse flux, while the forward flux is $J_0 + J$. The exchange flux may be very different for different stages of the process.

Let us see what part we may imagine the exchange current to play in crystal growth. Our aim is to grow a single, perfect crystal. Thus every atom has to end up in the right place. We cannot simply aim every atom at the right site and freeze it in place. We have to rely on there being an appreciable energy difference between a right site and a wrong site, so that wrongly placed atoms stand a greater chance of moving on than do rightly placed atoms. To increase the chance of a wrongly placed atom moving on, we must ensure that every atom explores a great many sites before becoming permanently integrated in the crystal. In other words, we require a high exchange current for some process or sequence of processes which brings about a rearrangement of the atoms on the surface. This process might be surface diffusion, or a sequence of process such as: a surface atom combines with the transporting agent, enters the vapour phase, re-adsorbs, and dissociates from the transporting agent. Thus we can have rearrangement of surface atoms while confined to the surface, or via the vapour phase.

The general form for the exchange current is, from Equation 5.4b:

$$J_0 = k_r \prod_i [A_i]^{\beta_i}$$

where $[A_i]$ is the concentration or partial pressure of species A_i, β_i is usually a small integer or quotient of small integers, and k_r is a kinetic coefficient which may be written:

$$k_r = k_0 \exp(-Q_r/RT)$$

Here Q_r is the activation energy for the reverse reaction, and k_0 depends on such quantities as the number of active surface sites, the vibration frequencies of adatoms, and on quantities which determine the bombardment rate from the vapour.

We can modify J_0 by several routes. We can change the $[A_i]$'s by changing the composition of the vapour. We can use temperature changes to modulate the exponential factor in k_r, and to some extent k_0. We can modulate k_0 by using such factors as surface orientation, surface contamination, defect and impurity concentrations. We can also change the activation energy Q_r by using a different surface orientation, or by changing the $[A_i]$'s, or by using a different transporting agent. Maximizing J_0 is, in short, an exercise in chemical kinetics. The root of the kinetic barrier may be in the structure of the surface, or in the structure of one or more vapour species, or some combination. For example, if the transport reaction produces volatile products at the crystal surface which are large complex molecules, e.g. $(Al\ Br_3)_2$ or As_4O_6 [235], these molecules may have a low probability of forming, while less complex molecules are strongly adsorbed and involatile.

The concept of exchange current is familiar to electrochemists [236], and we now pursue the parallel a step further.

Consider a typical vapour transport reaction at the crystal:

$$MX(g) + \tfrac{1}{2}H_2(g) \rightleftharpoons HX(g) + M(s)$$

$$K_p = \frac{p_{HX}\, a_M}{p_M \times p_{H_2}^{1/2}}$$

where M is the solid to be transported by transporting agent HX, and suppose that the deposition process goes mainly by the following steps:

(1) Adsorbtion of MX:

$$MX(g) \rightleftharpoons MX\cdot \quad K_1 = [MX\cdot]/p_{MX}$$

We use the notation $MX\cdot$ to denote a molecule adsorbed on the crystal surface.

(2) Adsorption of hydrogen:

$$\tfrac{1}{2}H_2(g) \rightleftharpoons H\cdot \quad K_2 = [H\cdot]/p_{H_2}^{1/2}$$

(3) Reaction between adsorbed MX and adsorbed hydrogen:

$$MX\cdot + H\cdot \rightleftharpoons M\cdot + HX\cdot \quad K_3 = [M\cdot][HX\cdot]/[MX\cdot][H\cdot]$$

(4) Desorption of the reaction product HX:

$$HX\cdot \rightleftharpoons HX(g) \quad K_4 = p_{HX}/[HX\cdot]$$

(5) Integration of M adatoms in the crystal:

$$M\cdot \rightleftharpoons M(s) \quad K_5 = a_M/[M\cdot]$$

Note that $K_p = K_1 K_2 K_3 K_4 K_5$

AREAS FOR FURTHER STUDY

We will consider now the extreme cases that arise when one of these steps is rate limiting, i.e. it requires far more overpotential than the other steps to be driven at a given rate. We assume that the reverse flux for the hindered step is small compared with the net rate, so that J is nearly equal to the forward flux for that step. We can now find out how the observed flux J depends on the partial pressures of reactants and products in each of the five possible cases. Thus if step (1) is rate-limiting,

$$J \simeq k_{f1}\, p_{MX} \quad \text{i.e.} \quad J \propto p_{MX}$$

where k_{f1} is the kinetic coefficient for the forward process of step (1).

If step (2) is rate-limiting:

$$J \simeq k_{f2}\, p_{H_2}^{1/2} \quad \text{i.e.} \quad J \propto \sqrt{p_{H_2}}$$

If step (3) is rate-limiting:

$$J \simeq k_{f3}\, [MX\cdot][H\cdot]$$

If only one step is significantly hindered, the other processes will achieve an equilibrium state to a good approximation, so that we may write:

$$J \simeq k_{f3} K_1 K_2\, p_{MX}\, p_{H_2}^{1/2}$$

Here J is bivariant in p_{MX} and $\sqrt{p_{H_2}}$.

If step (4) is rate-limiting:

$$J \simeq k_{f4}\, [HX\cdot]$$

$$= k_{f4}\, \frac{K_1 K_2 K_3 K_5}{a_M}\, p_{MX}\, p_{H_2}^{1/2}$$

$$= k_{f4}\, \frac{K_p}{K_4}\, \frac{p_{MX}\, p_{H_2}^{1/2}}{a_M}$$

where again we have assumed that all steps except (4) have achieved a near-equilibrium state. Again, J is bivariant in p_{MX} and $\sqrt{p_{H_2}}$.

If step (5) is rate-limiting:

$$J \simeq k_{f5}\, [M\cdot]$$

$$= k_{f5}\, \frac{K_p}{K_5}\, \frac{p_{MX}\, p_{H_2}^{1/2}}{p_{HX}}$$

In this case J depends on all three partial pressures. Thus we see that we may obtain some insight into the possible reaction mechanism by finding out how the net growth rate depends on the partial pressures over the crystal. The results are not always unambiguous, and there may be more than one series of steps which would explain the observed behaviour.

If we plot $\log J$ against the logarithm of the partial pressure of one of the reactants or products, we obtain a graph whose gradient at each point tells us the effective order of the surface reaction in that reactant or product. In general this may be the difference of comparable forward and reverse currents, but in a situation where the surface reaction is hindered almost entirely by a single slow step, we expect to obtain a gradient which is an integer or a quotient of small integers, reflecting the atomic steps involved in the reaction. For example, if step (3) were the hindered step in our previous example, a plot of $\log J$ versus $\log p_{MX}$ would have a gradient of unity over some range of p_{MX}, while a plot of $\log J$ versus $\log p_{H_2}$ would have a gradient of one half. If the partial pressure p_{MX} or p_{H_2} is reduced towards the equilibrium value, the net flux approaches zero, so that $\log J \to -\infty$. If the partial pressure is reduced yet further, the net flux becomes negative (i.e. the crystal is etched away by the vapour).

Consider Fig. 7.2, in which $\log J$ is plotted as the bold line as a

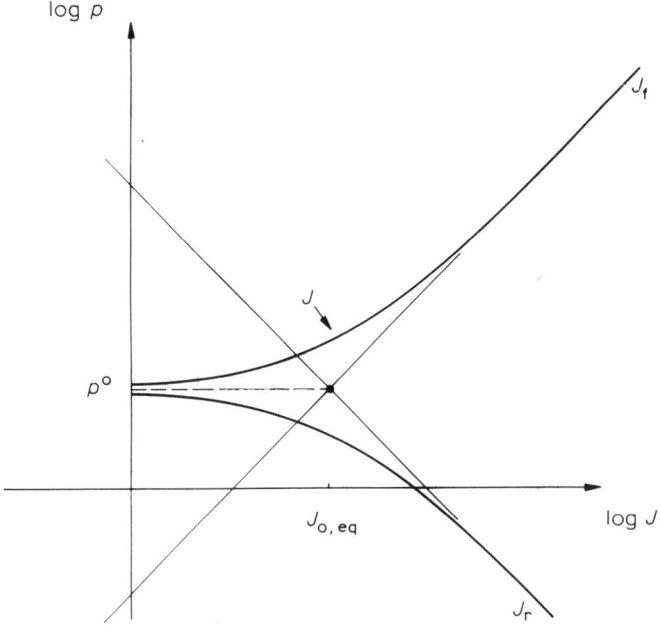

Fig. 7.2 Forward current J_f, reverse current J_r, and exchange current $J_{o,eq}$ in a chemical transport reaction.

AREAS FOR FURTHER STUDY

function of the logarithm of one partial pressure. At the equilibrium pressure p°, $\log J \to -\infty$. At very high pressures, J approaches the forward reaction rate J_f, while at very low pressures, J (which is now the etching rate) approaches the reverse reaction rate J_r. At equilibrium, J is zero, and the forward and reverse reaction rates are both equal to the equilibrium exchange current $J_{0,eq}$. This sort of diagram is equivalent to the Tafel plot of electrochemistry[236]. Note that $RT \log p$ is equivalent to a chemical potential, so that we are, in effect, plotting potential versus the logarithm of current. In general, J will depend on several partial pressures in a chemical vapour transport system, so that Fig. 7.2 is a section of a many-dimensional diagram.

Let us digress for a moment from chemical transport reactions and think what Fig. 7.2 would look like for a straightforward growth of a single-component solid from its vapour. The rate at which vapour atoms arrive at the surface is proportional to the vapour pressure, so that the forward current is plotted as a straight line of unit slope (Fig. 7.3). The evaporation rate is independent of pressure, to a good

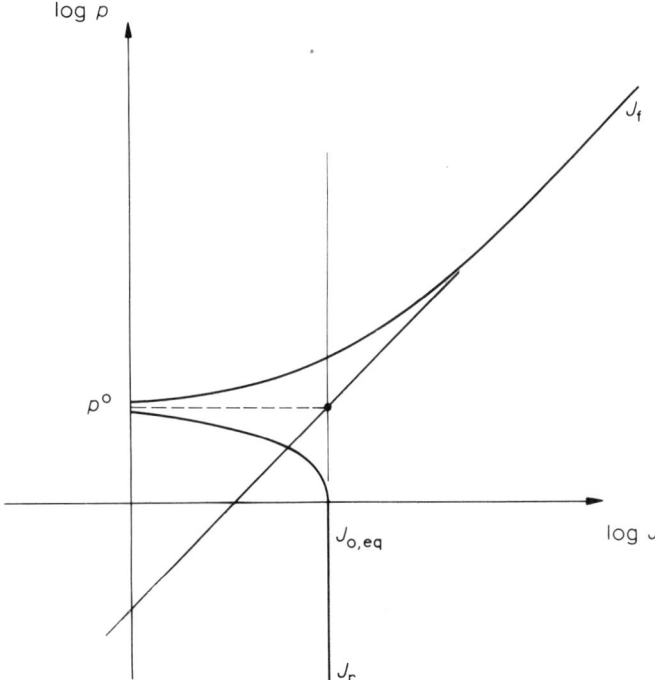

Fig. 7.3 Condensation current J_f, sublimation current J_r, and exchange current $J_{0,eq}$ in sublimation-condensation.

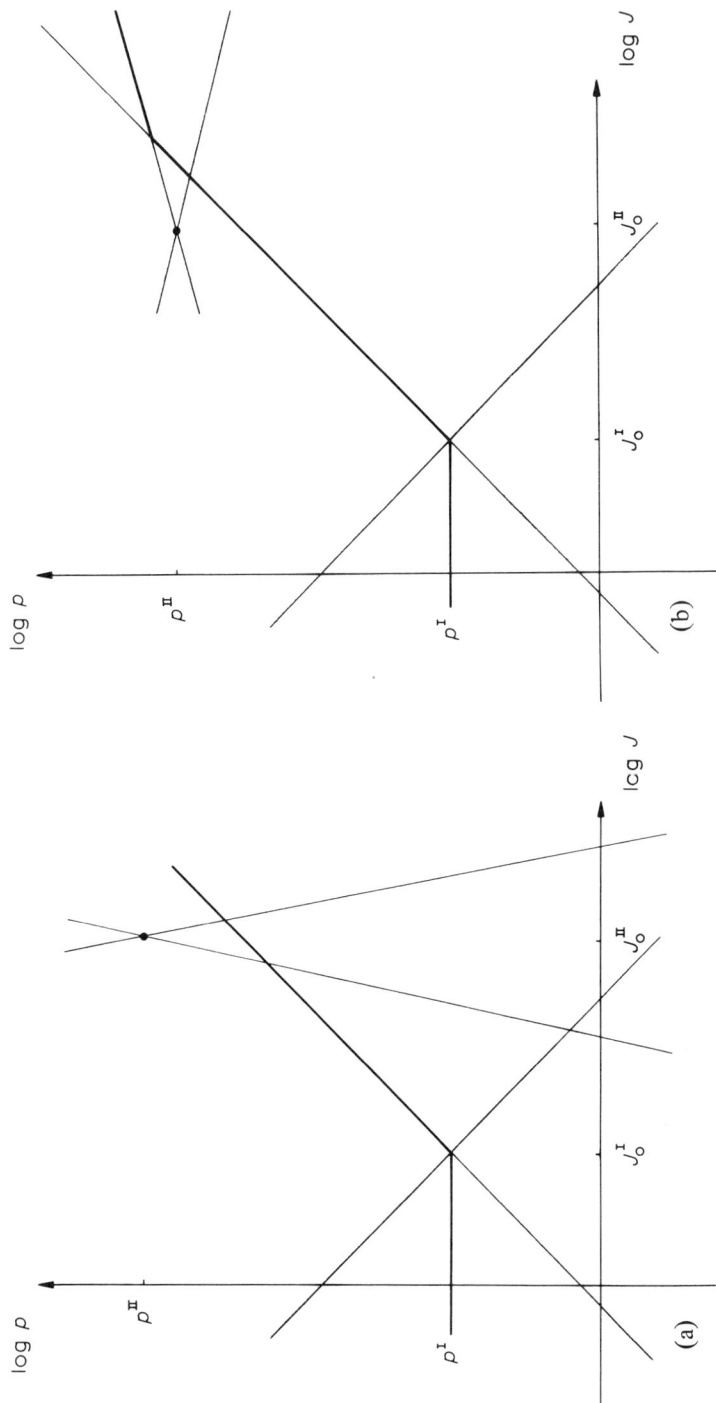

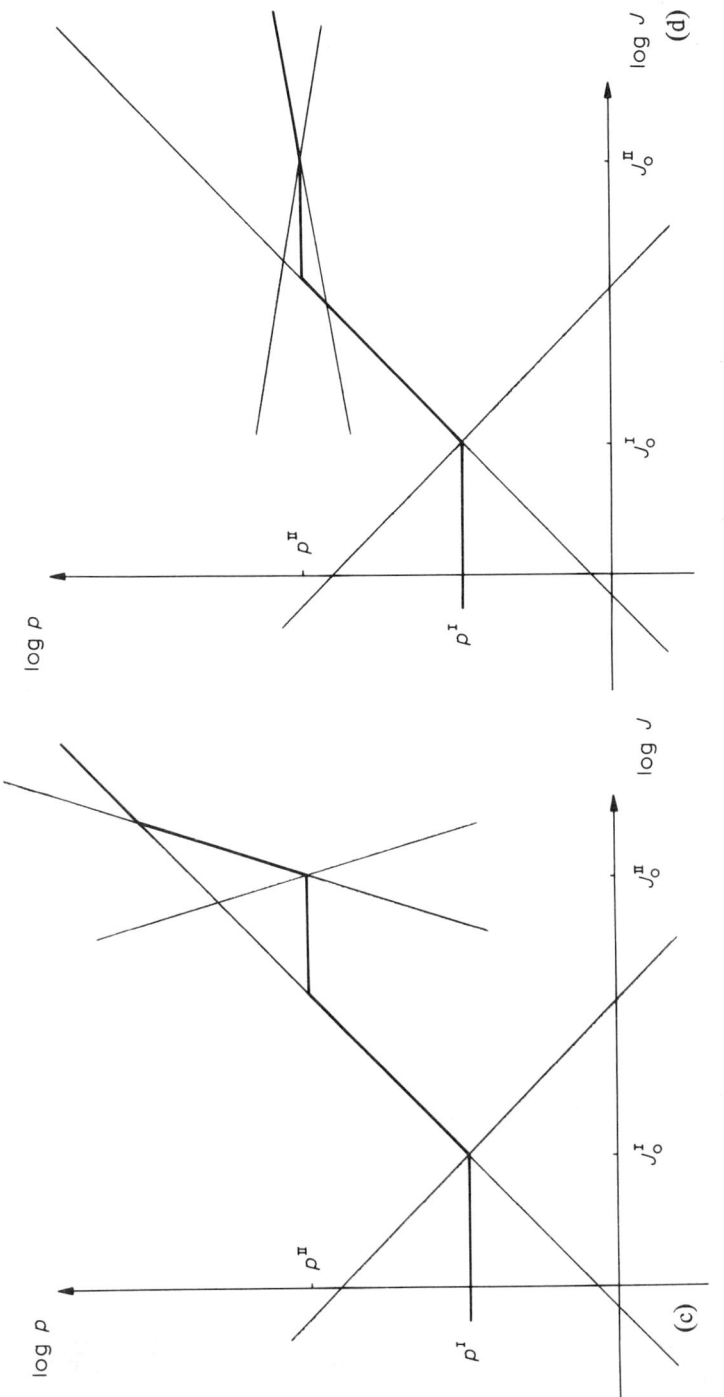

Fig. 7.4 Forward, reverse and exchange currents in a two-phase system.

approximation, and so is plotted as a vertical line. The net rate is shown as the bold line, and $p^°$ is the saturated vapour pressure. In this case, the exchange current J_0 during growth is independent of the vapour pressure.

We may make use of such diagrams to discuss the formation of metastable phases in crystal growth at high overpotentials. We include not only polycrystalline material and glassy material, but also second phases which might, for example, have a different composition or crystalline form and be stable phases under different conditions of temperature, pressure, and vapour composition. Let us consider a stable phase I with corresponding equilibrium pressure p^I and exchange current J_0^I and similarly for metastable phase II. On our 'Tafel' plot we can mark the points corresponding to (p^I, J_0^I) and (p^{II}, J_0^{II}). Four situations arise, according to the relative positions of these points and the dependence of the forward and reverse currents for the two phases on the partial pressure. These are indicated in Fig. 7.4. The net rates of growth of solid are indicated simply by straight lines with sharp kinks: in reality these kinks would be smoothed out.

In Fig. 7.4a we have a situation where solid II will not form at any potential, because the forward rate for production of solid II depends on the partial pressure to some lower power than does the forward rate for production of solid I. Also, J_0^{II} is to the left of the line representing the forward rate for production of solid I. When J_0^{II} is to the right of that line, the situation is as illustrated in Fig. 7.4c. Now solid II is produced over a limited range of partial pressures above p^{II}.

If the forward rate for production of solid II depends on the partial pressures to a higher power than does the forward rate for production of solid I, then solid II can always be produced at sufficiently high partial pressure (unless some other phase or rate-limitation intrudes). Thus in Fig. 7.4d, solid II is produced above p^{II}. In Fig. 7.4b, the rate of production of solid II is less than that of solid I until some pressure in excess of p^{II} is reached.

Above p^{II}, the net rate of production of both solids is positive, so that the solid produced will be a mixture in general, although it may be possible to make the rate of production for one solid orders of magnitude higher than for the other solid.

7.2 Maximizing the growth rate

The rate of growth of a crystal in a chemical vapour transport system is a function of many variables, of which a good number have been

AREAS FOR FURTHER STUDY

elaborated on in previous chapters. The crystal grower who is interested in growing large single crystals will want to make his transport system as efficient as possible. We have some observations to offer here which may be helpful. The crystal grower whose interest is in producing epitaxial layers will have problems of surface morphology, purity, stoichiometry, etc., uppermost in his mind, and is often embarrassed, rather than otherwise, by a rapid rate of growth.

Various authors have advanced theories of efficiency factors for vapour transport systems. These theories are alright so far as they go, but they fail to take account of more than a few of the factors that determine the rate of growth. This is not surprising. Crystal growth from the vapour is a topic for which general theories are not available.

We can dismiss some parameters from further consideration without much comment. The temperature difference between the source and the crystal, and the amount of inert gas, are parameters whose effect has been discussed quantitatively in Chapter 4.

If the crystal is growing with a negligible surface resistance, then the optimum conditions of growth are defined in terms of the operating temperature and vapour composition, and are determined by transport considerations. We want the temperature to be such that a large change in vapour composition between the source and the crystal may be brought about by a readily obtainable temperature difference. This aspect is also involved in the choice of a transporting agent, and has been discussed in Section 2.3.1.

The composition of the vapour is a variable, the effect of which was investigated in Chapter 4. We saw that in growth by dissociative sublimation, the maximum rate is achieved when the solid and the vapour have the same ratio of components. Here the composition of the vapour may be described by a single parameter, which we denoted by α in Chapter 4, with a second parameter if the growth system contains a significant amount of inert gas.

In chemical vapour transport, the number of composition parameters may be larger. In Section 4.4.4 we saw that again transport is most efficient if the vapour is stoichiometric, i.e., if it contains the components of the crystal in the same proportions as in the crystal. There are often other composition variables which enter the problem as well. We will not enter into a detailed investigation of this problem here.

If the growth rate of the crystal is limited not by the rate of transport but by the rate of some surface process, then the situation alters. Now the maximum growth rate may be achieved for some composition of the vapour which is considerably richer in some components of the crystal than in others. In addition, the operating temperature is best selected not only to provide a suitable change in equilibrium constant between the source and the crystal. There will be some advantage in increasing the temperature of the crystal, and of the source also, to make the surface reaction more facile, at the expense of optimum conditions for vapour transport.

To make these ideas a little more concrete, let us consider the growth of a binary compound AB by dissociative sublimation. The transport reaction is

$$AB(s) \rightleftharpoons A(g) + \frac{1}{m} B_m(g)$$

If the transport through the vapour were the process which determined the growth rate, we would obtain the maximum rate of growth by fulfilling the condition (see Section 4.3.3):

$$\alpha(l) = p_A(l)/p_{B_m}(l) = m$$

where (l) denotes conditions over the source. In fact, it is easy to show that if $\alpha(l) = m$, then $\alpha(0) = m$, i.e., the vapour over the crystal is also stoichiometric (no accumulation of either species). Suppose that the rate at which the crystal grows is limited by the rate of some surface process, such as:

$$B_m(g) \rightleftharpoons mB\cdot$$

or

$$B\cdot + A\cdot \rightleftharpoons AB(s)$$

where we use the notation $B\cdot$, etc, to denote atoms adsorbed on the surface. Obviously we can influence the forward rates of such processes by controlling the vapour composition. Thus, we can make the first process faster by increasing the partial pressure of B_m, by including some extra component B in the capsule or in a side-arm at a different temperature. Putting in extra component B will tend to depress the partial pressure of species A (at equilibrium, the product $p_A \cdot p_{B_m}^{1/m}$ over the solid is determined by the temperature only). If the partial pressure of species B_m is increased too much, the surface will become short of adsorbed A, and a different rate-limiting process may take over.

To make the second process go rapidly, we need to maximize the product of the concentrations of adsorbed A and B atoms. If we assume that the kinetics of this process are first-order in A and B adatoms, the forward rate is given by:

AREAS FOR FURTHER STUDY

$$J_f = k_f[A\cdot].[B\cdot]$$

The concentrations of A and B adatoms are related to the partial pressures of A and B_m in the vapour by the appropriate isotherms, which might be of the simple Langmuir form (Section 3.1):

$$\theta_A = \frac{p_A}{k_A + p_A}$$

$$\theta_B = \frac{p_{B_m}^{1/m}}{k_B + p_{B_m}^{1/m}}$$

where θ_A and θ_B are fractional coverages, and k_A and k_B are functions of temperature, enthalpy of adsorption, etc. The partial pressures p_A and p_{B_m} which maximize the product $[A\cdot].[B\cdot]$ (or $\theta_A.\theta_B$) will generally not be those given by considering the maximum rate of transport through the vapour. Clearly, similar conditions will apply to chemical vapour transport, with more components adsorbed on the surface, and more processes on the surface with rates to be maximized. We will not attempt to construct any theory for maximizing the rate of chemical vapour transport here. The theory developed in Chapters 4 and 5 enables one to write an equation relating the growth rate to transport and chemical kinetic variables. It is then a straightforward exercise in calculus[237] to find the conditions which make the growth rate maximum.

7.3 Growth by chemical vapour transport at high pressure

Growth of crystals from solution under high pressure is familiar, for example: the hydrothermal growth of quartz crystals. The high pressure is applied so that the temperature of the solution may be raised above its atmospheric boiling point, when the solubilities of the consituents of the crystal increase appreciably.

High pressure growth of crystals from the vapour is not so familiar, yet there are attractive features which make the idea worth considering. In this section, we discuss some of these features briefly.

As the pressure of a gas is increased to around 10^3 atm, its density approaches that of a liquid, and many of its properties change. For example, the ideal gas laws no longer apply because the volume occupied by the gas molecules is comparable with the total volume of the gas. Diffusion is no longer well described by a simple kinetic theory argument, such as that outlined in Section 4.1, but has become an activated process as in liquids and solids.

Of course, increasing the pressure in a chemical vapour transport system to this sort of level may cause one or more vapour components to liquify, so that the partial pressures of these components are then fixed by the temperature. The formation of more condensed phases may or may not be desirable. This point was discussed in Section 2.3.1. We will assume in the rest of this discussion that all components remain at pressures below their saturated vapour pressures.

Let us consider first of all the effect of increasing the total pressure on the equilibrium of a chemical vapour transport reaction. Le Chatelier's principle tells us that the increase in pressure will drive the reaction in the direction of the smaller molar volume i.e., fewer gas molecules. Consider the hypothetical reactions:

$$M(s) + HX(g) \rightleftharpoons MX(g) + \tfrac{1}{2}H_2(g) \tag{7.1}$$

$$M(s) + 2HX(g) \rightleftharpoons MX_2(g) + H_2(g) \tag{7.2}$$

$$M(s) + 3HX(g) \rightleftharpoons MX_3(g) + \tfrac{3}{2}H_2(g) \tag{7.3}$$

Here we could be considering a single element M with well defined oxidation states of 1, 2, and 3, or more likely, we would consider different elements M in Equations 7.1, 7.2, and 7.3.

The first reaction has an extra half a mole of vapour on the right, so that increasing the pressure would tend to shift the equilibrium to the left. The second reaction does not change the number of moles of vapour, and so an increase in total pressure would have little effect in moving the equilibrium one way or the other. The third reaction would be driven to the right by an increase in pressure. We observe, then, that an increase in pressure has the effect of stabilizing higher oxidation states.

Since we are discussing equilibrium here, we should note that the derivation of the expression for the equilibrium constant for a reaction, Equation 2.37, was derived on the assumption that the vapour was an ideal gas mixture. The chemical potential of each species is then a logarithmic function of its partial pressure (Equation 2.35):

$$\mu_i = \mu_i^0 + RT \ln p_i$$

At high pressures, this assumption is only roughly right. The way around this is to introduce the quantity 'fugacity', so defined that the chemical potential of each vapour component is given by:

$$\mu_i = \mu_i^0 + RT \ln f_i$$

where f_i is the fugacity of species i. Of course, at low pressures, f_i

AREAS FOR FURTHER STUDY

becomes indistinguishable from p_i. The equilibrium constant is now given by the equation:

$$K_f = \prod_i f_i^{\nu_i}$$

where ν_i is the stoichiometric coefficient of species i in the reaction. Both f_i and μ_i depend on all the components of the gas mixture. The ratio f_i/p_i is the fugacity coefficient. There are approximate methods for estimating fugacity coefficients, and the subject is well covered in the standard textbooks on chemical thermodynamics[4, 5]. We will not go into it any deeper here, except to note that the ratio of K_f and K_p is often not far from unity for pressures up to hundreds of atmospheres, so that as a rough approximation, it is often sufficient to ignore the difference.

If the transport reaction involves a decrease in the number of gas phase molecules, application of high pressure at constant temperature will shift the equilibrium to the right (i.e., increase the partial pressures of the products). There are situations where this is desirable in itself, to increase the efficiency of transport. We want the vapour over the source to consist mainly of the reactants, for efficient transport. There are situations where it is undesirable to change the operating temperature to bring this about. For example, one may not want to increase the temperature of operation if this would involve using different (more expensive or less easily fabricated) container materials. Or one might want to avoid lowering the temperature if this would make the surface reactions significantly less facile. If the transport reaction involves a decrease in the number of gas phase molecules, increase in temperature moves the equilibrium in the same direction as an increase in pressure if the reaction is endothermic, and in the opposite direction if the reaction is exothermic. In Section 4.2.3. we discussed the sort of reactions that might be endothermic while involving a decrease in the number of vapour molecules. Reactions such as:

$$M(s) + nHX(g) \rightleftharpoons (MX)_n(g) + \frac{n}{2} H_2(g)$$

are candidates. If the transport reaction which involves the monomer MX is endothermic, and if the polymeric species forms from the monomers with a positive or only slightly negative enthalpy change, then the transport reaction above will be endothermic. Operating at high pressure will then make the transport efficient at a lower temperature than would be necessary if one were operating at atmospheric pressure. This may often be a desirable feature in itself.

It should be realized that the reaction:

$$n\text{MX}(g) \rightleftharpoons (\text{MX})_n(g)$$

is driven to the right by increasing the pressure at constant temperature. Thus increasing the pressure will favour the transport reaction which proceeds via the polymer species.

We now consider the possible effect of high pressure on the surface reaction kinetics. With high pressures of the vapour components over the crystal surface, we expect the situation to approach that in solution growth. The crystal surfaces may well be covered by several atomic layers of liquid-like adsorbate. Note also that increasing the pressure of the vapour from say, 1 to 1000 atm decreases the translational entropy by 57.7 J mol^{-1} K^{-1}, according to the Sackur-Tetrode equation (Section 2.3.2). This means that the entropy change in going from solid to vapour is substantially reduced. According to Jackson's theory[92], this would have the effect of increasing the atomic roughness of the crystal surface by stabilizing atomic defects (adatoms and vacancies) on the surface. We may imagine that adsorption of vapour species which can form bonds to surface adatoms would also increase the surface roughness, by stabilizing adatoms. Increasing the pressure thus leads towards a situation in which the crystal can grow without the need for two-dimensional nucleation. Indeed, it may well be that the general success of chemical vapour transport as a way of growing crystals indicates that this situation is frequently achieved at atmospheric pressure or thereabouts. An increase in pressure should then bring this situation about at a lower temperature.

Finally, we should mention the effect of a large increase in pressure on the solid. In all cases we would expect a change in the equilibrium concentrations of defects – a decrease in the vacancy concentrations and an increase in interstitial atom concentrations. This will have a marked effect on, for example, the intrinsic conductivity of a semiconducting crystal, and also on the solubilities of impurities, as discussed in Section 2.3.3. In some substances we would find a more drastic change, in that a different phase may be stabilized by the high pressure. If a crystal is grown at high temperature and pressure, and cooled to room temperature without reducing the pressure, it may be possible to 'freeze in' the high pressure form so that it is preserved on reducing the pressure to atmospheric, particularly if the phase change would involve a substantial rearrangement of atoms into different structural units. The possibilities for producing hitherto unknown materials are considerable. High pressure synthesis has been well reviewed by Goodenough et al.[238], who provide a substantial list of references.

References

1. Laudise, R. A. (1970), *The Growth of Single Crystals*, Prentice-Hall Inc., New Jersey.
2. Schäfer, H. (1964), *Chemical Transport Reactions*, Academic Press, New York.
3. Powell, C. F., Oxley, J. H., Blocher, J. M., Jnr., (1966), *Vapour Deposition*, 2nd edition, John Wiley, New York.
4. Denbigh, K. (1971), *The Principles of Chemical Equilibrium*, 3rd edition, University Press, Cambridge.
5. Ives, D. J. G. (1971), *Chemical Thermodynamics*, MacDonald, London.
6. Kubaschewski, O., Evans, E. Ll., Alcock, C. B. (1967), *Metallurgical Thermochemistry*, 4th edition, Pergamon Press, Oxford.
7. Stull, D. R., Sinke, G. C. (1956), *Thermodynamic Properties of the Elements*, American Chemical Society.
8. J.A.N.A.F., Thermochemical Tables, (1971), 2nd edition, U.S. Department of Commerce, National Bureau of Standards.
9. *Selected Values of Chemical Thermodynamic Properties* (1949), U.S. Department of Commerce, National Bureau of Standards.
10. *Termodynamicheskie Svoistva Individualnykh Veschestv.*, Vols I and II. (*Thermodynamic Properties of Individual Compounds*), (1962), Akad. Nauk. S.S.S.R., Moscow.
11. Prigogine, I. (1955), *Introduction to Thermodynamics of Irreversible Processes*, Charles C. Thomas, Springfield, Illinois.
12. Logan, R. M., Hurle, D. T. J. (1971), *J. Phys. Chem. Solids*, 32. 1739.
13. Goldfinger, P., Jeunehomme, M. (1963), *Trans. Farad. Soc.* 59. 2857.
14. Phillips, F. C. (1963), *An Introduction to Crystallography*, 3rd edition, Longmans, London.
15. de Jong, W. F. (1959), *General Crystallography*, W. H. Freeman & Co., San Francisco.
16. Nye, J. F. (1967), *Physical Properties of Crystals*, Clarendon Press, Oxford.
17. Wooster, W. A. (1949), *A Text-book of Crystal Physics*, Cambridge University Press.
18. Mason, W. P. (1966), *Crystal Physics of Interaction Processes*, Academic Press, New York.

19. Goldschmidt, V. M. (1926), *Geochemische Verteilungsgesetze der Elemente*, in *Skrifter det Norske Videnskaps-Akad.*, Oslo 1. Matem – Naturvid Klasse.
20. Pauling, L. (1952), *The Nature of the Chemical Bond*, Oxford University Press (London).
21. Wells, A. F. (1950), *Structural Inorganic Chemistry*, 2nd edition, Clarendon Press, Oxford.
22. Náray-Szabó, I. (1969), *Inorganic Crystal Chemistry*, Akadémiai Kisdó, Budapest.
23. Greenwood, N. N. (1968), *Ionic Crystals, Lattice Defects and Non-stoichiometry*, Butterworths, London.
24. Roth, W. L. (1960), *Acta Cryst.* **13**, 140.
25. Cottrell, A. H. (1963), *Dislocations and Plastic Flow in Crystals*, International Series of Monographs on Physics, Clarendon Press, Oxford.
26. Frank, F. C. (1949), *Disc. Faraday Soc.*, No. 5. 48, 67.
27. Smallman, R. E. (1963), *Modern Physical Metallurgy*, Butterworths, London.
28. Bethge, H. (1970), *Proceedings of the 7th International Congress on Electron Microscopy*, Grenoble, France, 30th August, 1970, Soc. Française de Microscopie Electronique, Paris, France.
29. Masters, B. C. (1963), *Nature,* **200**, 254; (1965), *Phil Mag.*, **11**, 881;
    Bownie, M. E., Eyre, B. L. (1965), *Phil. Mag.*, **11**, 53;
    Meakin, J. D., Greenfield, I. G. (1965), *Phil. Mag.*, **11**, 277.
30. Ranganathan, S. (1968), *Field-ion Microscope Study of Interfaces*, in *Field-ion Microscopy*, edited by J. J. Hren and S. Ranganathan, Plenum Press, New York.
31. Brandon, D. G., Ralph, B., Ranganathan, S., Wald, M., *Acta. Met.* **12**, 813.
32. Hargreaves, F., Hills, R. J. (1929), *J. Inst. Metals*, **41**, 257.
33. Mott, N. F. (1948), *Proc. Phys. Soc.* (London) **60**, 394.
34. Bowden, P. B., Brandon, D. G. (1961), *Phil. Mag.*, **6**, 707; *Disc. Faraday Soc.*, **31**, 70; *Phil. Mag.*, **8**, 935; *J. Nucl. Materials*, **9**, 348.
35. Swalin, R. A. (1962), *Thermodynamics of Solids*, John Wiley & Sons Inc., New York.
36. Cottrell, A. H. (1964), *The Mechanical Properties of Matter*, John Wiley & Sons, Inc., New York.
37. Wallace, C. (1974), *J. Applied Crystallography*, (in press).
38. Wagner, R. S., Ellis, W. C. (1964), *Appl. Phys. Letters*, **4**, 89.
    Wagner, R. S. (1970), in *Whisker Technology*, ed. A. P. Levitt.
39. Green, J. M. (1970), *Metallurgical Transactions of the A.I.M.E.*, (1970), **1**, 647.
40. Jeffes, J. H. E. (1968), *J. Crystal Growth.* **3, 4**, 13.
41. Jeffes, J. H. E., Alcock, C. B. (1968), *J. Materials Sci.* **3**, 635.
42. Dasent, W. E. (1971), *Inorganic Energetics*, Penguin Educational Books, Harmondsworth.
43. Weiner, M. E. (1972), *J. Electrochem. Soc.* **119**, 496.
44. Garrett, I. (1972) unpublished work, internal P.O. Research Report.
45. Piacente, V., Malaspina, L. (1972), *J. Chem. Phys.*, **56**, 1780.

REFERENCES

46. Caveney, R. J. (1965), *Phil. Mag.*, **12**, 423.
    Caveney, R. J. (1968), *J. Phys. Chem. Solids*, **29**, 851.
47. Piacente, V., Bardi, G., Mancini, A., Desideri, A. (1971), *Rev. Int. Hautes. Tempér. et Réfract.* **8**, 237.
48. Bubnov, Yu. Z., Lur'e, M. S., Filaretov, G. A. (1971), *Russian J. Phys. Chem.*, 1715.
49. Glasstone, S. (1951), *Textbook of Physical Chemistry*, Macmillan and Co, London.
50. Partington, J. R. (1952), *Treatise on Physical Chemistry*, Vol. 3, Longmans, Green & Co., London.
51. Bondi, A. (1968), *Molecular Crystals, Liquids and Glasses*, John Wiley & Sons, Inc., New York.
52. Latimer, W. M., Buffington, J. (1926), *J. Amer. Chem. Soc.*, **48**, 2297.
53. Drozin, N. N. (1964), *Russ. J. Phys. Chem.*, **38**, 1386.
54. Partington, J. R. (1949), *An Advanced Treatise on Physical Chemistry*, Vol. 1, Longmans, Green & Co., London.
55. Johnson, D. A. (1968), *Some Thermodynamic Aspects of Inorganic Chemistry*, University Press, Cambridge.
56. Tosi, M. P., Fumi, F. G., *J. Phys. Chem Solids*, **23**, 359.
57. Ladd, M. F. C., Lee, W. H. (1964), *Lattice Energies and Related Topics*, in *Progress in Solid State Chemistry*, Vol. 1, edited by H. Reiss, Pergamon Press, Oxford.
58. Rao, K. J., Rao, G. V. S., Rao, C. N. R. (1967), *Trans. Farad. Soc.*, **63**, 1013.
    Rao, K. J., Rao, C. N. R. (1967), *Proc. Phys. Soc.* (London) **88**, 754.
59. Vervey, E. J. W., de Boer, J. H. (1940), *Rev. trav. chim*, **59**, 633.
60. Honig, A., Mandel, M., Stitch, M. L., Townes, C. A. (1954), *Phys. Rev.*, **96**, 629.
61. Moelwyn-Hughes, E. A. (1964), *Physical Chemistry*, 2nd edition, Pergamon Press, Oxford.
62. Waddington, T. C. (1959), *Advances in Inorganic Chemistry and Radio Chemistry*, Vol. 1, edited by H. J. Emeléus and A. G. Sharpe.
63. Faktor, M. M., Moss, R. H. (1968), Unpublished results. Internal P.O. Research Report.
64. Nesmeyanov, A. N. (1963), *Vapour Pressure of the Elements*, edited and translated by J. I. Carasso, Infosearch Ltd, London.
65. Vedeneyev, V. I., Gurvich, L. V., Kondrat'yev, V. N., Medvedev, V. A., Frankevich, Ye. L. (1966), *Bond Energies, Ionisation Potentials and Electron Affinities*, Edward Arnold, London.
66. Hume-Rothery, W. (1946), *Atomic Theory for Students of Metallurgy*, The Institute of Metals, London.
67. Thurmond, C. D., Struthers, J. D. (1953), *J. Phys. Chem.*, **57**, 831.
68. Reiss, H., Fuller, C. S., Morin, F. J. (1956), *Bell. Syst. Tech. J.*, **35**, 535.
69. Kröger, F. A. (1964), *The Chemistry of Imperfect Crystals*, North-Holland, Amsterdam.
69a. Kröger, F. A. (1970), *The Chemistry of Compound Semiconductors*, chapter 4, in *Physical Chemistry*, ed. W. Jost, Academic Press, New York.

70. Van Gool, W. (1966), *Principles of Defect Chemistry of Crystalline Solids*, Academic Press, New York.
71. de Boer, J. H. (1968), *The Dynamical Character of Adsorption*, 2nd edition, Clarendon Press, Oxford.
72. Todd, C. J. (1973), *Vacuum*, **23**, 195.
73. Ertl, G., Gerischer, H. (1970), *Semiconductor Surfaces*, chapter 7, in *Physical Chemistry*, Vol. 10, edited by W. Jost, Academic Press, New York.
74. Chupka, W. A., Inghram M. G. (1953), *J. Chem. Phys.*, **21**, 371 and 1313; (1955), *J. Phys. Chem.* **59**, 100.
75. Brewer, L., Bromley, L. A., Gilles, P. W., Lofgren, N. L. (1950), Papers 3, 4 and 7 in *Chemistry and Metallurgy of Miscellaneous Materials*, edited by L. S. Quill, National Nuclear Energy Series IV – 19 B, U.S.A.
76. Brewer, L. (1957), *Experimentia*, Supplement VII, 227.
77. Searcy, A. W. (1961), *High Temperature Inorganic Chemistry*, in *Progress in Inorganic Chemistry*, Vol. 3, ed. F. A. Cotton, Interscience, New York.
78. Roy, R. (1965), *Ionic Crystalline Solutions with "Massive" Concentrations of Point Defects*, in *Reactivity of Solids*, edited by G. M. Schwab, Elsevier, Amsterdam.
79. Anderson, J. S. (1946), *Proc. Roy. Soc.* **A185**, 69.
79a. Anderson, J. S. (1970), *The Thermodynamics and Theory of Nonstoichiometric Compounds*, chapter 1, in *Problems of Nonstoichiometry*, edited by A. Rabènau, North-Holland, Amsterdam.
80. Jeannin, Y. P. (1970), *Recent Progress in the Investigation of Nonstoichiometry*, chapter 2, in *Problems of Nonstoichiometry*, edited by A. Rabenau, North-Holland, Amsterdam.
81. Hyde, B. G., Bevan, D. J. M., Eyring, L. (1966), *Phil. Trans. Roy. Soc.* **A.259**, 583.
82. Libowitz, G. G., *Nonstoichiometry in Chemical Compounds*, in *Progress in Solid State Chemistry*, Vol. 2, edited by H. Reiss, Pergamon Press, Oxford.
83. Clark, L., Woods, J. (1968), *J. Crystal Growth*, **3, 4**, 127.
84. Frenkel, J. (1924), *Zeitschrift für Physik,* **26**, 287.
85. Langmuir, I. (1918), *J. Amer. Chem. Soc.* **40**, 1361.
86. Frank-Kamenetskii, D. A. (1969), *Diffusion and Heat Transfer in Chemical Kinetics*. Translation editor J. P. Appleton, Plenum Press, New York.
87. Brunauer, S., Emmett, P. H., Teller, E. (1938), *J. Amer. Chem Soc.*, **60**, 309.
88. Pickett, G. (1945), *J. Amer. Chem Soc.*, **67**, 1958.
89. Anderson, R. B. (1946), *J. Amer. Chem. Soc.*, **68**, 686.
90. Anderson, R. B., Hall, W. K. (1948), *J. Amer. Chem. Soc.*, **70**, 1727.
91. Huttig, G. F., Theimer, O., Mehlo, W. (1951), *Kolloid-Zeit.* **121**, 50.
92. Jackson, K. A. (1967), *Current Concepts in Crystal Growth from the Melt*, in *Progress in Solid State Chemistry*, Vol. 4, edited by H. Reiss, Pergamon Press, Oxford.
93. Buckley, H. E. (1951), *Crystal Growth*, John Wiley, New York.
94. Wilson, C. T. R. *Phil. Trans.* **189**, 265: *Proc. Roy. Soc.* **A 85**, 285; **A 87**, 277.

REFERENCES

95. Andres, R. P. (1969), *Homogeneous Nucleation in a Vapour*, chapter 2, in *Nucleation*, edited by A. C. Zettlemoyer, Marcel Dekker, New York.
96. Thompson, W., (1870), *Proc. Roy. Soc. Edinburgh,* **7**, 63; (1871), *Phil. Mag.,* **42**, 448.
97. Van Hook, A. (1961), *Crystallization Theory and Practice*, Reinhold, New York.
98. Zeldovich, J. B. (1942), *J. Exp. Theoret Phys.,* **12**, 525, (1943), *Acta Phys. Chem. USSR,* **18**, 1.
99. Becker, R., Doering, W. (1935), *Ann Physik,* **24**, 719.
100. Farkas, L. (1927), *Z. physik. Chem.* (Leipzig), **125**, 326.
101. Ostwald, W., (1897), *Z. physik. Chem.* **22**, 289; (1900), **34**, 493.
102. Pashley, D. W. (1965), *Advances in Physics,* **14**, 327.
103. Lewis, B. (1971), *Thin Solid Films,* **7**, 179; (1970), *Surface Science,* **21**, 273.
104. Walton, D. (1962), *J. Chem. Phys,* **37**, 2182.
105. Lewis, B., Campbell, D. S. (1967), *J. Vac. Sci. Tech.* **4**, 209.
106. Frankl, D. R., Venables, J. A. (1970), *Advances in Physics,* **19**, 409.
107. Venables, J. A., Ball, D. J. (1968), *J. Crystal Growth,* **3**, 180; Ball, D. J. Venables, J. A. (1969), *J. Vacuum Sci. Tech,* **6**, 468.
108. Kaschiev, D., Markov, I., (1972), *J. Crystal Growth,* **13/14**, 131.
109. Stowell, M. J., Hutchinson, T. E. (1971), *Thin Solid Films,* **8**, 411.
110. Bonzel, H. P. (1970), *Surf. Sci.* **21**, 45.
111. Lander, J. J. (1965), *Low-energy electron defraction and surface structural chemistry*, in *Progress in Solid State Chemistry*, Vol. 2, edited by H. Reiss, Pergamon Press, Oxford.
112. Fedak, D. G., Gjostein, N. A., (1967), *Acta Metallurgica,* **15**, 827.
113. Palmberg, P. W., Rhodin, T. N. (1968), *J. Chem. Phys.,* **49**, 134; Palmberg, P. W. (1969), *Auger Electron Spectroscopy in Low Energy Diffraction Systems*, in *The Structure and Chemistry of Solid Surfaces*, edited by G. A. Somorjai, John Wiley, New York.
114. Weyl, W. A. (1952) *Wetting of solids as influenced by the polarisability of surface ions*, in *Structure and Properties of Solid Surfaces*, edited by R. Gomer and C. W. Smith, University Chicago Press, Chicago.
115. Distler, G. I. (1971), *J. Crystal Growth,* **9**, 76; Distler, G. I., Lebedeva, V. N., Moskvin. V. V. (1971), *J. Crystal Growth,* **9**, 98.
116. Volmer, M., Estermann, I. (1921), *Z. Physik,* **7**, 13.
117. Alty, T., Clark, J. (1935), *Trans. Farad. Soc.,* **31**, 648.
118. Charlton, J., Semenoff, N., Schalinkoff, A. (1932), *Trans. Farad. Soc.,* **28**, 169.
119. Gjostein, N. A. (1963), *Surface Self-diffusion*, chapter 4, in *Metal Surfaces: Structure Energetics & Kinetics*, American Soc. for Metals, Cleveland.
120. Barrer, R. M. (1936), *J. Chem Soc.,* 1256, and 1261.
121. Lander, J. J., Morrison, J. (1962) *J. Chem Phys.,* **37**, 729.
122. Kossel, W. (1927), *Nachr. Akad. Wiss. Goettingen*, 135.
123. Stranski, I. N. (1928), *Z. physik. Chem.,* **136**, 259. (1931), *Z. physik. Chem.,* **11**, 421.

124. Burton, W. K., Cabrera, N., Frank, F. C. (1950), *Phil. Trans.* **A, 243**, 299.
125. Choi, J. Y., Sewmon, P. G. (1962), *Trans. AIME* **224**, 589.
126. Cho, A. Y. (1971), *J. Appl. Phys.* **42**, 2074.
127. Burton, W. K., Cabrera, N., Frank, F. C. (1949), *Nature*, **163**, 398.
128. Frenkel, J. (1945), *J. Phys. U.S.S.R.*, **9**, 392.
129. Burton, W. K., Cabrera, N. (1949), *Disc. Faraday Soc.* **5**, 33, 40.
130. Strickland-Constable, R. F. (1968), *Kinetics and Mechanism of Crystallization*, Academic Press, New York.
131. Hamilton, J. F., Brady, L. E. (1964), *J. Chem Phys.*, **35**, 414.
    Wagner, R. S., Ellis, W. C., Jackson, K. A., Arnold, S. M. (1964), *J. Chem. Phys.*, **35**, 2993.
132. Burmeister, J. (1971), *J. Crystal Growth*, **11**, 131.
133. Gjostein, N. A. (1963), *Acta Metall.*, **11**, 957.
134. Priestley, J. (1786), *Experiments and Observations on Natural Philosophy*, Birmingham, **3**, 390.
135. Jeans, Sir J., (1962), *An Introduction to the Kinetic Theory of Gases*, University Press, Cambridge.
136. Boltzmann, L. (1880), *Wien. Ber.* **81**, 117; (1881), *Wien. Ber.* **84**, 40 and 1230; *Collected Works* **2**.
137. Chapman, S., Cowling, T. G. (1970), *The Mathematical Theory of Non-Uniform Gases*, 3rd edition, University Press, Cambridge.
138. Enskog, D. (1911), *Phys. Zeit* **12**, 56 and 533; (1912), *Annalen. Phys.* **38**, 731.
139. van Liempt, J. A. M. (1931), *Z. anorg. u. allgem. Chem.*, **195**, 366; (1932), *Rec. trav. chim.* **51**, 114; (1938), **57**, 891.
140. Stefan, J. (1890), *Annalen der Physik und Chemie,* **17**, 550; (1890) *Annalen der Physik*, **41**, 725.
141. Faktor, M. M., Heckingbottom, R., Garrett, I. (1970), *J. Chem. Soc., A.,* 2657.
142. Reed, T. B., La Fleur, W. J. (1964), *Appl. Phys. Letters*, **5**, 191;
    Reed, T. B., La Fleur, W. J., Strauss, A. J. S. (1968), *J. Crystal Growth*, **3**, 115.
143. Faktor, M. M., Heckingbottom R., Garrett, I. (1971), *J. Chem. Soc.*, **A**, i.
144. Faktor, M. M., Garrett, I. (1971), *J. Chem. Soc.*, **A**, 934.
145. Weiner, M. E. (1972), *J. Electrochem. Soc.*, **119**, 496.
146. Faktor, M. M., Garrett, I., Moss, R. M. (1973), *J. Chem. Soc., Faraday I.* 1915.
147. Toyama, M. (1966), *Jap. J. Appl. Phys.*, **5**, 1204.
148. Woodbury, H. H. (1964), *Phys. Rev.*, **134**, A 492.
149. Somorjai, G. A. (1965), *J. Chem. Phys.*, **42**, 4140.
150. Pulliam, G. R. (1967), *J. Appl. Phys.*, **38**, 1120.
151. Partington, J. R. (1967), *General and Inorganic Chemistry*, 4th edition, Macmillan, London.
152. Wolfe, C. M., Stillman, G. E., (1971), *High Purity GaAs*, in *Gallium Arsenide and Related Compounds*, (Conference Series No. 9), Proceedings of the 3rd International Symposium, Aachen, 1970, Institute of Physics, London.

REFERENCES

153. Taylor, E. R. (1912), *Ind. Eng. Chem.*, **4**, 557.
154. Schäfer, H., Wiedermeier, H., (1958), *Z. anorg. u. allgem. Chem.*, **296**, 241.
155. Ban, V. S. (1971), *J. Electrochem. Soc.*, **118**, 1473; (1972), *J. Electrochem Soc.*, **119**, 761; (1973), *J. Phys. Chem. Solids*, **34**, 1119.
156. Shaw, D. W. (1970), *J. Electrochem Soc.*, **117**, 683; (1971), *J. Crystal Growth*, **10**, 251; (1972), *J. Crystal Growth*, **12**, 249.
157. Hurle, D. T. J., Mullin, J. B. (1967), *J. Phys. Chem. Solids*, Supplement No. 1., 241.
158. Day, G. F. *Heterojunction Device Concepts,* Varian Report No. 324-6Q, Air Force Contract No. AF 33 (615) - 1988, (Project No. 4460, Task No. 446006). Varian Associates, Palo Alto, California.
159. Kirwan, D. J. (1970), *J. Electrochem. Soc.*, **117**, 1572.
160. Rai-Choudhury, P. (1971), *J. Crystal Growth*, **11**, 113.
161. Frosch, C. J., Thurmond, C. D. (1962), *J. Phys. Chem.*, **66**, 877.
162. Battat, D., Faktor, M. M., Garrett, I., Moss, R. H. (1974), *J. Chem. Soc. Faraday Transactions I*, in press.
163. *Termecheskie Konstanti Veschestv,* Vol 5 (Thermal Constants of Substances) (1971), Acad. Nauk. U.S.S.R., Moscow.
164. Murray, J. J., Pupp, C., Potie, R. F. (1973), *J. Chem. Phys.*, **58**, 2569.
165. Poiseuille, J. L. M. (1840), *Compt Rend.* **11**, 961, 1041; (1841), *Compt. Rend..* **12**, 112; (1842), *Compt. Rend,* **15**, 1167; (1843), *Ann. Chim.* **7** 50; (1847), *Ann. Chim.*, **21**, 76; (1846), *Mem. div. sav.*, **9**, 433.
166. Sutherland, G. B. B. M. (1893), *Phil. Mag.* **36**, 507.
167. Reynolds, O. (1883), *Phil. Trans. Roy. Soc.* (London), **174**.
168. Eckert, E. R. G., Drake, R. M., Jnr., (1959), *Heat and Mass Transfer,* McGraw-Hill, New York.
169. Schubauer, G. B., Skramstad, H. K. (1947), *J. Research, Nat. Bur. Standards*, **38**, 281.
170. Prandtl, L., Tietjens, O. G. (1957), *Fundamentals of Hydro- and Aeromechanics*, Translated by L. Rosenhead., Dover, New York.
171. Bénard, H., (1900), *Rev. gén. sci.* **11**, 1261 and 1309.
172. Lord Rayleigh (J. W. Strutt), (1916), *Phil. Mag.* **32**, 529.
173. Shirtcliffe, T. G. L. (1969), *J. Fluid Mech. (GB)* **35**, 677.
    Gill, A. E. (1969), *J. Fluid Mech. (GB)* **37**, 289.
174. Eversteyn, F. C., Severin, P. J. W., Brekel, C. H. J. v.d., Peek, H. L., (1970), *J. Electrochem. Soc.*, **117**, 925.
175. Moss, T. S. (1959), *Optical Properties of Semiconductors*, Butterworths, London.
176. Ayers, S., Faktor, M. M., Marr, D., Stevenson, J. L. (1971), *J. Mater. Sci.* **7**, 31.
177. Stevenson, J. L., Ayers, S., Faktor, M. M. (1973), *J. Phys. Chem. Solids*, **34**, 235.
178. Berliner, J. F. T., May, O. E. (1925), *J. Amer. Chem. Soc.*, **47**, 2350.
179. Ubbelohde, A. R. (1965), *Melting and Crystal Structure*, Clarendon Press, Oxford.
180. Battat, D., Faktor, M. M., Garrett, I., Moss, R. H., *J. Chem. Soc. Faraday Trans I.* (in press).

181. Battat, D., Faktor, M. M., Garrett, I., Moss, R. H., *J. Chem. Soc. Faraday Trans I.* (in press).
182. Venables, J. A., Thomas, G. J. (1970), *Interstitials in f.c.c. metals*, in Proc. Int. Conf. on Vacancies and Interstitials in Metals, Julich, W. Germany, Sept 1968. North-Holland, Amsterdam.
183. Gibbs, J. W. (1928), *Collected Works of J. Willard Gibbs*, Yale University Press, New Haven.
184. Wulff, G. (1901), *Z. Krist.* **34**, 449.
185. Tiller, W. A. (1972), *On the energetics, kinetics and topography of interfaces*, in *Treatise on Materials Science and Technology*, edited by H. Herman, Acadmic Press, New York.
    Nason, D., Tiller, W. A. (1971), *J. Crystal Growth*, **10**, 117.
186. Sekerka, R. F. (1968), *J. Crystal Growth*, **3**, **4**, 71.
    Mullins, W. W., Sekerka, R. F. (1964), *J. Appl. Phys.*, **35**, 444; (1963), *J. Appl. Phys.*, **34**, 323.
187. Schäfer, H., Etzel, K., (1957), *Z. anorg. u. allgem Chem.*, **291**, 294.
188. Hildon, D. L., Gregory, N. W., (1972), *J. Phys. Chem.*, **76**, 1632.
189. Shelton, R. A. J., (1961), *Trans. Farad. Soc.*, **57**, 2113.
190. Guido, M., Balducci, G., Gigli, G., Spoliti, M., (1971), *J. Phys. Chem.*, **55**, 4566.
191. Tietjen, J. J., Enstrom, R. E., Richman, D. (1970), *R.C.A. Review*, **31**, 635.
192. Schäfer, H. (1972), *Chemical Transport as a Preparative Procedure*, in *Preparative Methods in Solid State Chemistry*, edited by P. Hagenmuller, Academic Press, New York.
193. Nitsche, R. (1971), *Kristallzucht aus der Gasphase*, in *Metallphysik*, edited by O. Madelung, Pergamon Press, Oxford.
194. Tanenbaum, M. (1959), *Semiconductor Crystal Growing* in *Semiconductors*, edited by N. B. Hannay, Reinhold, New York.
195. Scholz, H. (1967), *Philips Tech. Rev. (Netherlands)*, **28**, 316.
196. Bernal, J. D., Humphreys-Owen, S. P. F., private communication.
197. Honigmann, B. (1954), *Z. Elekrochem.*, **58**, 322.
198. Nitsche, R. (1967), *Fortschr. Miner*, **44**, 231.
199. Pizzarello, F. (1954), *J. Appl. Phys.*, **25**, 804.
200. Grimmeis, H. G., Rabenau, A., Koelmans, H. J. (1961), *J. Appl. Phys.*, **32**, 2123.
201. Kaldis, E., Widmer, R. (1965), *J. Phys. Chem. Solids*, **26**, 1697.
202. Hanak, J. J., Berman, H. S. (1966), *Proc. Int. Conf. Crystal Growth*, Boston, S. 249.
203. Reed, T. B. (1969), *MIT Report*, 1, 21.
204. Lord, G. W., Moss, R. H. (1970), *J. Phys. E. Sci. Instr.*, **3**, 177.
205. Kryzhanovskii, B. P. (1961), *Optics and Spetroscopy*, **10**, 359.
206. Bulakh, B. M., Pekar, G. S. (1970), *J. Crystal Growth*, **7**, 285.
207. Scholtz, H. (1965), *Chemie Ing. Technik*, **37**, 1173.
208. Kaldis, E. (1971), *J. Crystal Growth*, **9**, 281.
209. Piper, W. W., Polich, S. J. (1961), *J. Appl. Phys.*, **32**, 1278.
210. Prior, A. C. (1961), *J. Electrochem. Soc.*, **108**, 82.

REFERENCES

211. Nicholl, F. H. (1963), *J. Electrochem. Soc.*, **110**, 1165.
212. Robinson, P. (1963), *R. C. A. Review*, **24**, 574.
213. Gottlieb, G. E., Corboy, J. F. (1963), *R. C. A. Review*, **24**, 585.
214. Dryburgh, P., Private communication.
215. Schäeffer, P. S. (1965), *J. Amer. Ceram. Soc.*, **48**, 508.
216. Effer, D. (1965), *J. Electrochem. Soc.*, **112**, 1020.
217. Knight, J. R., Effer, D., Evans, P. R. (1965), *Solid State Electronics*, **8**, 178.
218. Shaw, D. W. (1971), *J. Crystal Growth*, **8**, 117.
219. Butcher, M. M., White E. A. D. (1965), *J. Amer. Ceram. Soc.*, **48**, 492.
220. Kuznetsov, F. A., Belyi, V. E. (1970), *J. Electrochem. Soc.*, **117**, 785.
221. Levitch, V. G. (1962), *Physicochemical Hydrodynamics*, Prentice-Hall, New Jersey.
222. Battat, D., Faktor, M. M., Sava, M. A., internal, P.O. Research Report.
223. Takahashi, R., Koga, Y., Sugawara, K. (1972), *J. Electrochem. Soc.*, **119**, 1406.
224. Bulakh, B. M., Pekar, G. S. (1971), *J. Crystal Growth*, **8**, 99.
225. Reisman, A., Landstein, J. E. (1971), *J. Electrochem. Soc.*, **118**, 1479.
     Reisman, A., Berkenbilt, B., Chan, S. A., Angilello, J. (1973), *J. Electronic Materials*, **2**, 177.
226. Manasevit, H. M., Erdmann, F. M., Simpson, W. I. (1971), *J. Electrochem. Soc.*, **118**, 1864; (1969), *J. Electrochem. Soc.*, **116**, 1725; (1970), *J. Electrochem. Soc.*, **117**, 196C; (1971), *J. Electrochem. Soc.*, **118**, 644; (1968), *Appl. Phys. Lett.*, **12**, 156.
227. Taylor, R. C., Sagadopan, V. (1971), *Appl. Phys. Lett.*, **19**, 361.
228. Takei, H., Takasu, S., (1964), *Jap. J. Applied Physics*, **3**, 175.
229. Nakado, I. (1973), British Association for Crystal Growth, Annual Conference, York, Sept. 1973. (unpublished).
230. Amelinckx, S., (1950), *J. Chim Phys.*, **47**, 213.
231. Grimley, R. T. (1967), *Mass Spectrometry*, in *The Characterisation of High-temperature Vapours*, edited by J. L. Margrave, John Wiley, New York.
232. Brebrick, R. F. (1968), 133rd meeting of the Electrochemical Society, Abstracts, Boston Mass., U.S.A., 5–9h May, 1968. Electochem Soc., New York. (1969), *J. Electrochem. Soc.*, **116**, 1274.
233. Kohlrausch, K. W. F., *Ramanspektren, Hand-und-Jahr-buch der chemischen Physik*, **9**, part 6;
     Herzberg, G. (1968), *Infra-red and Raman Spectra of Polyatomic Molecules*, Van Nostrand, New York.
234. Volmer, M. (1931), *Z. physik. Chem.*, Bodenstein Festband, 863;
     Paule, R. C., Margrave, J. L. (1967), *Free Evaporation and Effusion Techniques*, in *Characterisation of High Temperature Vapours*, edited by J. L. Margrave, John Wiley, New York.
235. Becker, K. A., Plieth, K., Stranski, I. N. (1962), *The Polymorphic Modifications of Arsenic Trioxide*, in *Progress in Inorganic Chemistry*, Vol. 4, edited by F. A. Cotton, Interscience Publishers, New York.
236. Bockris, J. O'M. (1954), *Electrode Kinetics*, in *Modern Aspects of*

Electrochemistry, No. 1, edited by J. O'M. Bockris and B. E. Conway, Butterworths, London.
237. Sokolnikoff, I. S., and Redheffer, R. M. (1958), *The Mathematics of Physics and Modern Engineering,* McGraw-Hill, New York.
238. Goodenough, J. B., Kafalas, J. A., Longo, J. M. (1972), *High Pressure Synthesis,* chapter 1, in *Preparative Methods in Solid State Chemistry,* edited by P. Hagenmuller, Academic Press, New York.
239. Shiloh, M., Gutman, J. (1973), *Mat. Res. Bull.* 8, 711.

Index

acceptors, 62
activation barrier to adsorption, 74
activation energy, 199
activation entropy, 199
active complex, 199, 200
active intermediate, 199
activity, 18
adatoms, 86, 87
adsorption, 71, 72
adsorption isotherm, 63
affinity, 17, 195, 252
Alcock, C. B., 15, 51, 55, 58, 139, 150, 158, 179
almost sealed capsule, 238
Alty, T., 92
Amelinckx, S., 260
Anderson, J. S., 69
Anderson, R. B., 78
Andres, R. P., 82
Angilello, J., 254
annealing of defects, 47
array of dislocations, 46
average diffusion coefficient, 202
axial planes, 27
Ayers, S., 184

Balducci, G., 220
Ball, D. J., 86
Ban, V. S., 153, 154, 256
Bardi, G., 54, 59, 60
Barrer, R. M., 94
Battat, D., 157, 202, 249, 250, 252, 266, 267
Becker, K. A., 270
Becker, R., 84
Belyi, V. E., 249

Bénard, H., 174
Bénard problem, 174, 177
Berkenbilt, B., 254
Berliner, J. F. T., 184
Berman, H. S., 229, 230
Bernal, J. D., 228
BET isotherm, 78
BET theory, 75
Bethge, H., 44, 260
Bevan, D. J. M., 69
binary diffusion coefficient, 201
birefringence, 35
Blocker, J. M. Jnr., 4
Bockris, J. O'M., 270, 273
de Boer, J. H., 57, 63, 71, 91, 161
Boltzmann, W., 103
Bondi, A., 55
bond strength, 56
Bonzel, H. P., 87, 95
Born–Haber cycle, 56, 58, 59, 67
Born–Landé model, 57
boundary layers, 167, 168, 169
Bourdon spoon gauge, 112
Bowden, P. B., 47
Bownie, M. E., 45
Brady, L. E., 98
Brandon, D. G., 46, 47
Brebrick, R. F., 265
Brekel, C. H. J. v.d., 177, 246, 248, 268
Brewer, L., 64
Bromley, L. A., 64
Brunauer, S., 75
Bubnov, Yu. Z., 54
Buckley, H. E., 79
Buffington, J., 55

Bulakh, B. M., 231, 232, 253
buoyancy, 266
Burmeister, J., 98, 207
Burton, W. K., 95, 96, 97, 98, 260
Butcher, M. M., 247

Cabrera, N., 95, 96, 97, 98, 260
cadmium sulphide, 58, 59, 123, 124, 125, 213
Campbell, D. S., 85
Caveney, R. J., 54
caesium chloride structure, 39
centre, 24
Chan, S. A., 254
Chapman, S., 103, 104, 108, 201
charge neutrality, 42, 68
Chariton, J., 92
Le Chatelier & Matignon rule, 55
Le Chatelier's principle, 120, 203, 280
chemical conductance, 197
chemical potential, 9, 10
chemical transport reactions, 119
chemical vapour deposition, CVD, 4
chemical vapour transport, CVT, 4, 51 64, 65, 70, 118, 136, 152
chemisorption, 71, 72
Cho, A. Y., 95
Choi, J. Y., 95
Chupka, W. A., 64
Clark, J., 92
Clark, L., 70, 239
Clausuis–Clapeyron equation, 22
close-spaced transport, 244
closed systems, 9
close-packed structures, 36
cluster, 85
cold-walled reactor, 246
complete condensation mechanism, 260
compositional convection, 173
condensation coefficient, 72, 74
configurational entropy, 13, 42, 61, 66
constitutional supercooling, 118, 208, 209
convection, 170, 180, 181
co-ordination, 35
co-ordination number, 60
Corboy, J. F., 245
Cottrell, A. H., 44, 45, 47
Coulombic forces, 61, 77
Cowling, T. G., 103, 104, 108, 201
critical cluster, 86

critical convective conditions, 175
critical growth rate, 213
critical nucleus, 84, 89, 200
critical temperature difference, 213
crowdions, 203
crystal imperfections, 41
crystallographic axes, 26
crystal structure, 24, 29
crystal symmetry, 24
crystal systems, 25
cubic close-packing, 36, 37
cubic system, 25, 30, 31

dangling bonds, 95
Doering, W., 84
Dasent, W. E., 51, 56, 58, 67
Day, G. F., 154
Debye law, 15
decoration of dislocations, 45
defect-sensitive properties, 69
degrees of freedom, 21
Denbigh, K., 8, 22, 281
depletion region, 92
Desideri, A., 54, 59, 60
diamond structure, 30
diffusion, 101
diffusion boundary layer, 177
diffusion in solids, 105
diffusion coefficient, 102
'diffusion only' approximation, 128
'diffusion only' model, 147
dihalides of metals, 67
dislocation cores, 45
dislocation node, 44
dislocations, 41, 42, 43, 44, 81, 86, 87
dissociative sublimation, 4, 118, 120, 124, 125, 129
Distler, G. I., 91
donors, 62
dopant, 53
doping by CVT, 217, 224
double refraction, 35
Drake, R. M. Jnr., 167, 168, 169, 173
Drozin, N. N., 55
Dryburgh, P., 246

Easterman, I., 92
Eckert, E. R. G., 167, 168, 169, 173
edge dislocation, 43
Effer, D., 247, 253, 254, 256
elastic strain, 61
electron affinity, 56, 67

electro-negativity, 58, 61
endothermic reactions, 67, 118
Ellis, W. C., 49
Emmett, P. H., 75
energy gap, 62
Enskog, D., 103, 201
Enstrom, R. E., 225, 226, 247, 253, 254, 255, 256
enthalpy, 10
enthalpy of formation, 15
enthalpy of solvation, 61
enthalpy of solution, 61
entropy, 9, 13, 52, 103
entropy of evaporation, 55
entropies of formation, 65
entropy of fusion, 55
entropy of sublimation, 55, 56
epitaxial deposits, 44
equilibrium constant, 16, 17, 18, 50, 53
Erdmann, F. M., 257
Ertl, G., 63
Etzel, K., 218
Evans, E. Ll., 15, 55, 58, 139, 150, 158, 179
Evans, P. R., 247
evaporation, 4, 22
Eversteyn, F. C., 177, 246, 248, 268
Eyre, B. L., 45
Eyring, L., 69
exchange current, 81, 82, 92, 100, 197, 269, 272, 273, 274, 275
experimental methods, 226
exothermic reactions, 118
extrinsic semiconductor, 62

face centred cubic structure, 26, 27, 29
Faktor, M. M., 59, 118, 120, 127, 152, 153, 155, 157, 184, 202, 210, 216, 221, 239, 242, 249, 250, 252, 266, 267
Farkas, L., 84
Fedak, D. G., 88
ferrous oxide, 43
Fick's law, 94, 102
Filaretov, G. A., 54
La Fleur, W. J., 118, 207, 208, 211
fluorite structure, 42, 68
form, 30
forward current, 272, 273
Frank, F. C., 44, 95, 96, 97, 98, 260
Frankevich, Ye. L., 59

Frank-Kamenetskii, D. A., 75, 201
Frenkel defects, 42, 67
Frankl, D. R., 85, 86, 92
Frenkl, J., 71, 96
Friedel's classes, 34
Frosch, C. J., 157
fugacity, 280
Fuller, C. S., 62
Fumi, F. G., 57

gallium arsenide, 241
Garrett, I., 53, 118, 120, 127, 152, 153, 155, 157, 202, 210, 239, 242, 252, 266, 267
Gerischer, H., 63
Gibbs free energy, 10, 12
Gibbs, J. W., 207
Gibbs–Thompson equation, 83
Gigli, G., 220
Gill, A. E., 176
Gilles, P. W., 64
Gjostein, N. A., 88, 94, 99, 207
Glasstone, S., 55
Goldfinger, P., 23, 59
Goldschmidt, V. M., 38
Goodenough, J. B., 282
Gottlieb, G. E., 245
grain boundary, 44, 45, 46, 47, 86
Grashof number, 173, 174, 177, 178
Green, J. M., 49
Greenfield, I. G., 45
Greenwood, N. N., 42, 69, 70, 206
Gregory, N. W., 220
Grimley, R. T., 265
Grimmers, H. G., 229
growth poison, 79
growth spirals, 97, 98
Guido, M., 220
Gurvich, L. V., 59
Gutman, J., 256

habit, 79
halides of metals, 67
halite structure, 39
Hall, W. K., 78
Hamilton, J. F., 98
Hanak, J. J., 229, 230
Hargreaves, F., 46
heat capacity, 13
heat of adsorption, 72
heats of atomisation, 67
heat flow, 178

heat transfer coefficients, 184
Heckingbottom, R., 118, 120, 210, 239
Hellbardt, G., 157
Helmholtz free energy, 10, 11
Henry's law, 18
Hermann–Mauguin notation, 30, 31
Herzberg, G., 265
hetero-epitaxy, 82
hexagonal close-packing, 36, 37
hexagonal system, 25, 30, 31
high pressure CVT, 279
Hildon, D. L., 220
Hills, R. J., 46
van't Hoff's equation, 19, 211
homo-epitaxy, 82
homogeneous nucleation, 82, 86
Honig, A., 57
Honigman, B., 228
Hume-Rothery, W., 61
Humphreys-Owen, S. P. F., 226
Hurle, D. T. J., 22, 61, 153
Hutchinson, T. E., 87
Huttig, G. F., 78
Hyde, B. G., 69

ideal gas mixture, 17
ideal solutions, 18
indices of direction, 29
indium oxide reflectors, 231
Inghram, M. G., 64
interfacial energy, 86
intermetallic compounds, 69
internal lattice energy, 57
interstitials, 42, 44, 47, 62
intrinsic carriers, 62
intrinsic conductivity, 41
ionic radii, 39
ionisation energy, 56
ionization potentials, 67
ionized impurities, 62
irreversible process, 9
isothermal growth, 184
isothermal transport, 187, 188, 189
Ives, D. J. G., 8, 281

Jackson, K. A., 78, 87, 88, 89, 261, 282
Jeans, Sir J., 103, 104, 161
Jeffes, J. H. E., 50, 51
Jeannin, Y. P., 69
Jeunehomme, M., 23, 59

Johnson, D. A., 56
de Jong, W. P., 24

Kafalas, J. A., 282
Kaldis, E., 229, 232, 233, 236, 268
Kappallo, W., 157
Von Karman trail, 170
Kashchiev, D., 87
kinematic viscosity, 168
kinetic theory, 102
Kirwan, D. J., 154
Knight, J. R., 247, 253
Knudsen cell, 264
Koelmans, H. J., 229
Koga, Y., 251
Kohlrausch, K. W. F., 265
Kondrat'yev, V. N., 59
Kossell, W., 95, 96
Kröger, F. A., 62, 134, 205
Kryzhanovskii, B. P., 231
Kuznetsov, F. A., 249
Kubaschewski, O., 15, 55, 56, 139, 150, 158, 179

Ladd, M. F. C., 57
laminar flow, 165, 249
Lander, J. J., 88, 94
Landstein, J. E., 254
Langmuir, I., 72, 73, 75
lattice, 29
lattice energy, 57
lattice mismatch, 44
lattice planes, 26, 29
lattice points, 25, 29
Latimer, W. M., 55
Laudise, R. A., 2, 226
Laue groups, 34
law of constancy of angle, 24
law of rational indices, 27
Lebedeva, V. N., 91
Lee, W. H., 57
Levitch, V. G., 249
Lewis, B., 85
Libowitz, G. G., 70
life-time, 100
Lofgren, N. L., 64
Logan, R. M., 22, 61
Longo, J. M., 282
Lord, G. W., 229
low-index planes, 29
Lur'e, M. S., 54
Madelung constant, 57, 58, 60

INDEX

Magneli phases, 70
Manasevit, H. M., 257
Malaspina, L., 54, 64
Mancini, A., 54, 59, 60
Mandel, M., 57
Margrave, J. L., 266
Markov, I., 87
Marr, D., 184
Mason, W. P., 34, 35
mass-spectrometry, 63
Masters, B. C., 45
Maxwell, C., 109
Maxwell–Boltzmann law, 106
Maxwell's relations, 119
May, O. E., 184
Meakin, J. D., 45
Medvedev, V. A., 59
mechanical turning, 47
mechanical strain, 44
Mehlo, W., 78
metanitroaniline, 184
metastable phases, 81
Michelitsch, M., 157
micro-facets, 98, 99, 100
Minden, H., 253, 254, 256
Miller–Bravais axes, 30
Miller indices, 26, 28
mirror planes, 25
misplaced atoms, 62
Moelwyn-Hughes, E. A., 57, 58
Mond process, 136, 138, 143
monohalides of metals, 67
monoclinic system, 25, 31
Morin, F. J., 62
morphological stability, 206
Morrison, J., 94
Moskvin, V. V., 91
Moss, R. H., 59, 118, 120, 157, 203, 229, 266, 267
Moss, T. S., 183
Mott, N. F., 46
Mullin, J. B., 153
Mullins, W. W., 216
multilayer adsorption, 75
multilayer film, 78, 80
Murray, J. J., 158

Nakado, I., 260
Naray-Szabo, I., 41
Nason, D., 216
native defects, 62
Nesmeyanov, A. N., 59, 113

Neumann's principle, 34, 35
Nicholl, F. H., 244
Nitsche, R., 226, 228
non-bridging ligands, 60
non-stoichiometry, 43
non-stoichiometric phases, 67
nucleation, 82
nucleation control, 228, 232
nucleation rate, 85
Nye, J. F., 24, 34, 35

open-flow systems, 223, 225, 245
open systems, 9
optical activity, 35
orthorhombic system, 25, 31
Ostwald, W., 84
oxidation state, 43
Oxley, J. H., 4

Palmberg, P. W., 88
parametral plane, 29
Partington, J. R., 55, 104, 144
Pashley, D., 85
Paule, R. C., 266
Pauling, L., 38, 56, 58, 61
Peck, H. L., 177, 246, 248, 268
Pekar, G. S., 231, 232, 253
perforated capsule, 239
phase rule, 20, 21, 22, 48, 136, 143
Pickett, G., 78
Phillips, F. C., 24, 30
physisorption, 71, 72
Piacente, V., 54, 59, 60, 64
piezoelectricity, 30, 35
pinning mechanism, 45
Piper, W. W., 238
Pizzarello, F., 229
place exchange, 43, 69
planar cluster, 45
planes of symmetry, 25
platelets, 79
Plieth, K., 270
Pockels effect, 35
point defects, 41, 42
point group, 24
Poiseuille, J. L. M., 160, 163
polarization, 39
polarization forces, 60
Polich, S. J., 238
Pottie, R. F., 158
Powell, C. F., 4
Prandtl, L., 170

Prandtl, number, 173, 174, 177
Priestley, J., 101
Prigogine, I., 17, 195
primitive cell, 25, 28
Prior, A. C., 239, 240
pseudo-symmetry, 46
Pulliam, G. R., 136, 257, 258
Pupp, C., 158
purification by CVT, 53, 217, 220
pyroelectricity, 30

Rabenau, A., 229
radius ratio rule, 38
Rai-Choudhury, P., 155
Ralph, B., 46
Raman spectroscopy, 265
Ranganathan, S., 46
Rao, C. N. R., 57
Rao, G. V. S., 57
Rao, K. J., 57
rate of adsorption, 72
rate of desorption, 72
Rayleigh, Lord, 174
reaction conductance, 197, 198
reaction eqilibrium, 16
reaction space, 198, 199
recrystallization, 47
Redheffer, R. M., 279
Reed, T. B., 118, 207, 208, 211, 229
re-entrant faces, 98
residence time, 71, 91
Reisman, A., 254
Reiss, H., 62
repulsion potential, 57
retrograde solubility, 61
reverse current, 272, 273, 274, 275
reversible process, 7
Reynold's number, 166, 167, 168, 174, 177
Reynolds, O., 165
Rhodin, T. N., 88
Richman, D., 225, 226, 247, 253, 254, 255, 256
Robinson, P., 244
rotational entropy, 51
rotation axes, 25
rotation–inversion axes, 25
rotary polarization, 35
Roth, W. L., 43
Roy, R., 68

Sackur–Tetrode equation, 55, 282

Sadagopan, V., 258
Savva, M. A., 249, 250
Schäfer, H., 2, 50, 70, 128, 219, 226
Schaeffer, P. S., 246, 254, 255, 268
Schalinkoff, A., 92
Schmidt number, 174, 177
Scholtz, H., 228, 233, 235
Schottky defect, 42, 67
Schubauer, G. B., 168
screw dislocation, 43
screw dislocation mechanism, 98
Searcy, A. W., 64
second law of thermodynamics, 7
Sekerka, R. F., 216
Semenoff, N., 92
separation by CVT, 217
sequential processes, 191, 194
Severin, P. J. W., 177, 246, 248, 268
Sewmon, P. G., 95
Shaw, D. W., 153, 194, 247, 251
shear structures, 69
Shelton, R. A. J., 220
Shiloh, M., 256
Shirtcliffe, T. G. L., 176
similarity theorem, 167
Simpson, W. I., 257
Sinke, G. C., 15
Skramstad, H. K., 168
small angle grain boundary, 46
Smallman, R. E., 44
Sokolnikoff, I. S., 279
solubility, 60, 61
solutal convection, 108, 173
Somorjai, G. A., 130
Soret coefficient, 108
sound velocity, 266
sphalerite structure, 38, 40, 42, 68
spiral gauge, 112
Spoliti, M., 220
stable clusters, 86
stacking fault, 41, 47, 48, 81, 86
stability, 50
Stefan flow velocity, 51, 65, 70, 109, 113, 116, 117, 118, 119, 139, 146, 147, 148, 150, 164, 171, 180, 181, 212, 218, 222
Stefan, J., 109
Stefan's equation, 110
Steno, Nicolaus, 24
Stevenson, J. L., 184
Stitch, M. L., 57
sticking coefficient, 82

INDEX

Stillman, G. E., 149
stoichiometry, 50
Stowell, M. J., 87
strain fields, 86
strain energy, 42, 43
Stranski, I. N., 95, 96, 270
Strauss, A. J. S., 208, 211
Strickland-Constable, R. F., 98
Struthers, J. D., 61
Stull, D. R., 15
sublimation, 4
sublimation of silver, 111
substitutional defect, 43
substitutional disorder, 67
substrate-induced twinning, 48
Sugawara, K., 251
superlattice, 61, 69
supersaturation, 83, 84, 96
surface diffusion, 91
surface free energy, 99
surface migration, 92
surface rearrangement, 87
surface roughness, 87
surface stoichiometry, 65
surface vacancies, 87
Sutherland, G. B. B. M., 162
Swalin, R. A., 61, 69

Tafel plot, 273, 276
Takahashi, R., 251
Takasu, S., 258
Takei, H., 258
Tanenbaum, M., 226
Taylor, E. R., 149
Taylor, R. C., 257
Teller, E., 75
template, 86
tensorial pyroelectricity, 30
terrace-ledge-kink model, 95
tetragonal system, 25, 31
tetrahedral co-ordination, 38
Theimer, O., 78
thermal conductivity, 184
thermal convection, 101, 108, 171, 173
thermal diffusion, 104, 108, 109
thermal diffusion ratio, 108
thermal entropy, 13
thermal mis-match, 47
thermochemical bond strength, 58
thermodynamics, 7
thermo-solutal convection, 176

third law of thermodynamics, 14
Thomas, G. J., 203
Thompson, W., 83
Thurmond, C. D., 61, 157
Tietjen, J. J., 225, 226, 247, 253, 254, 255, 256
Tietjens, O. G., 170
Tiller, W. A., 216
Todd, C. J., 63, 206
torsion effusion, 266
Tosi, M. P., 57
Townes, C. A., 57
Toyama, M., 128
translational entropy, 51
transport agent, 48, 49, 50, 51, 53, 63, 65, 66, 99
transport of carbon, 149
transport of gallium arsenide, 153
transport of nickel, 138, 142
transport of silver, 111, 114, 115
transport reactions, 50, 51
traps for adatoms, 95
triclinic system, 25, 31
trigonal system, 25, 31
trihalides of metals, 67
triple point, 20
Trouton's rule, 55
turbulent flow, 165
twin boundary, 46, 47
twinning, 45, 47, 98
twin planes, 45
two-dimensional cluster, 85
two-dimensional fluid, 85
two-dimensional gas, 62
two-dimensional nucleation, 91

Ubbelohde, A. R., 184
unique direction, 30
unit cell, 25

vacancy, 42, 44, 45, 47, 62
Van Gool, W., 62, 69, 206
Van Hook, A., 83, 97
Van Liempt, J. A. M., 105
Van't Hoff's equation, 19, 211
vapour–liquid–solid (VLS) mechanism, 49, 53, 78
vapour transport, 101
vapour transport conductance, 194, 198
variance, 21
Vedeneyev, V. I., 59

Venables, J. A., 85, 86, 92, 203
Verwey, E. J. W., 57
vibrational entropy, 42, 61
viscosity, 159
viscosity coefficient, 162, 164
viscous flow, 101, 108, 159
volatility, 50, 53, 60
Volmer, M., 92, 266
volume coefficient of expansion, 14

van der Waals force, 71, 72
Waddington, T. C., 58
Wagner, R. S., 49
Wallace, C., 48
Walton, D., 85
Weiner, M. E., 53, 113
Weiss zone law, 30
Wells, A. F., 41

Weyl, W. A., 90
White, E. A. D., 247
whiskers, 49, 79
Widmer, R., 229, 232, 233
Wilson, C. T. R., 82
Wolfe, C. M., 149
Woodbury, H. H., 130
Woods, J., 70, 239
Wooster, W. A., 34
Wulff, G., 207
Wulff's theorem, 207
wurtzite structure, 38, 40, 41, 42, 68

Zeldovich, J., 84
zone, 30
zone axis, 30
zero-point energy, 57